LAW AND JUSTICE IN A GLOBALIZED WORLD

PROCEEDINGS OF THE ASIA-PACIFIC RESEARCH IN SOCIAL SCIENCES AND HUMANITIES, DEPOK, INDONESIA, 7–9 NOVEMBER 2016: TOPICS IN LAW AND JUSTICE

Law and Justice in a Globalized World

Editors

Harkristuti Harkrisnowo, Hikmahanto Juwana & Yu Un Oppusunggu
Faculty of Law, Universitas Indonesia, Indonesia

LONDON AND NEW YORK

Routledge is an imprint of the Taylor & Francis Group, an informa business

© 2018 Taylor & Francis Group, London, UK

Typeset by V Publishing Solutions Pvt Ltd., Chennai, India

Although all care is taken to ensure integrity and the quality of this publication and the information herein, no responsibility is assumed by the publishers nor the author for any damage to the property or persons as a result of operation or use of this publication and/or the information contained herein.

The Open Access version of this book, available at www.tandfebooks.com, has been made available under a Creative Commons Attribution-Non Commercial-No Derivatives 4.0 license.

Published by: CRC Press/Balkema
Schipholweg 107C, 2316 XC Leiden, The Netherlands
e-mail: Pub.NL@taylorandfrancis.com
www.crcpress.com – www.taylorandfrancis.com

ISBN: 978-1-138-62667-6 (Hbk)
ISBN: 978-1-315-22329-2 (eBook)

Law and Justice in a Globalized World – Harkrisnowo et al. (Eds)
© 2018 Taylor & Francis Group, London, ISBN 978-1-138-62667-6

Table of contents

Foreword	ix
Organizing committee	xi
Keynote speech: On local wisdom and the pursuit of justice through criminal law reform: The Indonesian experience in deliberating the Bill of the penal code *H. Harkrisnowo*	1
Principle of balance in health services in Indonesia *Y.A. Mannas & E.L. Fakhriah*	9
Research report on the knowledge, experience, and attitude of consumers towards counterfeit medicine in Jakarta: A legal analysis *H. Marlyna & A. Sardjono*	15
Research report on protection of geographical indications in Indonesia: Advantages and challenges *M.M. Gabor & A. Sardjono*	23
The politics of laws in the establishment of the Human Rights Court in Indonesia *J. Saibih*	31
Indonesian pretrial: Can it fulfil the rights of the victims of an unfair trial to restoration? *F.M. Nelson*	39
The substance of good governance principles on government decisions in Indonesia *M.R. Bakry & A. Erliyana*	45
Does sanctity of contract exist in oil and gas contracts in Indonesia? *F.H. Ridwan, R. Agustina & J. Rizal*	55
Protection of personal information: The state's obligation to guarantee the right to privacy in Indonesia *N. Rianarizkiwati & J. Asshiddiqie*	63
The enforcement of payment of restitution in criminal proceedings as the base for filing of tort charges *S.L. Anindita & R. Agustina*	71
Repeat offence as aggravating circumstances in a criminal case: Differences in the definitions under the Criminal Code (KUHP) and the anti-corruption law *E. Elda & T. Santoso*	79
A new paradigm of the justice of outsourcing in Indonesia *I. Farida, S. Arinanto & J. Rizal*	85
Role of fiscal measures in financing responses to the financial crisis in Indonesia and their effects on fiscal sustainability *Y. Indrawati & A. Erliyana*	93

Abuse of substance, restorative justice and diversion S. Asa & S. Fitriasih	101
Waqf shares to create equitable economic distribution in Islam in relation to the Law Number 41 of 2004 on *Waqf* H.N. Lita, U. Hasanah & Y.S. Barlinti	109
Paradox of groundwater tax collection E.S. Sunarti & T. Hayati	119
The concept of a regulation of collateral under the *mudharabah* financing contract according to the Law No. 21 of 2008 on sharia banking in Indonesia M.Y. Harahap & U. Hasanah	127
Mediation as a means to provide *ishlah* (peace and harmony) in the resolution of *sharia* banking disputes in Indonesia Wirdyaningsih	133
Waqf banks under the Indonesian legal system A.K. Munthe & F. Prihatini	141
Presentation of suspects: The paradox of presumption of innocence A. Pangaribuan	149
The risk of joining the Trans-Pacific partnership for Indonesia: An investment perspective W. Setiawati	157
The establishment of small claims court in Indonesia: Expectation and challenge in encountering the globalisation era A. Afriana & E.L. Fakhriah	167
Causation in the context of environmental pollution crime A. Sofian & S. Fitriasih	173
Public official recruitment system: Towards creation of a rule of law based on *Pancasila* S. Anam & J. Asshiddiqie	181
A *waqf* and *musyarakah* implementation model in *takaful ijtima'i* as an alternative *sharia* insurance system: An experiment to maximise the realisation of the social justice principle in *sharia* insurance in Indonesia Z. Abdullah, U. Hasanah & G. Dewi	187
The implementation of Indonesia's penal sanctions to the increased illegal trade of endangered species A.G. Wibisana & W.P. Nuning	195
Analysis of the water resource regulation under the ordinance of the Greater Jakarta area R.I. Dewi	203
Review of the implementation of *murabahah* financing in *sharia* banks in the Greater Jakarta area A.R. Nurdin	211
Environmental damage and liability in Indonesia: Fancy words under conventional wisdom A.G. Wibisana & T.A. Dewaranu	221
Review of the principles of the criminal law on the ancient archipelago law book: The study of the Book of Kutara Manawa Darma Sastra of the Majapahit Kingdom and the Laws on Simbur Tjahaja of the Sultanate of Palembang in the context of criminal law development and condemnation in Indonesia I. Darmawan & H. Harkrisnowo	229

The existence of arbitration principles in commercial agreements: Lessons learned from an Indonesian court *Y.K. Dewi*	235
Cross-border child adoption: Protection and challenges in Indonesia *A.B. Cahyono*	241
Electronic identity management in ensuring national cyber security and resilience: Legal aspects of online identity and its secured transaction *E. Makarim & E.G. Pantouw*	249
The legal impact of the signing of the multilateral competent authority agreement on automatic exchange of financial account information: A banking law perspective *R.A.V. Napitupulu*	257
The unfair rules of intellectual property rights section of the trans-pacific partnership agreement *B.A. Prastyo & A. Sardjono*	265
Author index	271

Foreword

Law and Justice is one of the umbrella topics of the APRISH Conference held by Universitas Indonesia in November 2016. This academic undertaking is designed to provide a forum for researchers to disseminate and exchange their research findings and reports on this topic within and around Asia-Pacific countries.

It is well understood that there is a growing role of the Asia-Pacific region in the world, in particular in terms of global economy and politics. At this juncture, it is only natural that law plays a very significant role in the shaping and maybe transformation in each country, in the sub region and in the region itself. In addition, the number of Asia-Pacific countries, the diversity of countries and also the different historical background as well as their legal development and challenges provide a fertile avenue for critical analysis and research. It is the expectation of Universitas Indonesia that this conference would serve as a forum to reflect upon important developments in the region in various areas of law and justice.

It is important to note that in this book, the collection of information and analysis on the existence, clarity, and implementation of law on particular issues in one country is intended to provide detailed explanation to readers in conjunction with the specific issue addressed. It is the intention of the conference holder to facilitate a venue for interaction and interconnection among researchers, in particular to forge further collaboration in future legal research and publishing.

In this book readers would be able to find vast knowledge both on theoretical and empirical perspectives as to the working of laws aimed to achieve justice for the people. The role of law in the countries in this region has been demonstrated to shape not only the system of governance in each country, but also the well being of its people through the implementation of such law. As such, a number of articles in this book have also studied and revealed the historical antecedents on how legal rules and legal institutions were established in certain countries. Challenges found both in legislations and its implementation may be worth noting, for these could serve as lessons learned in other countries experiencing similar conditions.

Many fields of law are covered in this book which include not only public law, but also private law and transnational laws. Articles in public law vary from criminal law to criminal procedural law and environmental law. With regard to private laws, there are articles on inheritance law, adoption law, investment law and banking law, including those operated in line with Islamic law. In addition the importance of law in connection with other discipline is also discussed under the issues of human rights, good governance and democracy, which to some extent also indicated the degree of development of each of these three concept in the respective country. Since interconnection among countries in the region and worldwide has become a rule in today's globalized world, there are also articles on regional and transnational issues having significant parts not only in each country but also in the region, and maybe in the world as well.

It is to be noted however that most of the articles are related to legal development in Indonesia, a country which during the past two decades has experienced significant challenges in politics and social issues, where laws were repealed and amended, new laws were introduced, borrowing many conceptual frameworks from around the world, and adjusted to the Indonesian condition and structures.

May the readers find this book interesting and may it inspire new interests in various fields of law and justice.

The Editorial Board of the 1st APRISH Proceedings for Topics in Law and Justice

Prof. Harkristuti Harkrisnowo, Ph.D.
Prof. Hikmahanto Juwana, Ph.D.
Yu Un Opposunggu, Ph.D.
Faculty of Law, Universitas Indonesia

Organizing committee

STEERING COMMITTEE

Rosari Saleh, *Vice Rector of Research and Innovation, Universitas Indonesia, Indonesia*
Topo Santoso, *Dean, Faculty of Law, Universitas Indonesia, Indonesia*
Ari Kuncoro, *Dean, Faculty of Economics and Business, Universitas Indonesia, Indonesia*
Adrianus L., G. Waworuntu, *Dean, Faculty of Humanities, Universitas Indonesia, Indonesia*
Arie Setiabudi Soesilo, *Dean, Faculty of Social and Political Sciences, Universitas Indonesia, Indonesia*

INTERNATIONAL ADVISORY BOARD

Peter Newcombe, *University of Queensland, Australia*
Fred Piercy, *Virginia Tech University, Australia*
Frieda Mangunsong Siahaan, *Universitas Indonesia, Indonesia*
James Bartle, *University of New South Wales, Australia*
Elvia Sunityo Shauki, *University of South Australia, Australia*

SCIENTIFIC COMMITTEE

Manneke Budiman
Isbandi Rukminto Adi
Beta Yulianita Gitaharie
Surastini Fitriasih
Sri Hartati R. Suradijono
Elizabeth Kristi Poerwandari

CONFERENCE DIRECTOR

Tjut Rifameutia Umar Ali

CONFERENCE VICE-DIRECTOR

Turro Wongkaren

ORGANIZING COMMITTEE

Dewi Maulina, *Faculty of Psychology, Universitas Indonesia, Indonesia*
Intan Wardhani, *Faculty of Psychology, Universitas Indonesia, Indonesia*
Elok D. Malay, *Faculty of Psychology, Universitas Indonesia, Indonesia*

Josephine Rosa Marieta, *Faculty of Psychology, Universitas Indonesia, Indonesia*
Teraya Paramehta, *Faculty of Humanities, Universitas Indonesia, Indonesia*
Nila Ayu Utami, *Faculty of Humanities, Universitas Indonesia, Indonesia*
Priskila Pratita Penasthika, *Faculty of Law, Universitas Indonesia, Indonesia*
Efriyani Djuwita, *Faculty of Psychology, Universitas Indonesia, Indonesia*
Destri Widaya, *Faculty of Economics and Business, Universitas Indonesia, Indonesia*

WORKSHOP COMMITTEE

Corina D.S. Riantoputra, *Faculty of Psychology, Universitas Indonesia, Indonesia*
Fithra Faisal Hastiadi, *Faculty of Economics and Business, Universitas Indonesia, Indonesia*
Mirra Noormilla, *Faculty of Psychology, Universitas Indonesia, Indonesia*

TREASURERS

Robby Oka Yuwansa, *Faculty of Psychology, Universitas Indonesia, Indonesia*
Nurul Husnah, *Faculty of Economics and Business, Universitas Indonesia, Indonesia*

Keynote speech: On local wisdom and the pursuit of justice through criminal law reform: The Indonesian experience in deliberating the Bill of the penal code

H. Harkrisnowo
Faculty of Law, Universitas Indonesia, Depok, Indonesia

ABSTRACT: The problem of crime and punishment has existed for as long as human history. Nevertheless, it continues to be debated and remains controversial. The most significant question relevant to punishment is its philosophy or main purpose. Joining this debate, this paper elaborates on the undertakings of the legal drafters of the Indonesian Bill on the Criminal Code (or the Bill), who stipulated the aims of punishment in the Bill, while most penal code drafters in other countries appear to elude such stipulation. Furthermore, while the global trend is to consider only the aims developed in the Western hemisphere, such as retribution, deterrence, rehabilitation and restorative justice, the Bill's legal drafters also considered Indonesian local wisdom related to punishment. The incorporation of such local wisdom changes the paradigm of the existing penal code—which is the legacy of the colonial regime—into one that represents the core norms and values of the Indonesian people.

1 INTRODUCTION

Punishment has existed in all communities and nations worldwide, not only today but also in ancient times, as proven by old manuscripts and even various holy books addressing various misdeeds, religious or otherwise. Many would begin by theorising about why the state (or kingdom) is accorded the legal right to inflict punishment on the perpetrator of a crime and why it is permitted by moral judgement to treat such people in a different manner than those who have not violated a criminal law. Such is the importance of the social control mechanism in a society that Fyodor Dostoyevsky encapsulated it as follows: 'A society should be judged not by how it treats its outstanding citizens but by how it treats its worst criminals' (Dostoyevsky, 1915).

While the question of punishment has existed for as long as human history, its origin is difficult to trace. Rarely do we find any book depicting laws in any country at any time that does not address the issue of crime and punishment, although their conceptualisation might differ from one country to another and from one time to another. Despite its long history, punishment remains a debatable issue that generates vast volumes of literature on the aims of punishment.

Research on punishment is usually based on theories developed in the Western hemisphere, for the simple reason that Western literature is readily accessible. However, this study attempts to look at Indonesian local wisdom and ancient law books to enrich the understanding of our past and to examine how ancient patriarchs thought about, decreed and implemented punishments in their reigning periods. Discovering the content of old laws that once governed the land of Indonesia is an important task for the current Indonesian government and the parliament, who are deliberating the Bill on the Criminal Code to replace the existing criminal code, which is a legacy of the colonial regime.

This documentary research poses the following questions:

a. What are the purposes of punishment as found in the ancient law books in Indonesia, and how are they used in the drafting of the Bill on the Criminal Code today?
b. Which criminal principles found in such books could be incorporated into the Bill?
c. How different are the books from modern criminal law?

2 METHODOLOGY

Based on a qualitative study, this documentary research looked in depth at numerous documents and manuscripts. In addition to various books, journals and published research reports, the primary source examined for this study was ancient law books in Indonesia that have been translated into Indonesian.

3 DEFINING PUNISHMENT

From a legal point of view, punishment is the infliction of pain, deprivation or some other form of suffering and even the taking of a person's life. According to Foucault, punishment is just one manifestation of power. In essence he is of the opinion that punishment is a functioning of power that is exercised on the people being punished (in prison) by those supervising the punishment (Foucault, 1975). Naturally, such functioning of power carries legal attributes. David Garland, for example, underlined that punishment is a legally approved method designed to facilitate the task of the state in conducting crime control (Garland, 1990). With regards to the impact of punishment, Nietzsche (1956) believes that the broad effects of punishment in man and beast are the increase of fear, the sharpening of the sense of cunning and the mastery of the desires. Hence, punishment tames man, but does not make him better. In this context, Hart (1968) proposed that five elements could be found in the concept of punishment: 1) it must involve pain or other consequences normally considered as unpleasant; 2) it is imposed due to an offence against legal rules; 3) it must be imposed on an actual or supposed offender for his offence; 4) it must be intentionally administered by human beings other than the offender; and 5) it must be imposed and administered by an authority constituted by a legal system against which the offence is committed.

4 THEORETICAL BACKGROUND

Discourse on the theory of punishment addresses the rationale for why a person should be punished for violating a law. Also known as the philosophy of punishment, this discourse primarily aims to find a justification for the state to inflict suffering or pain on a person who has been proven guilty of committing a crime. Why is punishment considered a necessity in social interaction? Does the public need punishment against perpetrators of legal violations? Is it needed by the community to reassert societal values 'destroyed' by the perpetrator? In this regard, Jean-Paul Sartre (1948), the existentialist, argued that punishment does not have any benefit, but rather it is a limitation of freedom and a limitation on human rights. The earliest theory of punishment, known as the absolute theory, contended that a punishment is inflicted merely because a transgression of the law has been committed. In Latin, this is known as *punitur quia peccatum est*, which means 'a punishment is due for a crime committed'.

There are at least two major groups of theories relating to the justification of punishment: retributive and utilitarian. The retributive, retaliation or *lex talionis* school holds that an offender should be punished in payment for his or her transgression. In other words, this school of thought believes that there should be a moral connection between the criminal act and the punishment imposed. The emergence of social reaction to such transgressions is a reflection of disapproval, of morally blaming the perpetrator for the transgression, as well as a symbolic reaffirmation of long-held societal values. Hence, retributive justice is a refined form of the desire for vengeance against wrongdoing found in every human being. Nonetheless, the punishment must be proportional to the damage caused by the criminal act. This idea of 'just deserts' is heavily influenced by the philosophical thought of Immanuel Kant. The perpetrators should not profit from their crimes and, with the infliction of punishment, they cannot benefit from any gain originally expected from the commission of the crime.

Kant also asserted that punishment is a *kategorischer imperativ*, that is, a 'categorical imperative' or a logical consequence of the commission of a crime (Abel & Marsh, 1984). Hence, the infliction of punishment is a reaction demanded by ethical justice. In the same tone, Hugo Grotius formulated *mallum passionis propter malum actiones*, which means that an evil is to be inflicted because an evil has been committed (Rutherford, 1832).

In retributive theory, vengeance is not the only feature. There is also denouncement, a form of disapproval by society of such misbehaviour. As a symbolic reaffirmation of the norms and values upheld by society, it is expected that citizens will confirm these values. It is important also to observe that this theory emphasises the issue of proportionality; that is, the sentence imposed should be proportionate to the crime committed and not merely be a form of retaliation. The proportionality of a punishment, the main characteristic of retributivism, is supported today by Andrew von Hirsch, who believes that there are at least two advantages to this principle. First, it is ethically plausible and easy for the public to digest. Second, it provides guidance for decision-makers when determining the severity or leniency of a punishment. With this in mind, it is expected that both the legislature and the judiciary can determine a scale for crimes against punishments (von Hirsch, 1992).

The utilitarian school, on the other hand, challenges the view that there is a relationship between morality and punishment, because in its view punishment should only be imposed for future considerations instead of for metaphysical, backward-looking reasons. Jeremy Bentham (1948), for example, stated: '[A]ll punishment is in itself evil because it inflicts suffering and pain. But if it can be shown that the pain inflicted is in some way preventing or excluding some greater pain, thus it is useful in achieving greater aggregate of pleasure and happiness...'. In this regard, in the 17th century George Saville, 1st Marquis de Halifax, an English statesman of the House of Commons and later the House of Lords, made a fitting statement that 'men are not hanged for stealing horses, but that horses may not be stolen' (Brand, 1960). In essence, the utilitarian is of the opinion that criminal punishment shall, at the same time, benefit the state, the community and the perpetrators themselves. As such, punishment shall not merely serve as a 'payment' or retaliation for past wrongdoing and the damage caused by it. Instead, punishment shall have significant meaning for the totality of the law.

However, there are criticisms of this perspective. The utilitarian is deemed unable to provide any guidance. More than two centuries ago, Jeremy Bentham proposed that punishment must be weighed and determined to measure its deterrent effect against potential offenders. Of course, schemes of this kind are almost impossible due to the extensive wealth of data it would be necessary to collect in order to reveal the extent to which certain punishments produce deterrent effects. Andrew von Hirsch (1976) even boldly ensured that 'replacing deterrence with rehabilitation scarcely improves matters'. Fortunately, more modern utilitarianism later produced a number of theories of punishment, such as deterrence (special deterrence and general deterrence), incapacitation, rehabilitation and restorative justice. Deterrence theory basically assumes that punishment is imposed primarily to prevent the future perpetration of crime. There are two theoretical groupings from this forward-looking perspective: a) special deterrence theory, which holds that the objective of punishment is to deter or prevent a specific perpetrator from committing future crimes, and b) general deterrence theory, which focuses on the role of punishment in preventing future crimes by potential offenders among any members of the community.

Incapacitation theory differs from deterrence theory in that, according to this view, criminal punishment is meant to make offenders lose their capacity to commit future crimes. For example, in the past (though still applicable in some countries) this was done through the infliction of capital punishment or banishment. However, in modern times, when countries are slowly abandoning the use of capital punishment, the method of incapacitating perpetrators is through deprivation of liberty, by putting them in prisons or having them wear an electronic ankle bracelet that will alert law enforcement when they step beyond a legally allowed limit.

The rehabilitation perspective has emerged from medical science-influenced theorists who argue that crime is a 'disease', and that perpetrators are individuals who are suffering from this disease and need to be treated. Since people are different, it is possible that two persons committing the same criminal act would end up being punished differently due to differences in their psychological assessments conducted before a judicial decision. This approach has been criticised, mainly because it creates an indeterminate sentence, which is considered by some to reduce the certainty of law. Instead of inflicting a definite time for punishment, an indeterminate sentence does not state a specific period of time or release date (of imprisonment, for example), but just a range of time, such as three to five years. Hence the time to be served by the convicted person depends on the decision made by the prison officers.

Recent developments in punishment highlight a different view, no longer centred on the importance of crime and punishment. Restorative justice, essentially an approach involving all parties affected by a criminal act to jointly resolve and heal the condition. Basically, this is a process whereby the parties with a stake in a particular offence come together in the aftermath of the offence to collectively resolve it and its implications for the future (Marshall, 1999). Howard Zehr (1990) has described the mindset of restorative justice as follows: 'Crime is a violation of people and relationships. It creates obligations to the make things right. Justice involves the victim, the offender, and the community in a search for solutions to promote repair, reconciliation, and reassurance'.

5 FINDINGS

Various ancient law books found in Indonesia in the past are, among others, *The Book of Ciwasasana* or *Purwadhigama* (during the regime of King Dharmawangsa in the tenth century); *The Book of Gajamada* or *Kitab Gajamada*, which was named after the then prime minister of Majapahit in the middle of the fourteenth century; *The Book of Kutaramanawadharmasastra*, which is estimated to have been published in 1350; *The Book of Adhigama* or *Kitab Adhigama* (*Adhigama Kanaka*), which was made under the instruction of Prime Minister Kanaka in 1430; *The Book of Simbur Cahaya* or *Kitab Simbur Cahaya* in Palembang (circa 1630); *The Book of Kuntara Raja Niti* or *Kitab Kuntara Raja Niti* in Lampung (sixteenth century); *The Book of Lontara' ade'* or *Kitab Lontara' ade'* in South Sulawesi, which was still in force until the nineteenth century; *The Book of Patik Dohot Uhum ni Halak Batak* in North Sumatra, whose date of entry into force is unknown; and *The Book of Amana Gappa*, the Maritime Code of Makassar, South Sulawesi. These books are more the exception than the rule, since most of the old laws were unwritten. However, even if they were unwritten, the old legal stipulations could be found in ancient manuscripts, such as Kakawin (long narrative poems written in ancient Javanese language, poetry with specified rhythms and metres originating from Indian Sanskrit literature), Pepakem (charter, compilation of adat or customary laws or paragraphs taken from ancient books), Babad (historical texts), Jayapattra (decisions of the court), Prasasti (eulogy, usually carved on a stone), Lontara (stipulation of laws or other important issues in South Sulawesi, written on lontara leaves) and Awig-awig (laws of a Balinese Pakraman village, originally written on rontal leaves).

It is worth noting that, even today, Awig-awig continues to be acknowledged by the state, as demonstrated by Bali Province Local Regulation No. 3 of 2001 on Pakraman Village and Customary Institutions, which defines Awig-awig as a regulation made by a Pakraman village or banjar as a guideline for implementing Tri Hita Karana in accordance with the mawacara and Dharma Agama in the respective village (Sugiantari & Julianti, 2016).

6 LOCAL WISDOM ON THE AIMS OF PUNISHMENT

As described above, ancient Indonesian kingdoms and communities with customary laws had their own legal systems, although the organising and implementing mechanisms of the laws differed from what we have today. Several unique features of the old laws and legal systems are exemplified by the King holding the roles of executive, legislative and judiciary institutions at the same time. This absence of a separation of powers made it possible for the right to issue or adopt a law, to oversee its implementation and to pass judgement on its violation to rest in the hands of a king, the head of a community or the elderly of a community. Hence, the King must be a person of knowledge, literature, religion and justice. This is the very reason why, during the Majapahit era, the King inherited the role of king protector, as eloquently written in *Negarakertagama*, which is a chronicle of what was happening in the Kingdom of Majapahit at that time, although it has been assumed that some conditions were described in a better light than actually existed.

Another unique feature of the ancient law and other customary laws in Indonesia is the precedence of communitarianism over individualism, similar to Ferdinand Tonnies' *Gemeinschaft* ideal concept, in which personal relationships are very intimate and are defined and regulated on the basis of traditional social rules, and the division of labour is minimal. Furthermore, it is important to note that, during that period, no distinction or separation was made between criminal wrongs and civil wrongs, and many of these laws were unwritten (*ius non scriptum*). Both types of wrongdoings carried legal consequences, which were examined before and decided by the authorities, such as the King, the head of a village or the elderly of a community.

All ancient law books are considered to be the products of wise people, dominated by religious elders and usually endorsed by the King. These books stipulate different kinds of conduct, both public and private, civil and criminal. While 'sins' and 'crimes' are sometimes interconnected, specific religious wrongdoings have specific religious punishments, and wrongdoings against mundane norms have mundane punishments. Religious punishment could be inflicted in different forms, such as 1) expulsion from one's community for a definite or indefinite time; 2) prohibition on worshipping in one's place of worship; 3) prohibition on being helped by other members of one's society; 4) obligation to perform a specific religious ceremony; and 5) obligation to repent in a secluded place. Mundane punishments consisted of capital punishment, corporal punishment (such as mutilation, caning and sometimes torture), fines, compensation to the community, restitution to the victim, reprimands or admonishment and public shaming (Koesnoen, 1960). Despite the many forms of punishment, punishment by imprisonment—the dominant type of punishment today—is not found in any of the ancient law books or in any of the customary laws.

A closer look at the ancient law books reveals that, in ancient times in Indonesia, different levels of entities could execute punishments. First, a victim's clan or family could execute punishment by retaliation against (the clan of) the perpetrator. Second, there could be individual retaliation by a victim or the family of a victim against the perpetrator. Payment of peace offerings or blood money by the perpetrator to the victim or the victim's family or clan was also a form of punishment. To some extent, this condition appears to be similar to Beccaria's (1986) adage that 'punishment shall be certain, swift, and speedy' in order to achieve its efficacy and effectiveness.

Similar to modern theories of punishment, readings of ancient law books have indicated the existence of various aims of punishment, such as denunciation, prevention, incapacitation, restoring societal balance, restitution, conflict resolution and retribution. The aim of denunciation is found in Lontara Tribe Wajo in South Sulawesi, which reads, '[T]hey should not be killed unless it is their ills which kill them'. The rule that the King shall obey the law is stipulated in the law books, which states that 'the King shall follow all of the stipulations in the book in order to ensure the welfare of the country, since the primary trait (of the King) is to lead his people towards prosperity, which in turn would lead him to heaven as a result' (Muljana, 1967).

Denouncement as the aim of punishment sends a clear message to the public that whoever violates or breaks any proscription contained in the law books will receive a punishment, as the laws are the reflection of the people's collective values. Prevention as the aim of punishment can be found in Verse 93 of *Kutaramanawa Dharmasastra*, which stipulates, 'The intention of the King to impose fines is to control a person's desire so that this person will not be misguided, will not take the wrong path…'. The King, as the sole judge who could decide the severity or leniency of a fine, was to be guided by the law book and was not allowed to deviate from what had been written in it. Moreover, Verse 6 prescribes that no one could be exempted from the law if found guilty of certain crimes:

'Those who work for the King, even if he is a minister, if he commits one of the *astadusta* crimes which consist of taking a person's life], shall be put to death .... If he commits a theft, his misdeed is that of a thief; if the King's servants commit such crime, or a minister commits such crime, if he was murdered, such murder would not be punished, for those crimes are punished by death. Such is the law as made by the King'.

The incapacitation aim of punishment refers to the loss or restriction of an offender's liberty or ability to conduct normal community activities. This punishment was intended to prevent the recurrence of such wrongdoings by the offender. Restitution as the aim of punishment is found, for instance, in Verse 241: 'Whoever damages the property of others without intention, for example, breaking, cracking, due to the offender's throwing or hitting or due to war, is obliged to return goods twice the value of the damaged goods; he/she has to restore it as it was before, or replace with other goods to the owner'.

With regards to retribution, Verse 11 asserts: 'Whoever commits one of the above six crimes does not deserve to be pardoned by the King. If his/her guilt is proven, [he/she] must be executed without having to undergo any process'. Article 69 states that 'whoever steals gold, gems or fabric whose value is above 100, his hands shall be decapitated'.

Some of these expressed aims of punishment are incorporated into the Indonesian Bill on the Criminal Code (Ministry of Law and Human Rights, 2015), since they are, among others, depicting the norms and values embraced by the people, reflecting the close relationship of the people with their land and indicating the significant bond between law, community and morality, and religion.

At this juncture, the Indonesian legal drafters of the Bill on the Criminal Code apparently attempted to incorporate the aims of punishment; not only the ones derived from Western philosophy but also from the ancient law books. As such, they formulated the following aims of punishment in Article 55 of the Bill on the Criminal Code (Ministry of Law and Human Rights, 2015):

– Preventing of the commission of crime though upholding legal norms for the sake of societal protection,
– Reintegrating convicted offenders into society through rehabilitative treatment,
– Resolving conflict,
– Returning societal balance,
– Maintaining peace in society,
– Expiating offenders' guilty feelings.

The drafters of the Bill recognised the importance of such aims, although some of these are not found in modern, Western theory. This is also enhanced by the recognition of customary wrongdoings or the violations of customary laws that are still embraced in some regions. Yet, the application of such customary laws must not violate human rights and the international principles of law, thus the second paragraph of Article 55 of the Bill reflects the drafter's principle that punishment shall not constitute torture (in accordance with the Convention Against Torture, and Other Cruel, Inhuman and Degrading Treatment of Punishment ratified by Indonesia in 1998). Paragraph 2 of Article 55 asserts, 'Punishment is not meant to make people suffer or their dignity degraded'.

The Bill is in the process of becoming the first national criminal law that is in accordance with the national condition, local wisdom and the development of society, for basically this new law would replace the current criminal code, which is a legacy of the colonial regime in the nineteenth century. In a nutshell, according to the elucidation of the Bill, its objective is to produce a comprehensive National Criminal Code, with Indonesian characteristics, which respects customary and religious values and, at the same time, is modern, in accordance with international values, principles and standards (Ministry of Law and Human Rights, 2015, p. 133).

7   CONCLUSION: ON CRIME AND PUNISHMENT: THEN AND NOW

While many of the principles contained in the old law books are still relevant to today's criminal law, the development of society has led to the removal of a number of principles, crimes and punishments that were recognised in former days. The main difference, of course, lies in the fact that a separation of powers has become the norm in today's society, even if the head of state is a king, which carries many implications. First, most of the aims of punishment are still relevant in today's society. While retributivism is considered to be an outdated and primitive aim of punishment that is no longer morally appropriate in modern criminal law—and hence is not stipulated in the Bill of the Criminal Code—apparently, many judges continue to hold this as the main aim of punishment, as found in 79 court decisions. Second, modern criminal law has abandoned the use of corporal punishment, such as mutilation and other forms of torture, which is considered to be a punishment that is cruel and degrading to human dignity. In its place, deprivation of liberty has become the most common punishment in criminal law today. Third, capital punishment is used very sparingly, only for very serious and atrocious crimes. Fourth, punishment is inflicted only on the perpetrator, based on a fault requirement that only those found guilty should be punished; as such, the family members of a perpetrator cannot be punished. Fifth, the amount of compensation and restitution is still limited by the law but decided by the judiciary. Sixth, today, exculpation and excuses are much more elaborated. Seventh, stratification by status is no longer used to determine types of punishment. Eighth, today, criminal stipulations are separated from civil rules and regulations.

REFERENCES

Abel, C.F. & Marsh, F.H. (1984). *Punishment and restitution: A restitutionary approach to crime and the criminal*. London: Greenwood Press.

Beccaria, C. (1986). *On crime and punishments*. Translated by David Young. Indianapolis: Hackett Publishing Co.

Bentham, J. (1948). *Introduction to the principles of morals and legislation*. New York: Hafner Publishing Co.

Brand, G.E. (1960). The discipline of judges. *American Bar Association Journal, 46*(12), 1315–1318.

Dostoyevsky, Fyodor (1915). The House of the Dead, translated by Constance Garnett. London: Macmillan.

Foucault, M. (1975). *Discipline and punish: The birth of prison*. New York: Vintage Books.

Garland, D. (1990). *Punishment and modern society*. Chicago: University of Chicago Press.

Hart, H.L.A. (1968). *Punishment and responsibility*. Oxford: Clarendon Press.

Kant, I. (2010). *Metaphysics of morals: A critical guide*. Cambridge: Cambridge University Press.

Koesnoen, R.A. (1960). *Susunan Pidana dalam Negara Sosialis Indonesia.* (The Penal Structure in The Indonesian Socialist State Bandung, Sumur Bandung.

Marshall, T.F. (1999). *Restorative justice: An overview* (Report). London: Home Office Research Development and Statistics Directorate.

Ministry of Law and Human Rights of the Republic of Indonesia (2015). The Bill of the Criminal Code. (At the time of the writing of this article, the Bill is still under deliberation in the Indonesian Parliament).

Muljana, S. (1967). *Perundang-undangan Madjapahit*. (The Laws of Madjapahit) Jakarta, Bhratara.

Nietzsche, F.W. (1956). *The birth of tragedy and the genealogy of morals*. New York: Doubleday & Co.

Rutherford, T. (1832). *Course of lectures on Grotius De Jure Belli et Pacis*. Baltimore: William and Joseph Neal.

Sartre, J.P. (1948). *Existentialism and human emotions*. New York: Philosophical Library.

Sugiantari, A.A.P.W. & Julianti, L. (2016). Peranan awig-awig Desa Pakraman dalam mencegah tindak pidana pencurian benda sakral di Desa Pelaga Kecamatan Petang Kabupaten Badung. (The Role of Awig-awig – written customary law – in Preventing the Theft of Sacred artrifacts in the Village of Pelaga, Subregency of Petang, Badung Regency) *Jurnal Bhakti Saraswati, 5*(1), 60–69.

Von Hirsch, A. (1976). *Doing justice: The choice of punishments: Report of Committee for the Study of Incarceration*. New York: Hill & Wang.

Von Hirsch, A. (1992). Proportionality in the philosophy of punishment. In M. Tonry (Ed.), *Crime and justice: A review of research volume 16* (pp. 55–98). Chicago: The University of Chicago Press.

Zehr, H. (1990). *Changing lenses: A new focus for crime and justice*. Scottsdale PA: Herald Press.

Principle of balance in health services in Indonesia

Y.A. Mannas
Law Faculty, Andalas University, Padang, Indonesia

E.L. Fakhriah
Law Faculty, Padjadjaran University, Bandung, Indonesia

ABSTRACT: Balance manifests the function of law as a means of controlling social life by balancing interests that exist in the community or as a means of social control. The term balance in health care cannot be separated from the lack of regulation on legal relationship among parties. Laws have ruled in such a way on protection of parties involved in the health care, particularly doctors and patients, who are legal subjects in the therapeutic relationship. The emergence of many medical disputes led to the court has provoked a defensive form of treatment of most doctors due to fear of being prosecuted in providing treatment and a sense of absence of balance legal protection that in the field of health care. In principle, the positive law of Indonesia has set up the principle of balance in health care. Officers of law enforcement need understanding in handling cases of medical disputes, in term of distinguishing acts if omission or medical errors, with medical risks, as well as other non-law factors. Therefore, renewal is necessary in the field of health law by the particular court dealing with medical disputes, and both doctors and patients gain balance legal protection that is balance for both.

1 INTRODUCTION

Article 28H of the 1945 Constitution stipulates that anyone shall be entitled to health services. In an effort to realize health services in Indonesia, it is deemed necessary to issue a number of laws and regulations to ensure legal certainty and to attain health services that fulfill the fairness and justice among general public by determining the rights and obligations of patients, health paramedics, and hospitals. In health services, the importance of patient protection should be made the first priority of the operators of health services. Therefore, they have to respect such rights greatly. The Code of Ethics and the Hippocratic Oath also expressly provide for such obligations, particularly in Chapters I and II of the Council Code of Medical Ethics which provide for matters concerning general obligations and obligations of doctors to patients (Budianto & Utama, 2010).

For the sake of patient protection, it is expected that, in rendering services to patients, doctors should refer to and observe the services standards, professional standards, standard operating procedure so that the security and safety of the patients can be attained and maintained in an optimum manner and the actions taken should be in accordance with the said standards. Accordingly, these can serve as safety precaution for the prevention of malpractices. Patients have the right to obtain any information with respect to health services and medication patterns they are about to receive. The patients' right to get information is provided in a number of laws: Article 52 paragraph (1) of the Law Number 29 of 2004 on Medical Practices (hereinafter referred to as the Law on Medical Practices) stipulates that patients have the right to get full explanation regarding medical actions as described in Article 45 paragraph (3) of this Law on Medical Practices.

In addition to protection for patients, the law also stipulates protection for doctors when they perform their tasks and/or duties. Article 50 of the Law on Medical Practices stipulates that doctors get legal protection as long as they perform their tasks and/or duties in accordance with the professional standards and standard operating procedure. Doctors who perform their tasks and/or duties in accordance with the provisions of the three standards are entitled to legal protection in case of medical risks following medical actions. In this respect, the Indonesian positive law gives protection to both parties: doctors and patients, especially in the therapeutic relationship between them as both of them are in equal capacity as the subject of law entitled to legal protection.

Furthermore, the rights of patients are provided in Articles 4 through 8 of the Law Number 36 of 2009 on Health (hereinafter referred to as the Law on Health) which stipulates that anyone is entitled to health including access to resources in health sector, to get safe, quality and affordable health services, and to determine health services they require for and by themselves. Anyone is also entitled to balanced and accountable information and education on health, including information on their own health, the treatments and medications they have already taken and will receive from paramedics. Furthermore, Article 56 of the Law on Health also stipulates that anyone shall have the right to reject or accept part of whole supporting treatment after they have received and fully understood the information regarding such treatment. This indicates that information is essential and should be given to the patients before particular treatments are given. In Indonesian laws, protection for doctors is provided in Article 27 of the Law on Health which states that paramedics shall have the right to get reward and legal protection when performing their tasks and/or duties in accordance with the professional standards of paramedics. The rights and obligations of patients and doctors are also provided in Articles 50 through 53 of the Law on Medical Practices, which, in principle, contain similar provisions as those of the Law on Health.

The Law Number 44 of 2009 on Hospitals (hereinafter referred to as the Law on Hospitals) is the legal basis for legal obligations of hospitals in implementing patient safety. Article 34 of the Law on Hospitals expressly stipulates that hospitals shall apply patient safety standards. Such standards should be implemented by reporting incidents, analyzing and deciding solutions to the incidents. For reporting purpose, hospitals have to submit the same to the committee in charge of patient safety as designated by the minister. The Law on Hospitals also ensures that the legal responsibility regarding any negligence by paramedics is in the hands of the relevant hospital.

From the principle of relationship between doctor and patient (or other paramedics—hospital), there are therapeutic relationships or therapeutic transactions in which contracts or agreements occur (though not in writing) between patients and doctors in respect of medication or treatment of illnesses and between patients and hospitals in respect of health services by providing standardized facilities and infrastructures. Doctors and hospitals have to fulfill their obligation to provide medical services in accordance with the service standards, professional standards and standard operating procedure to patients, either upon request or otherwise, as in principle, from the therapeutic transaction, the health care provider and the health care receiver are equally the subject of law and have equal rights and obligations under the principle of equality before the law which are also stated in Article 1320 of the Indonesian Civil Code on legal requirements of agreements (Ratman, 2012).

In recent years, there have been occasions where the unsatisfied patients filed claims for compensation due to the mistake or negligence of doctors or paramedics in performing their tasks and/or duties. Various kinds of cases have been brought to court and have drawn serious attention of medical professionals and legal practitioners. The emergence of such cases indicates an increase in legal awareness among the public. The more they are aware of the law, the better their knowledge about their rights and obligations and the stronger and/or louder their voices when demanding that the law play its role in health sector more appropriately. The existence of the rights and obligations as the consequence of agreements between doctors and the relevant patients sooner or later may lead to medical disputes. This, of course, brings impacts on medical professionals. Doctors are worried and afraid of giving medical services to the patients. In a number of non-formal interviews the author has conducted with a number of doctors, it is found that this may result in another broader and alarming impact, that is doctors perform above the standard practices to avoid risks of being claimed and/or charged by the patients, which in turn leads to higher costs, brings about other impacts and adversely affects the public themselves as the consumers of health services. Essentially, in legal relations, both parties should be in a balanced position; however, in various disputes arising we frequently found imbalance in the relations between doctors and patients.

Based on the description above, the problem this writing addresses is how the principle of balance is implemented in the relationship between doctors and patients in respect of legal protection for doctors and patients as the parties involved in medical service relations.

2 DISCUSSION

The state regulates the life of its citizens to create fairness, order and peace in the public. The rules providing the life of citizens are contained in the prevailing laws and regulations based on the principles that underlie the creation of such laws and regulations. The relationship between the state and its citizens will create rights and obligations. Legal protection will become the right of the citizens and the obligation of the state. As such, legal protection will become an essential element of a state which is based on the rule of law as it aims to protect the citizens from injustice. The legal protection given is the impartial one. Accordingly, the state must implement and adopt the principle of balance to ensure that the interests of both parties are not adversely affected and legal order among the people as the objective of law can be attained and maintained.

In the legal theory generally adopted these days, it is admitted that a legal principle, in addition to laws and regulations, traditions, and court order, is also considered a source of law (Budiono, 2006). Besides having particular characteristics, the principle of balance consistently leads to logical truth and is adequately concrete in nature. Based on this consideration, the idea that develops that the principle of balance can be perceived as a reasonable, proper and just principle is later received as the basis of juridical bound. The principle of balance can be classified as principle of justice, given that the essence of such balance is also justice for the interests of the respective parties.

With respect to health services, in general, health services can be divided into two kinds, namely medical services and public health services. According to Leavell and Clark (1965), those two kinds of services have their own characteristics and/or features. Generally, it is said that medical services can be managed and/or administered individually with the main purpose to cure illnesses and/or diseases (curative) and to rehabilitate health (rehabilitative) and the main target is individuals, whereas public health services are, in general, managed and/or administered collectively in an organization, and in some occasion this involves the potential of general public and prevents illnesses and/or diseases and the main target is the public (Azrul, 1992).

Basically, the relationship between doctors and patients is included in the scope of the contract law. In the doctor–patient relationships or the so-called therapeutic relationship, there is a contract (though not in writing) between the doctor and the patient with respect to medication and treatment of illness and/or diseases, and between the patient and the hospital in respect of health services with facilities and infrastructures provided in accordance with the designated standards. Bound by an agreement, the relationship surely creates rights and obligations of both the doctor and patient. Both the health care provider and the health care receiver are the subject of law who have equitable obligations in accordance with the principle of equality before the law and this is provided in Article 1320 of the Indonesian Civil Code on the legal requirements of agreements. The performance of rights and obligations is likely to trigger or result in disputes between doctors and patients or the so-called medical disputes.

An effort to prevent such medical disputes from occurring is made by implementing the principle of balance in medical services. First, the principle of balance is implemented in the communication pattern of such therapeutic relationship. The communication pattern of mutual participation in this model is based on the idea that every person has equitable dignity and rights so that the patients will consciously be active and take part in their own medication.

As a matter of fact, this kind of relationship is rather difficult to be implemented on patients with poor educational and social backgrounds as well as on children or patients with a particular mental disorder. However, such kind of relationship should be developed both by doctors and patients (Azrul, 1992, p. 92).

The communication pattern of mutual participation is the one that respects the rights and obligations of both parties: doctors and patients. With this kind of relationship, the patients will give information necessary for the doctors to determine the best type of medication for the patients. Doctors also respect the rights of the patients by giving them the right, clear, and full information. Such information should be given by considering the background of the patients. Based on the information given to the patients, they are given the opportunity to choose the type of medication they will go through based on the information on the main points of the patients' diseases, and the doctors should supplement the same with the types of medication to be administered, prospect of success, other alternative medications and the potential risks. As such, the patients can decide the best medication for them as they have the right to reject or to accept the medication effort to be administered. Adoption of this communication pattern of mutual participation is the proof of implementation of the principle of balance in the therapeutic relationship.

Medical services can also mean a balance between the objectives and the facilities, between the facilities and the results, between the benefits and risks arising from the medical efforts taken. This principle is closely related to justice and/or fairness (Isfandyarie, 2006). The legal protection for the patients is expressly provided in laws. Legal protection for doctors is imperative in an effort to realize the operations of medical practices as mandated in the Law on Medical Practices. This also aims to attain a balance in the granting of legal protection for both parties. Article 2 of the Law on Medical Practices expressly stipulates that medical practices shall be conducted in accordance with some principles, including the principle of balance. Article 3 of this Law also stipulates that the regulation of medical practices provided in this law aims to give legal protection and certainty to the doctor and the patient as well.

Second, based on the principle of equality before the law, legal protection based on the principle of balance in giving medical services can be implemented by applying informed consent. Before performing medical treatment, doctors have to perform their obligation to give all information relating to such treatment including the main points of illnesses and/or diseases, prospect of success, other alternative medications and the potential risks, as well as other matters relating thereto. Based on the information from the doctors, the patients have the right to give consent or rejection to such treatment. This consent is the implementation of the principle of balance in the relationship between doctors and patients. If any later date, it is found that the patients are adversely affected by such treatment, either in the form of physical disabilities or death which later can be proven that it happened due to the mistake or negligence of doctors resulting from violations of professional standards, standard operating procedures or medical services standards, the relevant doctors will be subject to legal discharges. This aims to give legal protection to the patients. However, as long as doctors have done their work in accordance with the standards as described above, they are entitled to legal protection as well. In view of the foregoing, doctors and patients have equitable rights in getting legal protection.

The meaning of legal protection for doctors is not the elimination of criminal sanctions on them, as it stands to reason that doctors entitled to legal protection are those who have performed all of their obligations. Doctors are required to perform their medical practices in accordance with the professional standards, standard operating procedures or medical services standards. If in a medical dispute, it can be proven that there is a mistake or negligence on the part of the doctors in performing their profession, the protection will surely be given and enforced. Doctors must be accountable for all their medical mistakes and/or negligence if it is proved that there is a mistake and/or negligence adversely affecting the patient in the form of physical disabilities or death.

However, the frequent occurrences of malpractices that come to the surface, particularly those broadcast or blown up by the mass media have caused doctors to be defensive when giving medication. Law enforcers need a deeper understanding that medical disputes cannot be treated the same as other disputes in general. Not all medical practices that cause permanent disabilities or death are due to the mistakes or negligence of doctors as in the medical world there is a term called medical risks. It is this difference that should be understood by the law enforcers to enable them to enforce law in a just and impartial manner to both parties. A medical risk is information that should be conveyed or told to the patients or their family when an informed consent is being made and/or given. Worries and anxiety that make doctors feel insecure and restless when doing their work sooner or later will lower the quality of medical services in Indonesia. Doctors and patients are the subject of law in the therapeutic relationship; as such, the law should give equitable protection to both parties. Doctors and patients must fulfill their obligations and get their respective rights as mandate in the law to enable the state to give legal protection to both parties.

3 CONCLUSION

The implementation of the principle of balance in the relationship between doctors and patients should start from the relationship between both parties, by implementing a mutual participation pattern, or a relationship that respects the rights and obligations of both parties: doctors and patients. In this pattern, the patients give the information required by doctors to choose and determine the best medication for the patients. Doctors also respect the rights of the patients by giving them the right, clear, and full information. This mutual participation pattern is in line with the provisions of the Law on Medical Practices and other laws relating to medical services. The principle of balance must be applied when giving legal protection to doctors and patients. In the therapeutic relationship, doctors and patients have equitable capacity as the subject of law; therefore, the law must give equitable protection to both parties.

REFERENCES

Azrul, A. (1992) Menjaga mutu pelayanan rawat jalan [Maintaining The Quality of Service for Outpatient]. *Majalah Kesehatan Masyarakat Indonesia* [*Journal of the Indonesian Public Health Association*], XX (4), 196–201.

Budianto, A. & Utama, G.I. (2010) *Aspek Jasa Pelayanan Kesehatan dalam Perspektif Perlindungan Pasien*. Bandung, Karya Putra Darwati.

Budiono, H. (2006) *Asas Keseimbangan bagi Hukum Perjanjian di Indonesia*. Bandung, Citra Aditya Bakti.

Indonesia (1945) *Undang-undang Dasar 1945* [*The 1945 Constitution of Republic of Indonesia*].

Indonesia (2004) *Undang-undang No. 29 Tahun 2004 tentang Praktik Kedokteran* [*The Law Number 29 of 2004 regarding Medical Practices*].

Indonesia (2004) *Undang-undang No. 36 Tahun 2009 tentang Kesehatan* [*The Law Number 36 of 2009 regarding Health*].

Indonesia (2009) *Undang-undang No. 44 Tahun 2009 tentang Rumah Sakit* [*The Law Number 44 of 2009 regarding Hospitals*].

Isfandyarie, A. (2006) *Tanggung Jawab Hukum dan Sanksi Bagi Dokter Buku I*. Jakarta, Prestasi Pustaka.

Kitab Undang-Undang Hukum Perdata [*Burgerlijk Wetboek Voor Indonesie; Indonesian Civil Code*]. Translated by Subekti. Jakarta, Pradnya Paramita.

Leavell, H. & Clark, A.E. (1965) *Preventive Medicine for Doctors in the Community*. New York, McGraw-Hill.

Ratman, D. (2012) *Mediasi Non Litigasi terhadap Sengketa Medik dengan Konsep Win-Win Solution*. Jakarta, Elex Media Komputindo.

Suryani, B. (2013) *Panduan Yuridis Penyelenggaraan Praktik Kedokteran*. Jakarta, Dunia Cerdas.

Research report on the knowledge, experience, and attitude of consumers towards counterfeit medicine in Jakarta: A legal analysis

H. Marlyna & A. Sardjono
Faculty of Law, Universitas Indonesia, Depok, Indonesia

ABSTRACT: The issue of counterfeit medicines is a growing concern in developing countries, including Indonesia. In 2011, according to International Pharmaceutical Manufacturers Group (IPMG), 11% of medicines marketed in Indonesia were counterfeit products, causing 7.6 trillion rupiah in losses. This paper provides an analysis of consumer knowledge, experience, and attitutes towards counterfeit medicines viewed from the legal perspective. Based on the survey research method applied, questionnaires were used to collect data from 200 consumers above the age of 17 residing in Jakarta, who were selected using the quota sampling technique. This research concludes that consumers are incapable of protecting themselves from the harm of counterfeit medicines. Consumers' lack of knowledge concerning counterfeit medicines including the relevant regulations poses serious health risks to Indonesian consumers. Only a small number of consumers stated their experience of purchasing counterfeit medicines; however, it happened intentionally. This research recommends that BPOM and other government agencies as well as pharmaceutical companies cooperate to combat counterfeit medicines in order to protect consumers from the hazards to their health and safety.

1 INTRODUCTION

It is indisputable that counterfeit medicines may potentially harm consumer safety. Most counterfeit medicines contain toxic products that can have immediate and harmful effects on people exposed to them. Many counterfeit medicines use low cost material not intended for human consumption, and they are produced by using highly inappropriate non-sanitary machinery (Davison, 2011). Even though recent counterfeit medicines use less toxic materials, they are still potentially harmful to consumers causing them to spend more time and money to be cured from their illnesses. Such medicines can also cause death due to their inefficacy.

Counterfeit medicines are causing an increasing concern as they pose a threat to consumers' health and the reputation of pharmaceutical industry. In 2006 counterfeit medicines ranged from less than 1 percent of sales in developed countries, to over 10 percent in developing countries, depending on the geographical area (World Health Organization [WHO], 2016).

In Indonesia, based on the data obtained as the result of supervision by the National Agency for Drug and Food Control (Badan Pengawas Obat dan Makanan [BPOM]), in 2013–2015 counterfeit medicines found were dominated by medicines for erectile dysfunction (ED), antibiotics, antipyretic-analgesic, antihypertensive, and antimalarial medicines. Branded medicines of relatively higher prices tend to be more exposed to counterfeiting practices compared to generic medicines. The most frequently counterfeited branded medicines include Blopress, Cialis, Viagra, Ponstan, Bloppres, Incidal, ODI Diazepam, Anti-Tetanus Serum, and Nizoral. At the same time, based on the supervision conducted by BPOM during the period of January–June 2016, 17 branded illegal medicines including counterfeit medicines were found and dominated by vaccine, anti-tetanus serum, and erectile dysfunction medicines (BPOM, 2016).

Based on a study on the Economic Impact of Counterfeiting in Indonesia by LPEM FEUI and MIAP (Masyarakat Indonesia Anti Pemalsuan/Indonesian Anti Counterfeting Society) in 2013, out of 6 categories of most frequently counterfeited products (software, cosmetics, pharmaceutical, fashion, leather product, and food/beverages), 3.80% of counterfeit pharmaceutical products were sold in Indonesia. The said percentage is higher compared to the data found in the previous research conducted in 2010, namely 3.5%. (MIAP & LPEM FEUI, 2013).

Counterfeit medicines continue to pose threats to consumers in Indonesia. They are not only available at street peddlers, on the internet, or black market, but also in pharmacies. The research titled "Victory Project" conducted by the Faculty of Medicine of Universitas Indonesia/Cipto Mangunkusumo Hospital in 2011–2012 was aimed at quantifying the extent of counterfeit sildenafil in Indonesia. The study found that 45% of 100 mg sildenafil tablets in Indonesia were counterfeit and were widely distributed by street peddlers, drugstores, and pharmacies. Counterfeit sildenafil tablets were mostly found at "illegal" outlets, especially at street peddlers (100%) and in drugstores (56%). Pharmacies also had 13% counterfeit sildenafil. Among 12 sildenafil tablets purchased from the internet, 4 were counterfeit (Taher & Setiawati, 2013).

Indonesia has some regulations which can be used to combat counterfeit medicines as well as protecting consumers, including the Consumer Protection Law (the Law No. 8 of 1999), Health Law (the Law No. 36 of 2009), and Mark Law (the Law No. 20 of 2006). However, despite the existence of those regulations, the problem of counterfeit medicines continues to pose threats in Indonesia.

Despite the numerous studies on counterfeit medicines in Indonesia, there has been a limited number of research conducted to understand consumers' knowledge, experience, and attitude in purchasing and consuming counterfeit medicines particularly in Indonesia. This study has been constructed to explore and understand the relationship between consumers' knowledge, experience, and attitude towards counterfeit medicines in Jakarta. It is particularly aimed at examining consumers' knowledge on regulations and factors affecting the consumers' attitude towards pressing for legal action.

2 LITERATURE REVIEW

2.1 *Counterfeit medicines*

There is no universally accepted definition of 'counterfeit medicine'. The term 'counterfeit' that was firstly used as a matter of public health issue which is substandar medicine. The term has now become a legal term associated within intellectual property legislation especially trademark protection as in 1992 WHO provides definition on counterfeit medicine as:

> "one which is deliberately and fraudenlently mislabeled with respect to identity and/or source. Counterfeiting can apply to both branded and generic products and counterfeit products may include products with the correct ingredients or woth the wrong ingredients, without active ingredients, with insufficient (inadequate quantities of ingredient(s) or with fake packaging" (WHO, 2006).

The revision by International Medical Products Anti-Counterfeiting Taskforce (IMPACT) in 2008 has created a controversy regarding the concept of counterfeiting, raising the question as to whether it is an issue of intellectual property or public safety in relation to the quality, safety and efficacy of medicines. Some countries confront counterfeiting as an issue of intellectual property. They express a concern WHO's involvement in the enforcement of privately owned intellectual property rights without the endorsement of all member states in the WHO, by providing the secretariat for IMPACT. As WHO's mandate is to protect public health, WHO should combat substandard drugs (Clift, 2010).

The word counterfeiting has a literal meaning in the field of intellectual property, related to trademark infringement. The World Trade Organization (WTO) defines counterfeiting as:

"Unauthorized representation of a registered trademark carried on goods identical or similar to goods for which the trademark is registered, with a view to deceiving the purchaser into believing that he/she is buying the original goods" (WTO, n.d). Therefore counterfeiting is a trademark infringement, as an "unauthorized" use of a trademark that is "not only infringes a trademark owner's legitimate rights but does so in such an egregious way as to attract criminal penalties" (Clift, 2010).

In Indonesia, the definition of counterfeit medicines is stipulated in the Regulation of the Minister of Health No. 1010/MENKES/PER/XI/2008. Counterfeit medicines are medicines produced by unauthorized parties pursuant to the prevailing law or the production of medicines falsely indicating the identity of other registered medicines.

2.2 Consumer knowledge

Brucks (1986) defines knowledge as "a complicated construction characterized by the structure and the content of the information stored in the memory". Related to consumer's knowledge, the knowledge is "all the information related to the product and to the market which are stored in the long-term memory of the consumer allowing him to act on the market" (Korchia as cited by Bouzaabia & Salem, 2010).

There are three categories of consumer knowledge: "(a) subjective knowledge, perceptions which a person has of what he knows; (b) objective knowledge, a measure what an individual actually knows; and (c) prior experience, the amount of purchasing or usage experience a consumer has with the product" (Brucks, 1985).

2.3 Consumer experience

Most literatures use the term 'customer experience'. However, in this study the writer uses the term 'consumer experience' with the same understanding. A customer who is buying or consuming a service or product will have an experience "good, bad, or indifferent" and "a service always comes with an experience" (Carbone & Haeckel, 1994).

In this research, consumers' experience is merely defined by the actual living through an event, the real life as contrasted with the ideal or imaginary. It is what actually happened to him/her, or (still in the same sense) of his/her prolonged, cumulative life experience (Erlich, 2003).

2.4 Consumer attitude

In term of marketing, "an attitude is as general evaluation of a product or service formed over time" (Solomon, Russell-Bennet, & Previte, 2013). An attitude is a personal motive affecting a consumer's shopping and buying habits. Perner (2010) defines consumer attitude simply as "a composite of a consumer's beliefs, feelings, and behavioral intentions toward some object within the context of marketing".

Moreover, "attitude" is "...a learned predisposition to behave in a consistently favorable or unfavorable manner with respect to a given object" (Schiffman & Kanuk, 1997). An attitude correlates with one's intentions, which predict a behavior (Ajzen & Fishbein, 1980). Related to counterfeit products, if a person's attitude is considered as favorable the person would purchase a counterfeit product. On the other hand, if a person's attitude is considered as unfavorable, the person would not consider purchase a counterfeit product.

3 METHODS

The research was qualitative research to analyze the relations between variables. These variables were subsequently measured by using statistic procedures. The variables used were knowledge and experience with regard to the distribution of counterfeit medicines. Those variables were then linked to the attitude of the consumers of medicines towards the distribution

Table 1. Research sample.

Criteria	Data of DKI Jakarta in 2010	Total sample
No educational background	10%	4 respondents
Graduated from Elementary School	18%	7 respondents
Graduated from Secondary School	20%	8 respondents
Graduated from High School	38%	15 respondents
Graduated from University/College	18%	6 respondents

of counterfeit medicines in Jakarta. This cross-sectional research was conducted during the period of May-June 2016. It was based on a survey using questionnaires to examine a sample of a population. The targeted population was 200 medicine consumers in Jakarta, out of a total of 5 areas in Jakarta. The respondents were consumers live in Jakarta in the aged 17 years old and above. This study used the non-probability sampling design. The respondents were divided into 5 categories based on the percentage of educational background in Jakarta (Table 1).

4 RESULTS

4.1 Knowledge of counterfeit medicines

The majority of respondents (83.5%) did not have any knowledge to distinguish between original and counterfeit medicines. Only 16.5% had the knowledge to do so. Those respondents graduated from high school and university/college. Not suprisingly 100% of respondents having no educational background did not have the capability to distinguish between original and counterfeit medicines. Therefore, we can conclude that the higher the educational background, the better the consumers will be to differentiate between original and counterfeit medicines.

The low level of knowledge among consumers enabling them to distinguish between original and conterfeit medicines corresponds to the low level of knowledge of medicines in general, especially in relation to medicine classification. Only 24% of the respondents knew about medicine classification. In Indonesia, medicines are classified into four categories: over the counter, restricted over the counter, prescription, and psychotropic & narcotics medicines. Knowledge about medicine classification is useful to prevent consumers from buying counterfeit medicines. The classification tells consumers where they should buy the medicines. Therefore, for prescription medicines, consumers should only buy medicines at pharmacies. Prescription medicines bought at drugstores are at the risk of being counterfeited through illegal distribution.

The low level of knowledge to distinguish between original and counterfeit medicines is supported by the result that 87.5% of the respondents never received any information regarding how to prevent exposure to counterfeit medicines. Only 12.5% of the respondents received information either from the government or pharmaceutical companies. Based on the findings of this research, the low level of knowledge of the classification of medicines, as well as the inability to differenciate between original and counterfeit medicines, exposes consumers to the risk of purchasing counterfeit medicines.

A majority of the respondents (93%) stated that counterfeit medicines infringe the law. However, out of the 93%, 20% respondents stated that they did not know which specific law is infringed by counterfeit medicines. According to the respondents, counterfeit medicines infringe the mark law (23%), consumer protection law (47%), and health law (57%).

84.5% of the respondents understood that they should make a report after finding counterfeit medicines. Nevertheless, only 55% respondents know where they should file such report to, and they mentioned that more than one institutions deal with such report. A majority of the respondents (34%) mentioned that they should file the report to BPOM (24%), the police (24%), the National Consumer Protection Agency (Badan Perlindungan Konsumen Nasional/BPKN – 16%), non-governmental organizations (13.5%), sellers of medicine (13%), pharmaceutical companies (5.5%), Consumer Dispute Settlement Agency/Badan Penyelesaian Sengketa Konsumen (4.5%), the print media (3.5%), and the Ministry of Trade (1%).

4.2 Consumer experience

Most of the respondents stated that they never encountered counterfeit medicines (85%). However, 10.5% said they might have bought counterfeit medicines accidentally. In such case, they did not have the intention to do so and they did not realize that the medicines they had purchased were counterfeit. There were several reasons as to why they think that they might have purchased counterfeit medicines: the medicines did not cure their illnesses (67%), they experienced some side effects (38%), relatively low price (38%), suspicious packaging (33%), different taste (29%), and different shape (24%).

According to BPOM 2007, in Indonesia most counterfeit medicines are mislabelled with respect to the identity or the source and without an active ingredient, the packaging and label of generic medicines are replaced by branded medicine having similar active ingredients, and recycled medicine packaging is used from original medicines.

4.3 Consumer attitude

This research also reveals the factors being considered by consumers when purchasing medicines, namely the type of the medicines, trademark, the place to buy, and price. Interestingly, it has been found that consumers in Jakarta are not primarily motivated to purchase counterfeit medicines because they cannot afford to purchase legitimate medicines. In fact, price is the last factor they seem to consider in purchasing medicines. It supports the finding that consumers might have bought counterfeit medicines unintentionally. It is different from the finding of another research on Sudanese consumers showing that "non-accessibility and/or unaffordability of legitimate drugs may be the main contributors (strongest influence) encouraging purchase behaviour of counterfeit medicines in developing countries (Alfadi, Ibrahim, & Hassali, 2012). Therefore, "consumers who have weak accessibility and/or cannot afford the price of legitimate medicines are more likely to acquire stronger intent to purchase counterfeit medicines" (Alfadi, Ibrahim, & Hassali, 2012).

The respondents in this research were asked whether the news on counterfeit medicines has affected their attitude in purchasing medicines. A majority of the respondents (78%) would not purchase the medicines after learning from the news that medicines with certain trademark were found to have been counterfeited. They were afraid that the medicines would cause side effects to the consumers. However, 22% of the respondents stated that they would still purchase the medicines which have been proven to be effective in curing their illnesses and they would be confident when purchasing such medicines at a legitimate place.

Furthermore, respondents in this research were asked whether they had ever encountered counterfeit medicines and what action they took upon it. A majority of the respondents (52.4%) said that they would not take any legal action, 33.3% said they would inform others, while only 14.3% would file a report to the authorities. Most consumers who encountered counterfeit medicines (70%) reported their findings to the pharmacist or the seller of medicine, 20% reported it to the police, while 10% reported such matter to the BPOM. However, those who made the reports did not follow up, and no further action was taken.

This finding supports the previous study that majority of the victims are unlikely to take any legal actions. This result supports the previous study that "consumers lack knowledge and understanding of how and where to report cases of counterfeit medicines", therefore, consumers who encounter counterfeit medicines only tell pharmacists and others throw away the medicines and do nothing else (Bwemelo & Mashenene, 2015).

Consumers have at least 7 reasons for not taking any action after encountering counterfeit medicines because making a police report is costly (41%); they do not know where to report the case (35%); it is a waste of time (20%); they are afraid of the police or the authorities

(24%); a court procedure is time-consuming and complicated (24%); and there is a lack of confidence in the court system (12%).

The finding shows that consumers think that the legal procedure in the case of counterfeit medicines is not consumer friendly. According to Shofie (2009), consumers are reluctant to litigate since consumer protection norms are still unclear, and court procedures are complicated, time consuming, and costly. Consequently, consumers are likely to stay away from conflict, even though their rights have been infringed.

5 DISCUSSION

This study shows that consumers lack of knowledge of how to distinguish between counterfeit medicines and original medicines; the classification of medicines; and information on the regulations related to counterfeit medicines; and information on where to file a report upon encountering counterfeit medicines. In relation to consumers' experience related to counterfeit medicines, only 10.5% of the respondents have experienced buying counterfeit medicines. They became suspicious of the originality of the medicines based on certain factors, such as the occurrence of side effects; unreasonably low price, packaging, different taste, and different shape.

The results of this research confirm that in Jakarta price is not the most important factor for consumers when purchasing medicines. Therefore, price is not a driving factor in triggering the distribution of counterfeit medicines. It is evident from the findings of this research that consumers who encounter counterfeit medicines, although knowing that they should make a report when they find counterfeit medicines, they are unlikely to take legal action for some reasons, which support the fact that the legal procedure in the case of counterfeit medicines is not consumer friendly.

The research found that consumers are unable to protect themselves from counterfeit medicines. According to United Nations, "consumers often face imbalances in economic terms, educational levels, and bargaining power". On the other hand, consumers should have the right of access to non-hazardous products, as one of the consumers' basic rights is "the protection of consumers from hazards to their health and safety" (United Nations, 2003). This right has also been regulated in Indonesia in Law No. 8 of 1999 regarding Consumer Protection.

Therefore, it should be the obligation of the government and pharmaceutical manufacturers to protect consumers from the danger of counterfeit medicines. BPOM and other government authorities should cooperate with pharmaceutical companies to create dissemination programs involving consumers to prevent them from buying counterfeit medicines. Moreover, law enforcement should be prioritized by the Indonesian government in addition to legal action initiated by pharmaceutical companies to combat counterfeit medicines in cooperation with stakeholders and international organizations.

The research found that consumers' attitude in buying medicines in Indonesia is mostly influenced by the information they obtain through the media. Once they learn that a certain medicine is found to have been counterfeited, most of them would be unlikely to buy the type of medicine bearing the same trademark. Therefore, counterfeit medicines might harm the reputation as well as causing losses of revenues of pharmaceutical companies. As the trademark owners, pharmaceutical companies have the obligation to fight counterfeit medicines in order to protect their consumers as well as their own interest. The traditional justification of the trademark law is the protection of both consumers and trademark owners (McKenna, 2007).

6 CONCLUSION

This research concludes that consumers are unable to protect themselves from the harms of counterfeit medicines. This study shows that consumers do not have any knowledge of how to distinguish between counterfeit medicines and original medicines; the classification of medicines; information on the regulations relating to counterfeit medicines; and information on where to file a report upon encountering counterfeit medicines. Consumers' lack of knowledge concerning counterfeit medicines poses serious health risks to Indonesian consumers.

Although only a small number of consumers conveyed their experience in purchasing counterfeit medicines, they happened unintentionally. Moreover, even though consumers know that they should file a report when they find counterfeit medicines, they are unlikely to take legal action since the legal procedure in the case of counterfeit medicines is not consumer friendly.

Therefore, the research recommends that the BPOM and other governmental agencies as well as pharmaceutical companies cooperate to combat counterfeit medicines to protect consumers from hazards to their health and safety. Combating counterfeit medicines benefits consumers as well as the pharmaceutical companies as the tradewark owners.

REFERENCES

Ajzen, I. & Fishbein, M. (1980) *Understanding Attitudes and Predicting Social Behavior*. Englewood Cliffs, NJ, Prentice-Hall.
Alfadl, A.A., Ibrahim, M.I.M. & Hassali, M.A. (2013) Scale development on consumer behavior toward counterfeit drugs in a developing country: a quantitative study exploiting the tools of an evolving paradigm. *BMC Public Health*. [Online] 13 (1), 829. Available from: doi: 10.1186/1471-2458-13-829.
Badan Pengawas Obat dan Makanan (2016). *Badan POM terus perangi peredaran obat ilegal di Indonesia.* [Online] Available from: http://www.pom.go.id/new/index.php/ view/pers/316/.
Bouzaabia, R. & Salem, I. (2010) The relation between the consumer's knowledge and the browsing behavior. *Annals of "Dunarea de Jos" University of Galati Fascicle I: Economics and Applied Informatics*, 2, 59–70.
Brucks, M. (1985) The effects of product class knowledge on information search behavior. *Journal of Consumer Research*, 12 (1), 1–16.
Brucks, M. (1986) A typology of consumer knowledge content. In: Lutz, R.J. (ed.), *Advances in Consumer Research Volume 13*. Provo, Association of Consumer Research. pp. 58–63.
Bwemelo, G.S. & Mashenene, R.B. (2015) Consumers' response toward counterfeit medicines in Dar Es Salaam City, Tanzania: A comparative analysis. In: *Proceeding of the Second European Academic Research Conference on Global Business, Economics, Finance and Banking, 3–5 July 2015, Zurich, Switzerland*. Chennai, Global Business Research. p. Z546
Carbone, L.P. & Haeckel, S.H. (1994) Engineering customer experience. *Marketing Management*, 3 (3), 8–19.
Clift, C. (2010) Combating counterfeit, falsified and substandard medicines: defining the way forward? London, Chatham House. Technical Report.
Davison, M. (2011) *Pharmaceutical Anti-Counterfeiting: Combating the Real Danger from Fake Drugs*. New Jersey, Wiley.
Erlich, H.S. (2003) Experience-what is it? *The International Journal of Psychoanalysis*, 84 (5) 1127–1147.
Indonesia (1999) *Undang-undang No. 8 Tahun 1999 tentang Perlindungan Konsumen* [*The Law Number 8 of 1999 regarding Consumer Protection*].
Indonesia (2001) *Undang-undang No. 15 Tahun 2001 tentang Merek* [*The Law Number 15 of 2001 regarding Mark*].
Indonesia (2006) *Undang-undang No. 36 Tahun 2006 tentang Kesehatan* [*The Law Number 36 of 2006 regarding Health*].
Masyarakat Indonesia Anti Pemalsuan [MIAP] & Lembaga Penyelidikan Ekonomi dan Masyarakat Fakultas Ekonomi Universitas Indonesia [LPEM FEUI] (2010) *Dampak Ekonomi dari Kegiatan Pemalsuan di Indonesia*. Jakarta, LPEM FEUI. Report.
McKenna, M.P. (2007) The normative foundations of trademark law. *Notre Dame Law Review*, 82 (5), 1839–1857.
Perner, L. (2010) *Consumer behavior: the psychology of marketing*. [Online] Available from: http://www.consumerpsychologist.com/.
Schiffman, L.G. & Kanuk, L.L. (1997) *Consumer Behavior,* 6th edition. Englewood Cliffs, NJ, Prentice Hall.
Shofie, Y. (2009) *Kapita Selekta Hukum Perlindungan Konsumen di Indonesia,* 3rd edition. Bandung, Citra Aditya Bakti.
Solomon, M.R, Russel-Bennett., R. & Previte, J. (2013) *Consumer Behavior: Buying, Having, Being*, 10th eds. Frenchs Forest, Pearson Australia.

Taher, A. & Setiawati, A. (2013) Victory project: a study of counterfeit PDE5 inhibitor (sildenafil) in Indonesia. *Acta Medica Indonesiana*, 45 (4), 290–294.

United Nations. (2003) UN: United Nations guidelines for consumer protection (as expanded in 1999) Resolution 39/248. New York, United Nations. Available from: http://www.un.org/esa/sustdev/publications/consumption_en.pdf.

World Health Organization (2006) WHO: Counterfeit Medicines: an update on estimates. Available from http://www.who.int/medicines/services/counterfeit/impact/TheNewEstimates.pdf.

World Health Organization (2016) WHO: Substandard, spurious, falsely labeled, falsified and counterfeit (SSFFC) medical products. Fact Sheet Number 275. Available from http://www.who.int/mediacentre/factsheets/fs275/en/.

World Trade Organization (n.d) WTO: *Glossary Term*. [Online] Available from: https://www.wto.org/english/thewto_e/glossary_e/counterfeit_e.htm.

Research report on protection of geographical indications in Indonesia: Advantages and challenges

M.M. Gabor & A. Sardjono
Faculty of Law, Universitas Indonesia, Depok, Indonesia

ABSTRACT: The protection of Geographical Indications (GI), which has its traditional and cultural roots in European countries, is being accepted in Asian countries. The concept and legal protection of GIs were introduced into the Indonesia's legal system following Indonesia's ratification of the WTO Agreement in 1995. Indonesia provided for GI protection in the Trade Mark Law No. 15 Year 2001, followed by the Government Regulation No. 51 of 2007 concerning GI registration. Almost nine years later, there is a need to assess the advantages and challenges related to the implementation of Indonesia's GI protection, namely as follows: first, does the concept of GI actually promote Indonesia's interest; second, what is the reason behind the relatively low number of GI registrations; and third, how can law be used to promote more effective implementation? The article sets the ground for a more in-depth analysis of the issue at a conceptual and theoretical level. The analysis is based on a combination of socio-legal research, philosophical reflection with a critical reflective approach, with a particular focus on historical, conceptual, paradigm development. Legal and non-legal contributing factors along with the pragmatic implications of the same are also considered.

1 INTRODUCTION

The concept of geographical indications (hereinafter referred to as GI) as a form of intellectual property rights (hereinafter referred to as IPR) is a relatively new concept in Indonesia's legal system. Other more traditional forms of IPR, particularly trade mark and copyright laws, date from pre-independence era. The legal protection of GI was adopted in Indonesian national law after Indonesia ratified the Agreement Establishing the World Trade Organization (the Law of the Republic of Indonesia Number 7 Year 1994 concerning the Ratification of the Agreement Establishing the World Trade Organization (Persetujuan Pembentukan Organisasi Perdagangan Dunia), the State Gazette Year 1994 Number 57, the Supplement to State Gazette Number 3564) and the Trade Related Aspects of Intellectual Property Rights (hereinafter referred to as TRIPs) as one of its attachments which became effective as from January 1, 1995. To fulfill Indonesia's obligation under TRIPs, several of its existing IPR laws were amended. The protection of GIs was integrated into the Law Number 15 Year 2001 concerning Trade Marks (the State Gazette Year 2001 Number 110, the Supplement to State Gazette Number 4131) (hereinafter referred to as Trade Mark Law). The Trade Mark Law adopts the constitutive approach requiring GI registration (Art. 56 paragraph (2)), mandating the issuance of a government regulation for the registration procedure and mechanism (Art. 56 paragraph (9)). The Government Regulation Number 51 concerning Geographical Indications (hereinafter referred to as GR 51/2007) was issued seven years later, on September 4, 2007.

Despite the government's efforts, foreign assistance, and high expectations particularly among academicians, the comprehensive provisions of GR 51/2007 fell short of optimally achieving the Trade Mark Law's objective of providing effective GI protection in Indonesia, as evident from the relatively low number of GI registrations occurring thereafter. A total of

46 GIs have been registered to date, and 6 out of the total number are foreign GIs (Ramli, 2016), in comparison to India which has a total of 272 GIs registered through March 2016, and only 10 out of which are foreign (Intellectual Property India, 2016). Referring to the familiar aphorism, 'Too often, people complain that they have good laws, but they are poorly implemented (Seidman & Nalin, 2001), there is a need to assess the roots of relatively slow increase in the number of GIs registered even after rules and regulations have been put into place, and after many years of implementation.

This article aims to answer three specific research questions. First, does the existing legislation provide adequate basis for effective GI protection in Indonesia? Second, is the implementation of existing GI legislation adequately supported by GI consciousness at various levels of stakeholders? And third, how can law be used to promote more effective implementation of GI protection?

2 LITERATURE REVIEW

As several authors point out, the roots of the concept and definition of GI as we know it today under the TRIPs Agreement can be traced back to wine regulations in the 18th century France in Europe (Munsie, 2002). Commentators have argued that in the history of the privileges of French wine growers, as illustrated by 'Bordeaux' and 'Champagne', one finds the combination of elements of rural policy, guarantees of authenticity, and the search for competitive advantage, which marks the current WTO debate about global regime on GIs (Das, 2010). The influence of the AOC and the resulting GI system are thus closely related to economic, political, and socio-cultural factors. The AOC influence transpires throughout international agreements and conventions such as the 1958 Lisbon Agreement, EEC/UE regulations, and ultimately the WTO/TRIPs Agreement, to which Indonesia is a signatory.

2.1 *GI in Asian countries: 'norm migration' theory*

Wang speaks of the 'migration of the GI norm' and three levels of GI 'legal consciousness' in Asian countries, namely first, passive acceptance of the 'transplanted' IG norm; second, developing an awareness of their own GI interest; and third, taking a stance in the ongoing North-North discussion on the implementation of TRIPs provisions on GI. Wang's perspective is influenced by 'norm migration' theories associated with the concept of 'epistemic community', exploring actors involved in the process of norm migration and their interests (Wang, 2006).

2.2 *GI: Global struggle of concepts and paradigms*

The current GI definition and provisions under TRIPs have been the result of a compromise between the sui generis protection paradigm advocated by EC vis-à-vis the trade mark protection paradigm supported by the U.S. (Ayu, 2009). This explains in part the lack of a uniform concept and inconsistency in GI terminology, particularly in view of the scope and a higher level of GI protection as recognized by many commentators (Panizzon, 2006). As some commentators note, neither of these paradigms are free from conceptual difficulties. On the one hand, the U.S. proposed trade mark protection paradigm creates contradiction at the conceptual level due to the fundamental difference between the concept of GI, basically a collective right, and trade mark which is basically an individual right. On the other hand, the EU sui generis paradigm treats GI as a separate IP category and thus deviates from the existing IP system (Wang, 2013).

Gervais refers to it as 'irreconcilable differences' between the 'common-law approach to protecting certain GIs as trademarks and the (currently) mostly European approach of using a sui generis system to protect GI's' (Gervais, 2015). At the same time, Raustiala and Munzer argue that the international GI debate is 'primarily driven not by philosophical arguments but by political interests' pointing out that European governments are pushing forward for a higher standard of GI protection at the international level 'in an effort to protect traditional producers from increasing competition from abroad' (Raustiala & Munzer, 2007).

As the 'struggle' between the sui generis and trade mark regime GI protection at the global level continues, there are an increasing number of countries referred to as "Friends of GIs" led by India, the Checz Republic, and Swiss which support the idea of expanding a higher level of protection to GI products other than wines and spirits (Wang, 2006). Munzer and Raustiala argue against it, taking the position that there is a need to 'harmonize downward to the general TRIPS standard rather than upward to the absolute protection standard' (Raustiala, Munzer, 2007).

Their position is based on the argument that while recognizing GI as a particular type of IPR and as such deserving some protection in international law, the existing level of protection under TRIPs and the current demands of the European Union for even greater protection are unjustified. Their main argument is that at the core of the justification of GI protection, as that of trademarks, it is the consumer-based rationale; in other words, the protection of GIs, just as that of trademarks, is aimed at reducing confusion and limiting consumers' search costs in the marketplace (Raustiala & Munzer, 2007).

2.3 *GI and IP social justice theory*

While GI issues at the conceptual level evolve primarily around the approaches taken by the two main legal traditions, namely common law and European continental law, a majority of developing countries are increasingly seeking their own justification of GI protection (Kongolo, 2008). As pointed out by Mtima (2015) 'equitable access, inclusion, and empowerment are the essential and extrinsic values of every enterprise,' but they stand in contrast with 'IP law and economics' (Landes & Posner, 1989) which promote and perpetuate social inefficiencies in the IP protection apparatus. Through IP empowerment, IP social justice seeks to realign such a mechanism, thus recouping social losses and maximizing efficiency and the productive impact/output of the overall IP system (Mtima, 2015).

2.4 *GI protection in Indonesia: Proposed approaches*

Commenting on IP in general, Sardjono (2011) points out that "there is a generally prevailing view among common [Indonesian] people that IPR is an abstract, altogether strange concept", "a concept they do not seem to understand", conceding that the Government and legislators tend to take a formalistic stance when it comes to IPRs, viewing them "as a single and functional concept". The various attempts to explain the justification for GI protection in Indonesia include, among others, the cultural and rights approach by Sardjono (2011) and Ayu (2009) respectively; the national policy approach by Sasongko (2005); the consumer protection approach by Rudita (2011); and the effectiveness approach by Gabor (2012) in research undertaken by them respectively. They, as some other commentators, agree that there is a need for a sui generis GI law for effective GI protection in Indonesia.

3 RESEARCH TYPE AND METHODOLOGY

The complex nature of GI protection in Indonesia calls for research exploring its essence with a historical, philosophical, and doctrinal approach leading to a more in-depth understanding of la raîson d'être, the reason or justification for existence of GI protection in Indonesia. This is in line with the method typically applied in socio-legal studies, which consider law in the context of broader social and political theories. Socio-legal studies look at, among other things, whether and how law is implemented and enforced. By exploring law's connection with broader social and political forces—domestic and international—a perspective on ideology, culture, identity, and social life is gained. Socio-legal research involves interdisciplinary and multi-disciplinary approaches (University of Bristol Law School, Socio-Legal Research).

However, due to the limited scope of this article, the current research merely lays the ground for more in-depth and exhaustive doctrinal research, using comprehensive primary and secondary sources of data. Primary sources of information include statutory materials, laws, and regulations, international conventions and international agreements related to GI protection. Secondary sources of information include textbooks, reference books, studies, thesis, and dissertation written by other authors on the subject, legal periodicals, parliamentary debates, government reports, seminar and conference materials and reports, speeches, statistical data on GI, and excerpts from GI registry from relevant authorities. Such research will combine the historical, comparative, and trans-disciplinary research model with a philosophical approach typical of socio-legal research.

The current discussion puts an emphasis on two particular aspects, namely the challenges and advantages of GI rules in Indonesia, while setting the ground for exploring further the ideal basis for GI rules in Indonesia.

4 FINDINGS AND DISCUSSION

The strategic role and importance of GIs as the instrument of economic growth and national development have been acknowledged at the top national level, as well as by GI producers. Following is a brief overview of the advantages and challenges of GI protection in Indonesia from the vantage of GI producers and communities and the Indonesian government (DGIP, TCF, Future of Geographical Indications in Indonesia, 2016).

4.1 *GI producers and communities*

As GI success stories indicate, there have been considerable proven advantages of GIs, while certain challenges still remain. The success stories include, among other things, Preanger Java Tea registered GI, Sumbawa Honey registered GI, Lada Putih Muntok (Muntok White Pepper), and Arabica Coffee Gayo. In all instances, the respective communities have stated that they have enjoyed considerable benefits and advantages after obtaining a GI registration certificate which includes 'economic benefit', 'manifold price increases', 'revitalization of the industry', 'improved social welfare of both producers and traders', 'positive environmental impact', 'huge market demand', 'improved legal certainty'.

On the other hand, challenges include 'GI control' and 'dealing with misappropriation cases', 'lack of government support and financial assistance, particularly in the initial phases of GI development', as well as the 'local government's lack of sense of ownership'.

The words of the Chairman of Indonesian Geographical Association, Slamet Bangsadikusumah (2016), represent the overall view of GIs in Indonesia. After "having heard, seen, and felt the benefit of GI", there is no doubt that every effort needs to be made in order to ensure that GIs are sustainable and effective. GI guarantees originality and quality, and it brings negotiating power. GI protection is in line with the national agenda of Nawa Cita, which refers to the nine-priority agenda spelled out in Joko Widodo-Yusuf Kalla Presidential Campaign with the title: The Way of Change for Indonesia which is Sovereign, Autonomous, and Possesses Character; Vision Missions and Action Program [Jalan Perubahan untuk Indonesia yang Berdaulat, Mandiri dan Berkepribadian; Visi Misi dan Program Aksi]. GI requires technical skills, capital, quality awareness, synergy; human and financial resources as well as coordination at the national and regional levels. There is a need for a strategic policy to ensure that Indonesian GIs are truly effective. Clearly a formulated law/government regulation/local government regulation is also required in order to ensure all of those, including adequate allocations in the National and Regional Budget for effective implementation. There is a need for cooperation with academicians as well as with the bureaucracy, particularly by involving the local bureaucracy.

4.2 *The Indonesian government*

From the government's perspective, the statement of the Director for Trademark and Geographical Indication of DGIP sums up the current and future issues related to GI in

Indonesia. First, GIs are yet to be understood as an intangible asset. Second, GI supporting structure is still rather weak in terms of funding, human resources, and there is also a need for strengthening the role of Regional Governments as GI stakeholders. Third, there are challenges related to marketing both domestically as well as overseas, due to, among other things, lack of information. Fourth, consumers still lack GI awareness; hence, they are often not prepared to pay a higher price for GI products. Fifth, the regulatory framework requires further strengthening.

4.3 *Indonesia's stance in global GI*

As affirmed by the representative of the Ministry of Foreign Affairs, Indonesia belongs to the group of WTO member states supporting the idea of expanding GI protection to products other than wine and spirits. While waiting for the continuation of WIPO negotiations, Indonesia is studying the possibility to become a party to the Lisbon Agreement.

4.4 *Draft amendment to the Trade Mark Law*

The draft amendment to the Trade Mark Law currently being finalized by the Working Committee of the House of Representatives of the Republic of Indonesia (DPR-RI) includes more detailed GI provisions as compared to the GI provisions under the currently applicable Trade Mark Law (Damarsasongko, 2016).

4.5 *GI rule in Indonesia: The way forward*

The following is an analysis of GI protection in Indonesia based on three stages of GI legal consciousness as expressed by Wang (2007).

4.5.1 *1st stage of GI consciousness*
In the first stage of its GI consciousness, Indonesia adopted GI provisions under its Trade Mark Law separately from Collective Marks and Certification Marks. Thus, it recognized GI as a particular type of IPR. However, it failed to accommodate the specific character and the most prominent distinguishing features of GI vis-à-vis trade mark, namely the collective nature and cultural aspect of GI.

Following a seven year vacuum pending the issuance of implementing a government regulation, GR 51/2007 was issued. It presented a more comprehensive, sui generis type approach to GI protection, perhaps in an attempt to fill the gaps left by the previously existing GI provisions in the Trade Mark Law, particularly in terms of definitions. However, by doing so, it also went beyond its mandated scope as the implementing regulation for the registration procedure and mechanism (Trade Mark Law, Art. 56 paragraph (2)). In such sense, GR 51/2007 failed to bring a conclusive solution to GI protection at the normative level in Indonesia.

4.5.2 *2nd stage of GI consciousness*
Despite the above mentioned conceptual and normative problems, GR 51/2007 has been an important starting point in the process of developing Indonesia's second level of GI consciousness. Apart from facilitating a participatory process by involving the local community in the formulation of GR 51/2007 parallel to implementing the pilot project of Kintamani Arabica Coffee GI as the first GI registration in Indonesia in 2007, it also triggered the growth of GI awareness among local producers.

The case of the Arabica Coffee Gayo is an excellent example to illustrate this point. Starting off with non-existent GI consciousness, the local Gayo community developed a keen interest in GI in reaction to misappropriation overseas. After successfully completing the GI registration process at home, they ended up as the first Indonesian GI ever to be registered in the EU.

Based on the findings of this research, the following challenges have been identified: first, a gap of awareness as a result of lack of dissemination in general; second, gap between the growing aspiration of local communities to obtain GI protection on the one hand, and the constraints they are faced with in obtaining them on the other. Such constraints are related to the internal capacity of local communities caused by lack of information, capacity, and opportunity to have their GIs registered, maintain the quality of their products, and obtain access to markets. On the other hand, such constraints are related to the government's capacity to provide GI protection, such as the lack of a strong regulatory and an institutional framework for implementation and enforcement, budgetary and human resources, coordination across agencies, and a stronger involvement of local governments.

GIs are gaining significant momentum in Indonesia at this particular time. However, the government's strong political will is yet to be translated into policy and concrete measures based on a law that recognizes the special nature of GI, the need for GI protection, while being deeply rooted in the values and culture of Indonesian society. Based on the foregoing, Indonesia's currently applicable GI rules still appear to be 'trapped' in the 'struggle of paradigms'. While recognizing the need for a change in the law, Indonesian legislators appear to be reluctant to afford GI protection the approach merited by its nature, namely a 'one of its kind' or a sui generis approach.

4.5.3 *3rd stage of GI consciousness*

Upon entering the 3rd stage of GI consciousness, Indonesia needs to release itself from the 'self-imposed exogenous pressure' of following, or feeling compelled to reject, a particular protection paradigm. After all, as described above, such paradigms have been the result of compromises at many different levels influenced by many different non-law factors. By understanding the nature of the GI concept as applicable in the Indonesian context, translating them into effectively implementable provisions of the law and implementing them wholeheartedly, Indonesia will be able to gradually develop its own GI concept and norms that work best not only in the Indonesian context, but also in its relations with the rest of the world.

5 CONCLUSION

In sum, although initially adopted in Indonesia's legal system as a foreign concept, GI is not contradictory to the fundamental principles of the Indonesian people as articulated in the state philosophy Pancasila and the 1945 Constitution. GI rules in Indonesia were not initially created in response to the genuine need of the Indonesian people; rather than that, it was a result of 'self-imposed exogenous pressure' when Indonesian signed the WTO/TRIPs Agreement. The legal consequence arising from such gap of consciousness was the absence of a solid legislative basis for effective implementation and subsequently the relatively slow development of GI registrations, thus leaving Indonesian GIs with international reputation exposed to misappropriation, particularly in the context of international trade. However, GI consciousness has developed steadily.

In answering the above stated research questions, the following can be stated. First, the concept of GI protection is gaining an increasing ground in Indonesia's legislation as evident from the currently proposed amendment to the existing Trade Mark Law. GI provisions in the new law offer a more solid legal basis for GI protection. However, there still appears to be reluctance on the part of legislators to take GI provisions out of the Trade Mark Law and provide for them under a sui generis system. Second, there have been increasing recognition and GI consciousness among stakeholders at all levels. In fact, GI producers increasingly demand stronger GI protection. And third, there is a need for a law capable of accommodating such consciousness and demand for stronger GI protection. An effectively implementable sui generis GI law with a combination of the top-down and bottom up approach could make Indonesian GI protection a success, primarily to the advantage of local communities, producers, and consumers, while developing the reputation and confidence needed for access to the international market.

Finally, Indonesia needs to be able to create its own GI rules building on pre-existing models; however, it is free from the 'self-imposed exogenous pressure' to follow, or rigidly reject any of them. Apart from effective legislative drafting techniques, Indonesian legislators should primarily seek inspiration from the spirit of the Founding Fathers, demonstrating profound understanding, wisdom, and vision, upholding first and foremost the genuinely 'Indonesian spirit', values and principles that define and make the Indonesian nation great. The resulting effective GI protection will thus transcend economic advantages to Indonesia.

REFERENCES

Ayu, M.R. (2009) *Geographical Indications Protection in Indonesia based on Cultural Rights Approach*. Jakarta, Nagara.

Bangsadikusumah, S. (2016) Asosiasi indikasi geografis Indonesia. In: *Joint Seminar EU-Indonesia Trade Cooperation Facility (TCF) and Indonesian Directorate General of Intellectual Property (DGIP RI): Masa Depan Indikasi Geografis di Indonesia on August 29, 2016*. [Indonesian] Jakarta, DGIP RI & EU-Indonesian TCF.

Damarsasongko, A. (2016) New rules related to the draft of the Amendment of Law on Trademark, Industrial Design, and Patent. In: Indonesian Intellectual Property Academy (IIPA) and The Center for Continuing Legal Education—Fakultas Hukum Universitas Indonesia (CLE-FHUI) and World Intellectual Property Organization (WIPO)—Singapore *Advance Course Recent Issues on Intellectual Property Law, 30 Agustus 2016, Depok, Indonesia*. Depok, CLE-FHUI.

Das, K. (2010) *Geographical Indications at the WTO: An Unfinished Agenda* [Online]. Available from: doi:10.2139/ssrn.1597091.

Gabor, M.M. (2012) *Efektivitas Perlindungan Hukum Indikasi Geografis di Indonesia*. Depok, Universitas Indonesia. Disertasi Doktor.

Gervais, D.J. (2015) Irreconcilable differences? The Geneva act of the Lisbon agreement and the common law. *Houston Law Review*. [Online] 53(2), 339–371. Available from: https://works.bepress.com/daniel_gervais/51/.

India Ministry of Commerce & Industry (2002). *The Geographical Indications of Goods (Registration and Protection) Rules March 8, 2002*. New Delhi, India Government.

India Parliament. (1999) The geographical indications of goods (registration and protection) act. *Gazette of India Extraordinary No.48 December, 30 1999*. New Delhi, India Parliament.

Indonesia (1945) Undang-undang Dasar 1945 [*The 1945 Constitution of the Republic of Indonesia*].

Indonesia (1994). *Undang-Undang Nomor 7 Tahun 1994 tentang Persetujuan Pembentukan Organisasi Perdagangan Dunia. Lembaran Negara Republik Indonesia Tahun 1994 Nomor 57. Tambahan Lembaran Negara Republik Indonesia Nomor 3546* [*The Law Number 7 of 1994 regarding Ratification of Agreement Establishing The WTO. The State Gazette Year 1994 Number 57. The Supplement to State Gazette Number 3564*].

Indonesia (2001). *Undang-Undang Nomor 15 Tahun 2001 tentang Merek. Lembaran Negara Republik Indonesia Tahun 2001 Nomor 110. Tambahan Lembaran Negara Republik Indonesia Nomor 4131* [*The Law Number 15 of 2001 regarding Marks. The State Gazette Year 2001 Number 110. The Supplement to State Gazette Number 4131*].

Indonesia (2007) *Peraturan Pemerintah Nomor 51 Tahun 2007 tentang Indikasi-Geografis* [*Government Regulation Number 51 of 2007 regarding Geographical Indications*].

Komisi Pemilihan Umum [KPU] Republik Indonesia (2014) *Jalan perubahan untuk Indonesia yang berdaulat, mandiri dan berkepribadian: visi misi dan program aksi*. [Online] Jakarta, KPU RI. Available from: http://kpu.go.id/koleksigambar/VISI_MISI_Jokowi-JK.pdf.

Kongolo, T. (2008) *Unsettled International Property Issues*. Alphen aan den Rijn: Kluwer Law International B.V.

Landes, W.M. & Posner, R.A. (1989) An economic analysis of copyright law. *Journal of Legal Studies*, 18 (2) 325–363.

Mtima, L. (2015) From swords to ploughshares: towards a unified theory of intellectual property social justice. In: *Intellectual Property, Entrepreneurship, and Social Justice: From Swords to Ploughshares*. 1st edition. Northampton, Edward Elgar Publishing.

Munsie, J. (2002) *A Brief History of the International Regulation of Wine Production*. Harvard Law School. Available from: https://dash.harvard.edu/bitstream/handle/1/8944668/Munsie.html?sequence=2.

Panizzon, M. & Cottier, T. (2006) Traditional Knowledge and Geographical Indications: Foundations, Interests and Negotiating Positions. [Online] Available from: doi:10.2139/ssrn.1090861.

Ramli, A.M. (2016) Indikasi Geografi di Indonesia. *Joint Seminar EU-Indonesia Trade Cooperation Facility (TCF) and Indonesian Directorate General of Intellectual Property (DGIP RI): Masa Depan Indikasi Geografis di Indonesia on August 29, 2016*. Jakarta, DGIP RI & EU-Indonesian TCF. Speech.

Raustiala, K. & Munzer, S.R. (2007) The Global Struggle over Geographic Indications. *The European Journal of International Law EJIL.* [Online] 18(2), 337–365. Available from: doi:10.1093/ejil/chm016.

Rudita, L. (2011) Hak Kekayaan Intelektual & Perlindungan Konsumen: Studi tentang Indikasi Geografis dari Perspektif Kepentingan Konsumen. Depok, Universitas Indonesia. Disertasi Doktor.

Sardjono, A. (2011) Culture and intellectual property development in Indonesia. *Indonesia Law Review*, 1(3), 237–252.

Sasongko, W. (2005) *Indikasi Geografis: Studi tentang Kesiapan Indonesia Memberikan Perlindungan Hukum terhadap Produk Nasional.* Bandar Lampung, Penerbit Universitas Lampung.

Seidman, A., Robert B.S. & Nalin A. (2001) *Legislative Drafting for Democratic Social Change, A Manual for Drafters.* Dordrecht: Kluwer Law International.

Wang, M.C. (2006) The Asian Consciousness and Interests in Geographical Indications. *The Trademark Reporter.* [Online] 96 (4), 906–942. Retrieve from http://www.inta.org/TMR/Documents/Volume%2096/vol96_no4_03.pdf.

Wang, S.Y. (2013) *Geographical Indications as Intellectual Property: In Search of Explanations of Taiwan's GI Conundrum.* [Online] Newcastle, The University of Newcastle. Ph.D. eTheses. Available from: https://theses.ncl.ac.uk/dspace/items-by-author?author=Wang%2C+Szu-Yuan.

World Intellectual Property Organization [WIPO] (1958) Lisbon Agreement for the Protection of Appellations of Origin and their International Registration. Stockholm, WIPO.

World Trade Organization [WTO] (1994). Agreement on Trade-Related Aspects of Intellectual Property Rights (TRIPs). Geneva, WTO.

The politics of laws in the establishment of the Human Rights Court in Indonesia

J. Saibih
Faculty of Law, Universitas Indonesia, Depok, Indonesia

ABSTRACT: In spite of the existence of international human rights laws and the inclusion of similar laws at the national level, serious human rights violations have continued to occur. In the case of Indonesia, severe violations of human rights have been committed systematically as a means of maintaining the government's status quo. After a long discussion, the House of Representative (Dewan Perwakilan Rakyat or hereinafter so called as DPR) passed the bill as a definitive law. The enactment of the Human Rights Law on 23 November 2000 implies a new hope for the enforcement of human rights in Indonesia. In this law, it is stipulated that the Human Rights Court has the authority to try severe human rights violations, including crimes of genocide and crimes against humanity. In the explanation, it is stated that the severe human rights violations stipulated under this law are in accordance with the Rome Statute of the International Criminal Court. This paper will depict the substance of the law on the Human Rights Court and the enactment process, more specifically concerning the debate in parliament while discussing the substance of the law. Hence, this paper will focus on the debates on the legislative process in the DPR by presenting views from different factions in the parliamentary debate in the legislative process. Depicting the views of some of the factions in the parliamentary debates will give an idea to the readers about the importance of the law on the Human Rights Court in relation to the settlement of some severe past violations of human rights.

1 INTRODUCTION

Considering the significant number of cases of human rights violations, the government decided to establish a Human Rights Court to try these cases. The Law Number 26/2000 established this Human Rights Court. Before that, the President, according to his authority, issued a Government Regulation in Lieu of the Law (PERPU) Number 1/1999 regarding the Human Rights Court. The reform of the court in the context of law enforcement was a significant step in bringing justice to those who seek it. However, the DPR voted down the PERPU, and this situation compelled the government to revoke the submission of the PERPU.

The President submitted the draft law on 5 April 2000, along with the cover letter number R.08/PU/IV/2000. Based on this letter, the DPR accepted the submission of the Draft Law on the Human Rights Court, and discussed or reviewed it together with the government. The submission of the draft law replacing the PERPU that was rejected compelled by the DPR (the DPR plenary session, 13 March 2000) not more than a month before was a very quick response, and it shows the urgency of having this kind of law that promulgates the establishment of the Human Rights Court in Indonesia. Thus, the submission of the Draft Law on the Human Rights Court was based on the consideration of its substance, as set forth in PERPU, with some amendments to some articles that were basically necessary for the implementation of Article 104 paragraph (2) of the Law No. 39 of 1999 on Human Rights. Some of the improvements to the draft law submitted to the DPR were to provide justice and legal certainty for the people.

For the first time in history, the legislation contained provisions that regarded the retroactive aspect of the Draft Law on the Human Rights Court as a different specification.

Discussions regarding the implementation of this retroactive aspect against gross human rights violations were carried out in depth by the development team to capture the growing aspirations of the people. The variety of political streaming also shows different perspectives on the establishment of the Human Rights Court, so it is important to show political views in relation to the establishment of the Human Rights Court.

The writer will depict the substance of the law on the Human Rights Court and the enactment process, more specifically concerning the debate in parliament while discussing the substance of the law. Hence, this paper will focus on the debates on the legislative process in the House of Representatives by presenting views from different factions in the parliamentary debate on the legislative process. Depicting the views of some of the factions in the parliamentary debates will give an idea to the readers about the importance of the law on the Human Rights Court in relation to the settlement of some severe past violations of human rights.

2 METHOD

This paper aims to develop a model that rethinks transitional justice in Indonesia, which in the interim period has established the Human Rights Court to gain public trust in the government. This article also shows the responses to the victims' expectations through the political movement as these victims join political parties, in addition to the acknowledgement of the incidents that they experienced before. This study has gained information about transitional justice responses that were implemented by the Indonesian government, perpetrators and victims in settling their case through peaceful agreement and any initiatives that can prevent such abuses in the future. To obtain specific information about people's experiences and transitional justice responses to specific human rights violations, the narrative theory[1] (Moyle, 2002) and methods have been chosen.

The writer intends to explore people's stories and their responses to the efforts of the government in bringing the perpetrators to trial and to achieve reparation for them. In writing this paper as part of his thesis, the writer has interviewed people, by locating documents containing comments or responses created by the Indonesian government, human rights advocates and religious groups, and by finding other related documents. The aforementioned research has located a significant amount of information, which is not limited to academic literature, but also includes legislation, interviews, documents containing comments or responses created by Indonesian government agencies, human rights advocates and religious groups, articles or news reports from newspapers and magazines, court case decisions and many other related documents.

Understanding the social realm is not simple. The social sphere requires elasticity in reactions to it. Another reason why narrative theory is used is because the nature of history in relation to social and political events is flexible and incomplete (Moyle, 2002). Narrative theory allows researchers to identify and analyse the social contexts for the transitional justice responses, especially the reparation process. The use of narrative theory also makes it possible for the identification of the relationships between systems and social structures (Mumby, 1993).

3 PARLIAMENTARY DEBATE BEFORE THE ENACTMENT OF THE LAWS

3.1 *Blocs in the parliamentary session after the 1999 general election*

Suharto had begun to lose people's confidence, and the movement to oust Suharto from his presidency was getting stronger all over the country (Soemardjan, 1999). The reforms in the political field were related to the general election, the organisation and the position of the

---

1. There are many definitions of narrative theory. A narrative is a story. Narrative theory is used in many subjects, such as psychology, sociology, linguistics, history and culture.

Table 1. The composition of blocs in the DPR 1999–2004.

Fraction (parties members)	Members sitting
Indonesia Democratic Party in Struggle (PDI-P)	153
Functional Groups Party (Golkar)	120
Development Unity Party (PPP)	58
National Awakening Party (PKB)	51
Reform Bloc	41
National Mandate Party (PAN)	34
Justice Party (PK)	7
Crescent Moon and Star Party (PBB)	13
Indonesian National Unity Bloc	12
Unity and Justice Party (PKPI)	4
Indonesia Democratic Party (PDI)	2
Unity Party (PP)	1
Indonesian National Party-Marhaen Front	1
Indonesia National Party (PNI)	1
Indonesian Unity in Diversity Party (PBI)	1
IPKI	1
Democratic Catholic Party (PKD)	1
Union of Muslim Sovereignty Bloc (PDU)	9
Nahdlatul Ulama Party (PNU)	5
Islamic People's Awakening Party (PKU)	1
Islamic People's Party (PUI)	1
Indonesia Muslim Association Party (PSII)	1
Masyumi Indonesian Muslim Political Party (PPIIM-Masyumi)	1
Nation Love Democracy Party (PDKB)	5
Military/Police (TNI/POLRI)	38
Total Members	500

Source: (http://www.ndi.org/files/1079_id_preselect.pdf.)

People's Consultative Assembly (MPR) and also the House of Representatives (DPR) and Council (Sato, 2003). Since 1999, the Electoral Law has been accommodating aspirations of political ideas. All elements were involved in the general election and were given the authority to oversee the exercise of democracy at the polling stations (TPS). Forty-eight political parties participated in this election. This is far less than the number of existing parties registered at the Ministry of Justice and Human Rights, which is 141 parties.

The General Election Commission has successfully organized the general election and elect the members of DPR and MPR as result of general election 1999. The elected member of MPR then will conduct the general assembly meeting to decide some of the decision in relation with the guidelines of state policy, elect the Speaker of the house and also President as a head of state and government. The General Assembly was held between 1 June and 21 October 1999. In the General Assembly, Amien Rais was voted as the Speaker of the MPR and Akbar Tanjung as the Speaker of the DPR.

After the election of 1999 (the first election after the Reform), there were 500 members of the DPR, the legislative body. They consisted of 462 members elected at the 7 June general election and 38 who were appointed from the military and the police.

From the 48 parties that contested the general election, only 21 parties were eligible to have members sitting in the DPR. Only six of those parties, which had a minimum of ten members sitting in the DPR or 2% of the electoral threshold, were eligible for the 2004 election. These six parties were PDI-P, Golkar, PPP, PKB, PAN and PBB. From the total number of elected seats, those six parties won a total 429 of the 462 elected seats or 93% of elected seats. As the six parties had their own bloc, then 15 parties had to merge with others in forming a bloc. By having blocs, they have a right to nominate candidates for speaker, assign members to commissions, and make speeches on the floor. From 8 of the 21 parties represented in the DPR, nine blocs were formed (Table 1). The military is the tenth, separate bloc[2].

3.2 *Islamic factions' views on past severe violations of human rights cases*

Islamic political parties in the House of Representatives can be categorised into blocs, as mentioned above. There were the Union of Muslim Sovereignty (PDU) Bloc, the National Awakening Party Bloc (hereinafter, FKB), the Crescent Moon and Star Party Bloc (hereinafter, FPBB) and the Development Unity Party Bloc. Their views in the discussion of the promulgation of the Human Rights Court were important for this thesis; of particular importance were their views of some cases of the violation of human rights in the Soeharto era. Their voices led to the importance of the prosecution of human rights violations that happened in the Soeharto era.

The FPBB, in delivering their views on the draft law, asserted that the need for the completion of cases of human rights violations that had occurred throughout Indonesia was due to the demands of the people of Indonesia and the global community. Moreover, the FKB articulated that several human rights violations that occurred in the past, in addition to lowering the degree of civilisation of Indonesia as a nation, would also have negative effects for the development of democracy in Indonesia. In fact, the world would highlight the little or no commitment to the enforcement of human rights in Indonesia, which in turn would result in a threat to Indonesia's economic and trade relations with developed countries.

The FPBB, in their opening statement on the draft law, highlighted that, in the multidimensional crisis in Indonesia, the enactment of the Human Rights Court law should be followed by reform in the tribunal administration, which still faced corruption issues. The FPBB asserted that the enactment of the Human Rights Court law was needed to deal with human rights violations not only after the enactment but also those that had occurred in the past. Thus, the law on the Human Rights Court should apply retrospectively without expiration to prosecute the crimes. In addition to that, the FPBB also claimed that the establishment of the Truth and Reconciliation Commission was very important to settle the past human rights violations.

Another Islamic party bloc, the FPP, believed that the reform of the law was primarily concerned with resolving cases of violations of human rights, which had truly become an urgent necessity at that time. The FPP also claimed that the Indonesian people were increasingly aware of the importance of resolving the issues related to the human rights abuses in Indonesia that had occurred in the past, as well as preventing them happening in the future. If the aspirations and demands continued to strengthen on these issues, but they were not responded to quickly, wisely and in a prudent manner, then the threat of the disintegration of the nation-state of Indonesia could come true.

The need for a Human Rights Court is very important and urgent, and it will reveal and settle various events that indicate the occurrence of human rights violations, as mandated by

---

2. For those nine blocs, five were formed in the name of the party and the rest of the blocs were formed jointly by two or more parties. PAN and PK, for instance, formed a coalition named The Reform Bloc; these parties had substantial support from urban Muslim voters. Some of the nationalist parties formed their own bloc, named The Indonesian National Unity (KKI) Bloc. Another five traditional Muslim parties had a joint bloc, named The Union of Muslim Sovereignty (PDU), consisting of five Muslim parties. It concludes that the Islamic parties' bloc consisted of nine parties that were members of five different blocs. They were Partai Umat Islam (Islamic People's Party or PUI), Partai Kebangkitan Umat (Islamic People's Awakening Party or PKU), Partai Persatuan Pembangunan (Development Unity Party or PPP), Partai Syarikat Islam Indonesia 1905 (1905 Indonesia Muslim Association Party or PSII 1905), Partai Politik Islam Indonesia Masyumi (Masyumi Indonesian Muslim Political Party or PPIIM), Partai Kebangkitan Bangsa (National Awakening Party), Partai Bulan Bintang (Crescent Moon and Star Party or PBB), Partai Keadilan (Justice Party or PK) and Partai Nahdlatul Umat (Nahdlatul Umat Party or PNU). Those blocs won one third of the elected seats. Hence, the voice of these parties in the DPR was substantially significant in deciding some important decisions in the House.

Law No. 39 of 1999 on Human Rights. Various efforts made by the National Human Rights Commission over human rights violations had become 'sterile' because they were not able to be brought to court in the absence of the regulation on the Human Rights Court, which regulates the investigation and prosecution. Consequently, cases such as Bloody Incident in Tanjung Priok 1984; Warsidi incident in Lampung; Application of Military Operation Region (Daerah Operasi Militer so called as DOM) in Aceh; the incidents that caused many casualties in the eastern province, Irian Jaya, Bloody Trisakti Incident, Bloody Semanggi I–II Incidents and other human rights violations, were just a pile of documents that were difficult to process legally.

The preparation of legal instruments concerning the Human Rights Court in Indonesia needs to achieve international credibility and independence of the judiciary, so the law on the Human Rights Court must be adapted to the standards of international law. Convincingly proven human rights violations can be tried in the International Criminal Court, except where the relevant country will earnestly prosecute violators of human rights. The human rights court law is one of the law that govern about the trial for serious crimes that meet with international laws standard it will be a means by which we will protect Indonesian citizens, against those who violated human rights. The perpetrators of human rights should just be tried in Indonesia, which itself also gets international legitimacy.

In the draft law there was visible dominance or government intervention in the ad hoc tribunals for the process of human rights violations that occurred prior to the draft law. It is stipulated in Article 37 of the draft law that the President of the House of Representatives, based on the proposal, will establish an ad hoc court for human rights violations that occurred before this draft law. This article will provide an opportunity for the elite government (the executive) and the legislative body to determine matters for which an ad hoc body may be formed. This opportunity can be used by both of them (the government and the legislative body) to protect the perpetrators at other times, and then The Supreme Court will establish the ad hoc court. This is because it was the Supreme Court that had the authority to determine such matters. The Supreme Court needed to increase public confidence in law enforcement, in particular human rights, and also to strengthen the consistency and independence of the judiciary from the influence of any intervention for the purpose of realising the principles of the rule of law.

With regards to the scope of its authority, the Human Rights Court is only limited to human rights violations (Chapters 4 and 5 of the draft law). If the Court is only limited to human rights violations, then in the future the work of the Human Rights Court will not be effective, because of case scarcity in the violation of human rights. This is based on the idea that the shift pattern and the future government system could change the types of violations of human rights. Thus, the FKB proposed that the scope of the authority of the Human Rights Court could also reach the breach of Covenant on Civil and Political Rights and the Covenant on Economic, Social and Cultural Rights, as part of Law No. 39 of 1999 on Human Rights.

The FKB, in their response to the draft law, also argued that most of the crimes against humanity were committed by state institutions, and that at the time the crimes were committed there was moral and legal justification for the actions by the state, as they were committed in order to maintain integration, political stability and security. Moreover, the FKB also asserted that the human rights violations, which happened during Soeharto's government, were committed in order to save the foundation of the state. The crimes against humanity that accompanied it were not able to be prosecuted under the ordinary Criminal Code (Indonesian Criminal Code, hereinafter KUHP) but should be provided for by laws that specifically regulate human rights crimes, since the Indonesian Criminal Code only covers ordinary crimes. Therefore, the FKB suggested that the scope of the authority of the draft law submitted by the government should cover crimes against humanity. This means that, for crimes committed by the state, this draft law should cover the institution that makes the policy. Crimes committed by the direct action of the state can be ongoing crimes against humanity that are both broad and systematic (Extraordinary crimes).

In the view of the FKB, the existence of the Human Rights Court and the Truth and Reconciliation Commission as institutions can guarantee social justice on human rights violations. Both institutions can learn from other countries, such as South Africa and South Korea, to resolve the problem of human rights violations, and still maintain their relations among the fellow citizens of the nation and preserve the national integration and justice for the people. The settlement of human rights violations through the Truth and Reconciliation Commission will accelerate us in terms of political and security stabilisation, while simultaneously providing conducive conditions for economic recovery and accelerating the democratisation process in the life of the nation.

The FPBB supported the views of the FKB that the incidents in Ambon, Poso (Central Sulawesi), Sambas (Central Kalimantan), Tanjung Priok incident, the 27 July incident, Trisakti's incident, DOM incidents in Aceh and other incidents need to be resolved through law enforcement, either through the court or through draft laws on the Human Rights Commission and the Truth and Reconciliation Commission that will soon became the law.

3.3 *Military faction views on the debate in the parliamentary session*

The military faction in the DPR asserted that the people's aspirations on the protection and respect for human rights have been accommodated in the MPR Decree No. XVII/MPR/1999. Furthermore, the military faction articulated that the Decree of MPR-RI also set restrictions on the use of the fundamental rights of human beings, for instance, in Articles 34 and 36 of the MPR Decree it states that every person shall respect the human rights of others in the social life of the nation. In addition, to ensure their rights and freedoms, every person shall be subject to the restrictions that are established by the law with the sole purpose being to ensure recognition and respect for the rights and freedoms of others and to meet fair demands in accordance to considerate social security and public order in the people's democracy. The restriction on the implementation of human rights set out in the specific law on human rights is not a new provision, because previously it had been set out in the Charter of the United Nations Universal Declaration of Human Rights 1948.

In responding to the draft law submitted by the government, the military faction outlined some issues related to the provisions in the draft law, as follows:

1. Certainty and legal protection and justice for everyone.
2. Draft law on the Human Rights Court should adhere to the principles of applicable law and be more comprehensive than Government Regulation in lieu of Law No. 1 of 1999 on the Human Rights Court.
3. When it has been passed into law, the draft law should be able to ensure the establishment of national unity and the future of the nation.

The military faction also criticises the definition of the violation of human rights in the draft law, even though the definition was taken from Law No. 39 of 1999 on Human Rights. Such provision defines a violation of human rights as that which causes serious bodily or mental injury, or material or immaterial losses, and the resulting insecurity both to individuals and society. In addition to that, the definition is very subjective because it depends on the different conditions of both individuals and situations. Such a definition has not made it clear whether the offence is a serious human rights violation or a criminal offence that can be dealt with by the ordinary Criminal Code. Therefore, a human rights violation referred to in this draft law is a violation in an extraordinary crime that has broad impacts both nationally and internationally. The definition of what qualifies as a violation of human rights should be very clear in defining the elements of the crimes, as this will impact on the judicial process in proving the criminal elements.

The military faction argued that the formulation of Article 37 paragraph (I) in the draft law on the retroactivity principle was also contrary to the purpose of the establishment of the law, namely the establishment of the rule of law and justice for everyone. Providing the retrospective provision without placing a time limit on the prosecution of crimes that happened in the past, indicates that the draft law does not provide legal certainty and justice.

The importance of providing a time limit for past human rights violations is because the determination of whether a case can be tried by an ad hoc Human Rights Court will be very dependent on the assessment of the investigator, public prosecutor and the ad hoc Human Rights Court. Lastly, the military/police faction argued that, since the violation of human rights is set out in the draft law as a serious crime, the provision on the time limit for prosecution might be exempted from the provisions of KUHP. However, the time limit has been firmly secured for legal certainty, as has been stated.

3.4 *Government views in responding to the aspirations in the parliamentary debate*

In responding to the aspirations of the blocs in the parliamentary session, the Minister of Law and Human Rights, on behalf of the government, delivered a general explanation especially on the retroactive principle. Regarding such issues, the government claimed that the implementation of the retroactivity principle will be determined only by the ad hoc Human Rights Court that is established by a Presidential Decree upon the recommendation of the DPR. In addition to determining to which events the examination can be applied retroactively, it all depends on the DPR as an institution authorised by the (draft) legislation to propose the establishment of an ad hoc Human Rights Court.

Thus, if the DPR proposes to the President that the case of Westerling in 1948 (The Jakarta Post, 2010) should be reopened, then the President will establish a special ad hoc Human Rights Court for that case only. Moreover, the government highlighted that the draft law should not be only about the establishment of an ad hoc Human Rights Court, but that it should, most importantly, establish a permanent Human Rights Court to adjudicate the violation of human rights that happened after the enactment of the Human Rights Court law. The retroactivity principle could not be applied to adjudicate past human rights violations through a permanent Human Rights Court. However, the specific cases that occurred in the past, before the enactment of the law on the Human Rights Court, will open the possibility to form an ad hoc Human Rights Court specifically for that case only, not for other cases.

Thus, a Human Rights Court is similar to the ad hoc tribunals, such as the International Tribunal for Rwanda or the former Yugoslavia, which were created specifically to prosecute certain cases that occurred at certain times in both regions. A tribunal was not intended to be a permanent institution to prosecute all cases of human rights violations that have occurred in the past. Various practices occurring internationally show that the retroactive principle was only applied to certain cases that occurred in the past, not for all cases of violations of human rights that have not yet happened. Concerning the opportunity to protect the perpetrators of political elites, this can be understood, but in the end the issue is up to the goodwill of all parties who want to find a fair resolution of the issues that need to be resolved.

4 CONCLUSION

The establishment of a Human Rights Court in Indonesia could not be separated from the political circumstances at that time. Firstly, the government's transition from the New Order Era to the Reform Era was the first trigger that urged the establishment of a Human Rights Court. The transition of the government has claimed many victims of violations of human rights that suffered from the New Order government. The pursuit by the victims of the establishment of a Human Rights Court has brought about the enactment of the MPR's resolution for the settlement of various violations of human rights that happened in the New Order Era.

Secondly, the transition of the government also led to the holding of the first democratic general election since 1955, which many political parties contested as participants in the election. The general election gave the opportunity to many former political prisoners of the New Order government to become legislative candidates, including the victims of Tanjung Priok incident, such as A.M. Fatwa (National Mandate Party/PAN) and Abdul Qodir Djaelani (Moon and Crescent Party/PBB). Later on A.M. Fatwa became a Vice Speaker of the House of Representatives for the period 1999–2004. A.M. Fatwa believed that his becoming a member of parliament brought many changes in the meaning of reform and in the establishment of a Human Rights Court, especially the recognition of Tanjung Priok incident as a past severe violation of human rights.

Thirdly, the transition of the government brought about a wind of change in three regulations on politics, such as the general election law, social and political organisations law and the law on the composition of the MPR/DPR. The changes in the law on social and political organisation have led to the establishment of many political parties in various political streams, such as the nationalist stream and the Islamic stream. This change also brought back former political parties that had not existed for a long time after the fusion policy of the New Order government. The Islamic parties in parliament brought many changes, in addition to the recognition of Tanjung Priok incident as a past severe violation of human rights.

Lastly, the reform of the military faction in the DPR has also brought some changes in the views of the military bloc in the DPR, especially on the issue of human rights. The minutes of the meeting of the DPR debate on the establishment of a Human Rights Court show that the military agreed to the establishment of a Human Rights Court but that the military bloc wanted to make sure that equality before the law will be applied to anyone involved in severe violations of human rights, which needs to be defined clearly.

REFERENCES

Indonesia. (1999). *Undang-Undang Nomor 39 Tahun 1999 tentang Hak Asasi Manusia* [*The Law Number 29 of 1999 regarding Human Rights*]. Jakarta: Direktorat Jenderal Peraturan Perundang-undangan, Kementrian Hukum dan Hak Asasi Manusia.

Indonesia. (1999). *Peraturan Pemerintah Pengganti Undang-undang No. 1 Tahun 1999 tentang Pengadilan Hak Asasi Manusia. Lembaran Negara Republik Indonesia Tahun 1999 Nomor 191. Tambahan Lembaran Negara No.3911* [*Government regulation in lieu of law regarding Human Rights Court. The State Gazette Year 1999 Number 191. The supplement to State Gazette Number 3911*]. Jakarta: Direktorat Jenderal Peraturan Perundang-undangan, Kementrian Hukum dan Hak Asasi Manusia.

Indonesia. (2000). *Undang-Undang Nomor 26 Tahun 2000 tentang Pengadilan Hak Asasi Manusia. Lembaran Negara Republik Indonesia Tahun 2000 Nomor 208. Tambahan Lembaran Negara Republik Indonesia Nomor 4026* [*The Law Number 26 of 2000 regarding Human Rights Court. The State Gazette Year 2000 Number 208. The supplement to State Gazette Number 4026*]. Jakarta: Direktorat Jenderal Peraturan Perundang-undangan, Kementrian Hukum dan Hak Asasi Manusia.

Indonesia. (2000). *The People's Consultative Assembly (MPR)'s Resolution Number V/MPR/2000 on the Consolidation of National Unity*. Jakarta: Sekretaris Jenderal MPR.

Moyle, K. (2002). *Digital technologies in Australian public schools: A narrative study of government policies* (PhD. Thesis). Swinburne University of Technology, Hawthorn.

Mumby, D. K. (1993). *Narrative and social control: Critical perspective Volume 21*. California: Sage Publication.

Sato, Y. (2003). *Democratizing Indonesia: Reformasi period in historical perspective*. IDE research paper. Chiba, The Institute of Developing Economies.

Soemardjan, S. (Ed.). (1999). *Kisah Perjuangan Reformasi (The story of the struggle for Reformasi)*. Jakarta: Pustaka Sinar Harapan.

The Indonesian National Commission on Human Rights. (2000). *Annual Report 1999*. Jakarta: The Indonesian National Commission on Human Rights.

The Jakarta Post. (2010, May 19). Westering's War. *The Jakarta Post*.

United Nations. (1998). *Basic facts about the United Nations*. New York: Department of Public Information United Nations.

Indonesian pretrial: Can it fulfil the rights of the victims of an unfair trial to restoration?

F.M. Nelson
Faculty of Law, Universitas Indonesia, Depok, Indonesia

ABSTRACT: Unfair trial can be defined as a process in which the judicial procedures are not conducted in a fair manner. A balance system is needed in an effort to evaluate such conduct in the criminal procedure. In this context, pretrial has a role as a balance system to ensure the implementation of the principle of fair trial. In the Indonesian Code of Criminal Procedure (KUHAP), pretrial is one of the powers that a district court has in order to examine and decide the legality of an arrest or detention; the legality of the termination of an investigation or prosecution; and requests for compensation or rehabilitation (Article 1 Paragraph 10 of the Law No. 8 of 1981). However, there are several problems related to pretrial—one of which is regarding the implementation of the procedures in restoring the rights of the victims of an unfair trial. This paper is based on empirical legal research done by exploring the victims', law enforcers' and other related stakeholders' experiences.

1 INTRODUCTION

As one of the institutions that is most relied upon in upholding justice, the judicial institution is an important element of a country that bases itself upon the rule of law. The judicial institution is held in such high regard that it is given the authority to 'violate' human rights under certain circumstances in order to uphold them, which would include coercive actions in the investigation and prosecution phase. It is of the utmost importance that these actions are carefully and cautiously thought out before they are implemented. This has to be done in order to prevent authorised parties from abusing or misappropriating the authority to conduct coercive actions.

There have been various cases in which the investigation and prosecution were not properly conducted. A few examples of these cases are: JJ Rizal's case in 2009 and the war hero's widow's case in 2010. Deviations from proper investigation and prosecution could also be committed in the victimisation of susceptible citizens, such as Deden in the Ahmadiyah case in 2011 and the college students' case in 2011. From these cases, it can be seen that a person suspected of a criminal act could also be the victim of the supposed criminal act. The 'victims' in this context are not those resulting from a criminal act, but rather the ones who were the product of an unfair trial.

These so-called victims of unfair trials have remained forgotten in most studies concerning the attempts to fulfil human rights. The existence of pretrial as a mechanism that facilitates the claims of victims of unfair trial is deemed to be insufficient to fulfil the victim's rights. This is mostly due to the following facts: that a pretrial will only award compensation for specific types of requesters, i.e. victims of illegal coercive actions; the complicated compensation process (procedural and administrative) by the civil proceeding mechanism (Articles 77 to 83 the Law No. 8 of 1981); and the extremely limited amount of compensation that victims of unfair trial are able to request.

Based on the abovementioned facts, this paper will discuss the implementation of the procedures for restoring the rights of unfair trial victims, the effectiveness of pretrial as an institution in restoring the rights of unfair trial victims and alternative measures that can be undertaken in an attempt to restore the rights of unfair trial victims.

This research is conducted using the empirical legal research method. It was determined and built through a series of discussions and workshops that involved experts. From these workshops, the issues of the research and the methods used throughout the research were determined. This study also involved using a literature review, determining informants, developing the questionnaire, training for interviewers, sampling and also drafting the study report.

2 RIGHT TO RESTORATION FOR VICTIMS OF UNFAIR TRIAL UNDER THE INDONESIAN LAW ON CRIMINAL PROCEDURE

The right to restoration for victims of unfair trial is provided in Chapter XII in the Indonesian Code of Criminal Procedure (hereinafter referred to as KUHAP) on Compensation and Rehabilitation, which consists of three Articles—Articles 95, 96 and 97. The right to receive compensation and rehabilitation is made up of two principles (Rukmini, 2003, p.105):

1. Citizens' rights to receive compensation (in the form of money) and rehabilitation by means of restoration; and
2. The obligation of law enforcers to always remain accountable for their actions during the pre-adjudication phase.

This proposition is based on the term *habeas corpus* (Garrett, 2012, p.61), which states that not a single person can be arrested, be deprived of his property, be exiled, or have his rights reduced, except as the law so prescribes, and in accordance with the procedures that it has laid down (Eddyono et al., 2014, p.11). In Indonesia, *habeas corpus* takes the form of a pre-trial institution that is derived from the need for intense judicial scrutiny against all actions that deprive an individual of their liberty.

In addition to that, KUHAP defines compensation as the right of a person to have his requests granted in the form of the payment of an amount of money for having been arrested, detained, prosecuted or tried without any reason based on law or due to an error regarding the person or the law, which is applied as regulated in KUHAP (Article 1 Paragraph 22 of the Law No. 8 of 1981). In addition, rehabilitation is the right of a person to have his rights restored to their capacity, status, dignity and integrity, which is granted at the stage of investigation, prosecution or adjudication for having been arrested, detained, prosecuted or tried without any reason based on law or due to an error regarding the person or the law, which is applied as regulated in KUHAP (Article 23 Paragraph 10 of the Law No. 8 of 1981). Even so, KUHAP gives a limit of only 7 (seven) excuses that can be used to request for rehabilitation, as referred to in Article 97 paragraphs (1) and (3) of KUHAP.

3 MEANS TO RESTORE THE RIGHTS OF THE VICTIMS OF UNFAIR TRIAL IN THE INDONESIAN CIVIL CODE

Apart from the criminal procedure, Article 1365 of the Indonesian Civil Code (hereinafter referred to as KUHPer) can be used as the legal grounds for unfair trial victims to claim for compensation. Regarding this issue, Rosa Agustina quotes Mariam Darus Badrulzaman's statement that 'even though there are laws that specifically regulate means of claims against unlawful acts, the provision in Article 1365 of KUHPer will continue to apply due to its importance in emphasizing and formulating natural laws' (Agustina, 2003, p.283). Indeed, an attempt to restore rights by the use of such provision will be exercised through civil proceedings in a district court.

Later, the court will examine whether the officers' act was unlawful. Hoffman argues that to state that an act is unlawful, it must fulfil the prerequisites or elements that are provided in Article 1365 of KUHPer, as follows: 1) There must be an act, whether positive or negative— which means it could either be a commission or omission; 2) Such an act must be against the law; 3) Damages ensued; 4) There is a causation between such an illegal act and the damages that incurred; 5) Such an act is a fault that must be accountable.

Even though the amount of compensation is not strictly regulated in KUHPer, Article 1371 Paragraph (2) of KUHPer provides guidance on such matters, by stating that: '...compensation for damage caused by the injury...shall also be evaluated based upon the status and the financial condition of the individuals involved, and upon the circumstances' (Article 1371 Paragraph 2 of the Civil Code of Indonesia). Therefore, the amount of compensation must be assessed case by case. A judge, in deciding upon the amount of compensation, is not bound by laws and regulations. This particular matter has been the subject of a case law from Indonesia's Supreme Court following its decision on 23 May 1970, which states that a judge, in ruling on the amount of compensation, must do so by virtue of justice (*ex aequo et bono; naar redelijkheid en billijkheid* atau *in goede justitie*) (Agustina, 2003, p.81).

4 A PORTRAIT OF INDONESIA'S PRETRIAL

The five cases that I examined here are based on the Supreme Court's decisions from 2002 to 2011. I found five interesting facts that could portray Indonesian pretrials, which will be the main issues of discussion in this subsection.

First of all, I discovered that a request for a pretrial can be submitted by individuals or legal entities that have access to legal aid. From those five decisions, three of them (*Arief Fikriansyah and Deden Sura Agung case*, *Philips Fernando Sinay case* and *Abu Bakar Ba'asyir case*) were accompanied by an attorney from a law firm, whereas the other two (*Hasan Basri case* and *Lina case*) were accompanied by an attorney from a Legal Aid Foundation.

Secondly, one of the people interviewed during this research claimed that he did not know about the pretrial procedure. This explains why he did not submit a request for such a procedure. This, in my opinion, is a problem on its own, because, even 23 years after KUHAP was enacted, there are still people who do not know about it, including those people who are currently dealing with legal problems. This also shows that access to legal aid is still limited.

Moreover, the requests for a pretrial are mostly submitted on specific criminal cases, such as corruption and narcotics. Requests for a pretrial on alleged theft cases are mostly submitted by Jakarta Legal Aid Institute (Lembaga Bantuan Hukum Jakarta [LBH Jakarta])—whose priority is to help those who are poor, have little understanding of the law and who are suppressed (http://bantuanhukum.or.id.); thus, a pretrial is one of the options in advocating for individuals who fall into the above categories.

Another finding is that families have an important role in the request for a pretrial. In the pretrial request of Syahri Ramadhan Burhanuddin, a child suspected of theft and who was arbitrarily arrested, he submitted a pretrial request to Cibinong District Court through his legal counsel from Jakarta Legal Aid Institute, in which a single judge decided that the request would not be granted because it had already been submitted to the District Court and had gone through the examination up to the main trial of the supposed case. At the time, Syahri, as a victim of unlawful arrest, was not aware of the existence of pretrial provision and its procedures. In the interview, he explained that: 'I believe that there is such a thing, after the cassation, the Legal Aid informed my mother of that matter, but I personally don't really know anything about the procedures, all I know is that it has been resolved, that's it'. This is similar to the *Hasan Basri case*, in which the person who was actively involved in requesting a pretrial was his wife, Khotimah.

Furthermore, the detention of the suspect/accused further complicates the request for a pretrial. Even though KUHAP has already anticipated such matters by allowing the request for a pretrial to be submitted by the family or legal counsel of the detainee, with the suspect/accused being detained, access to and communication with the suspect/accused becomes extremely limited. If the suspect is detained throughout the pretrial process, it will undoubtedly increase risks and induce fear for the suspect. This may happen because the suspect is being detained in a police/prosecutor detention room throughout the pretrial process.

5 FACTORS THAT INFLUENCE THE INEFFECTIVENESS OF A PRETRIAL IN RESTORING THE RIGHTS OF THE VICTIMS OF UNFAIR TRIAL

There are four obstacles that influence the ineffectiveness of a pretrial. Firstly, several law enforcement agencies do not have a budget that is specially allocated for pretrial compensation. This is mainly because the budget of every agency is determined before the fiscal year begins. In addition, several agencies do not even include pretrial compensation into their budget planning—or only allocate an extremely small amount for it. This is chiefly due to the state budget policy, which uses budget absorption as one of the indicators of the agency's performance. If by the end of the fiscal year the compensation budget pot is not fully utilised, then it will be assumed that the agency has performed poorly. On the other hand, if the amount of compensation paid exceeds the amount assigned to the budget pot, it means that the agency is forced to use the 'tactical fund' or the law enforcement officer's personal fund, *which is prone to be interpreted as corruption*.

Consequently, this budgeting policy affects the compensation paid in two ways: (1) compensation for the victims is not paid or (2) the compensation that is paid comes from the law enforcement officer's personal fund. If the second condition were to happen, it would not result in a liability for the agency. It is, however, an effective deterrent tool for law enforcers who illegally conduct an arrest and detention. It can now clearly be seen that the problem is not only in the budgeting phase but also in the execution phase.

Secondly, the victims and those who represented the victims of illegal arrests and detentions often suffered from a pretrial manipulation by law enforcers. The pretrial institution would then be subject to a manipulation performed by none other than the law enforcers themselves, by accelerating the submission of the letter bringing the action to the district court, which causes the pretrial request to automatically fail (Article 82 of the Law No. 8 of 1981).

The pretrial of Syahri Ramadhan would be a good example of this. The request was submitted by a legal counsel from Jakarta Legal Aid Institute on 30 June 2009, while the police submitted the letter bringing the action to the prosecutor's office on 29 June. What would then be considered as odd is the fact that the prosecutor immediately submitted the bill of indictment to the court on 30 June 2009. The pretrial hearing then commenced on 15 July 2009, for the reason that Cibinong District Court needed two weeks to give a trial notification to Jakarta Legal Aid Institute, which is domiciled in Jakarta. On the other hand, the primary case trial started on 9 July 2009, so hence the pretrial hearing on 15 July 2009 automatically failed, with reference to Article 82 of KUHAP. What is even odder was that during the first day of the pretrial hearing, the judge promptly proceeded to read out the neatly typed decision and handed it out to both parties. This practice differed greatly from the common practice of trial in Indonesia, yet when the decision was being read out, the judges had not yet prepared the complete (typed) version of the decision, and the parties would have to wait for it to be completed in order to obtain a copy of the decision.

This fact is often ignored by the judge that examines a pretrial request, because judges are bound by Article 82 of KUHAP, which regulates that once the primary case is submitted to the court, a pretrial request will abort accordingly. This results in a lack of correction for law enforcers and the victim's rights being drowned in the trial scheme, where the victim is positioned as a person accused of a criminal act. Concerning this problem, one of the interviewees, Dr. M. Naser, a member of the National Police Commission, stated that this is one of the main reasons why pretrial is considered ineffective.

Thirdly, illegal arrests and detentions are often followed by the torture of the arrested suspect/accused. Jakarta Legal Aid Institute revealed that 74.66% of arrestees suffered torture and that 78.57% of them also experienced torture during the examination process (LBH Jakarta, 2008, p.43). Similar conduct is also engaged in against juveniles. The torture is often vicious and affects the victim's psychological condition. The absence of psychology/psychiatry services to restore the victim's condition, and the fact that the torture instilled trauma in the victim, would greatly discourage them from submitting a request against the law enforcers that have arrested/detained them. Aside from the victim, their family is also the target of intimidation by the law enforcers, such as the intimidation experienced by Khotimah, Hasan

Basri's wife. Another grave concern is when the victim is being detained in a detention room in a police station, while the police from the police station are being tried in a pretrial for their illegal conduct. Under these conditions the victim is extremely vulnerable to intimidation, because he is physically under the 'authority' of the officers who have illegally detained him.

Fourthly, poor access to legal aid also contributes to victims being unaware of the pretrial mechanism, or, in cases where they do know about it, it is still unreasonably difficult to access a legal counsel that is willing to represent a detainee, for instance, in the Hasan Basri case. These findings about the ineffectiveness of pretrial lead to public distrust of the entire pretrial mechanism. This is further asserted by the fact that most court decisions dismiss the pretrial mechanism.

6 PRETRIAL: WHAT IS AND WHAT OUGHT TO BE (*DAS SEIN* AND *DAS SOLLEN*)

Historically, pretrial was half-heartedly established—it was none other than a mediocre remedy for the authority to exercise coercive actions that were given to the police. However, exercising such authority without a proper judicial process would inevitably violate human rights, the philosophy of an independent judiciary and the presumption of innocence principle. The gap between the "what is" and "what ought to be" (*das sein* and *das sollen*) of a pretrial results in the ineffectiveness of pretrial as a correctional mechanism. This can be examined through KUHAP's enactment process, which shifts the concept of *habeas corpus* to only deal with administrative matters. For example, in the case where an investigator has already issued a decision and a notification following a coercive action, then it can be automatically concluded that such coercive action will be considered as a legal act (Pangaribuan, 2014, p.9).

In addition to that, one of the people interviewed for this study, Narendra Jatna—who at that time was Head of the District Attorney's Office in Cibadak, West Java, Indonesia—argued as an expert in one of the workshops that essentially the argument behind the provisions on pretrial is incorrect. He then proceeded to refer to the principles that are internationally acknowledged and applied, whereas within those said principles, none of them supports the exercise of coercive actions by an investigator. This mistake, according to him, contributes to an ineffective pretrial, and if the circumstances do not change anytime soon, then violations in the form of illegal coercive actions will continue to happen.

He also stated that based on his research on the government's responsibility for compensating a suspect's right after being violated by the officials, it is showed that every agency in other countries has its own budget post of such compensation—which is specially established in order to execute court decisions, including compensation. Such a concept makes the compensation mechanism clear. Meanwhile, until now, Indonesia has not implemented such a concept, which makes the whole compensation mechanism unclear.

Furthermore, the scope of pretrial as referred to in Article 77 of KUHAP is limited to deciding whether an arrest, detention, investigation and prosecution is legal or illegal. Nonetheless, as has been discussed above, other issues beyond the current pretrial scope arise.

For the above reasons, the National Law Commission (Komisi Hukum Nasional [KHN]) argued that the scope of the authority of pretrial institutions as a supervisory instrument needs to be extended so that it would also cover law enforcers' abuse of authority, in order to protect the rights of individuals that seek to obtain justice. KHN recommended the following '... to determine whether the preliminary evidence is sufficient, it must first be requested by a commissioner judge before the commission of a coercive action, including the authority of the commissioner judge to examine and decide: (i) an examination without legal aid; and (ii) an arrest must be made after the request for such action is granted by the judge' (Komisi Hukum Nasional, 2007).

The maximum amount of compensation is limited to Rp 3,000,000.00, as referred to in the Government Regulation No. 9 of 1983 on the Explanation of KUHAP, which has not yet been amended. Such a limitation on the amount of possible compensation means that these victims will have to resort to a civil claim in order to be fully compensated.

Narendra Jatna also stated (N. Jatna, personal communication, 2013) that the pretrials that have been conducted to date have not functioned effectively. One of the reasons for this is that there exists a problem with compensation for a suspect in cases where a pretrial request is granted by the judge. He stated that: 'if the agencies try to compensate, then they wouldn't be able to give a full compensation due to the non-existence of such budget. If they do try to give a compensation, then it would be considered as corruption'. Concerning this problem, he proposed a solution, which is that the budgeting item for accommodating pretrial compensation should always exist under the 'unexpected budget' item, and that it is imperative to adjust the country's current finances in order to accommodate the pretrial mechanism.

7 CONCLUSION

From the above analysis, it can be concluded that 1) the mechanism to restore the rights of the victims of unfair trial have been provided in KUHAP on pretrial mechanism, compensation and rehabilitation. Other than that, it is also possible to use the unlawful act as referred to in Article 1365 of the Indonesian Civil Code. 2) Currently, Indonesia's pretrial institution is still considered to be ineffective in restoring unfair trial victims' rights, due to four factors and the flaws of its concept, which include: an insufficient budget for executing compensations; the difficulty in gaining access to legal aid; a limited scope of authority; the limited amount of compensation that may be requested; the short period of the request submission; the utilisation of civil proceedings to examine material matters; limited legal remedy options; and law enforcers' rejection of pretrial. 3) The alternative option to restore unfair trial victims' rights by submitting an unlawful conduct claim is less desirable because it would take an unreasonable amount of time and money, and, aside from that, the evidentiary process would only examine whether an arrest and detention warrant on the supposed case exists.

Thus, an amendment to KUHAP and its implementing regulations is imperative. In addition, there needs to be full support from the government in the attempt to prepare a budget for compensation, which stems from a legal process, and this must be supported by the improvement of the country's financial system, in which the country, through its ministries, should arrange a budget for compensation and restoration for unfair trial victims. There should also be a Supreme Court Regulation that regulates pretrial proceedings with more clarity, in order to avoid discretions in its implementation.

REFERENCES

Agustina, R. (2003). *Perbuatan Melawan Hukum* [Tort Law] (1st ed.). Jakarta: Program Pascasarjan Fakultas Hukum Universitas Indonesia.

Eddyono, S. W., Djafar, W., Sufriyadi, Napitupulu, E. A. T. & Sriyana. (2014). *Praperadilan di Indonesia: Teori, Sejarah, dan Praktiknya* [Pretrial in Indonesia: Theory, History and Practice] (1st ed.). Jakarta: Institute for Criminal Justice Reform.

Garrett, L. B. (2012). Habeas corpus and due process. *Cornell Law Review*, 98(1), 48–123.

Kitab Undang-undang Hukum Perdata [Indonesian Civil Code] 1847 (Id.).

Komisi Hukum Nasional. (2007). *Penyalahgunaan wewenang dalam penyidikan oleh polisi dan penuntutan oleh jaksa dalam proses peradilan pidana* [Abuse of Power within the Investigations by the Police and the Prosecutions by the Prosecutors in Criminal Procedure Process]. Jakarta: Komisi Hukum Nasional.

Lembaga Bantuan Hukum [LBH] Jakarta. (2008). *Mengungkap Kejahatan dengan Kejahatan, Survei Penyiksaan di Tingkat Kepolisian Wilayah Jakarta Tahun 2008* [Revealing Crime with a crime, Survey on Torture in Jakarta Special Region Police]. Jakarta: LBH Jakarta.

Pangaribuan, L. M. P. (2014). Hakim Pemeriksa Pendahuluan (HPP) dalam rancangan sistem peradilan pidana di Indonesia [Preliminary Examining Judge in The Design of The Criminal Justice System in Indonesia]. *Teropong*, (1), 2–21.

Rukmini, M. (2003). *Perlindungan HAM melalui Azaz Praduga Tidak Bersalah dan Azaz Persamaan Kedudukan dalam Hukum pada Sistem Peradilan Pidana Indonesia* [Human Rights Protection Through The Principle Presumption of Innocence and The Principle Equality Before the Law in The Criminal Justice System in Indonesia] (1st ed.), Bandung: PT Alumni.

Undang-undang No.8 Tahun 1981 tentang Hukum Acara Pidana [Law Number 8 of 1981 regarding The Code of Criminal Procedure] 1981 (Id.).

The substance of good governance principles on government decisions in Indonesia

M.R. Bakry & A. Erliyana
Faculty of Law, Universitas Indonesia, Depok, Indonesia

ABSTRACT: Good Governance (GG) principles are an advanced instrument for making dynamic government decisions. This article analyses the urgency level on the principles of GG, based on the substance principles, on government decisions in Indonesia. The focus of the article is the setting of the principles of GG in the Netherlands, international organisations and some of the laws and regulations in Indonesia. Comparisons were made to assess the suitability of the substances, so that the urgency level on each principle of GG could be determined, especially on government decisions in Indonesia. The conclusion is that there are three substances on the principles of GG in Indonesia, namely fundamental, micro and macro. The urgency degree of each GG principle on government decisions in Indonesia is determined according to the level of the substance: (1) fundamental, (2) micro, and (3) macro. Principles regarding the procedures and the truth of the facts on government decision are grouped and fit into the substance of the fundamental principles and micro substance principles. Subsequently, the macro principles of GG substance are used as the parameter principles in the publication of government decisions in Indonesia.

1 INTRODUCTION

The trend towards globalisation that is occurring in various aspects of life, especially in developing countries, including Indonesia, has shifted the paradigm of government from 'government' to 'governance' (Nugraha, 2006). Nugraha (2006) puts forward the argument that the shift of the government paradigm is closely related to the condition of past governments, which held the top level of governance and other functions within society with absolute power.

According to Durcker, in response to this situation, the people desired a government that does not only govern and regulate, but that can also govern in a good and proper manner (as cited in Osborne & Gaebler, 1993). The function of ensuring and protecting development, which is the core objective of every modern country, essentially depends on whether or not government decisions can respond to the changes and development in society.

Addink et al. (2010) are of the view that Good Governance (GG) today can potentially be one of the three main pillar of any modern state other than rule of law or democratic state. In theory, GG embodies principles that are essentially aimed towards creating a good government. Ideally, each of the principles of GG has a different urgency in forming dynamic government decisions, based on the substantive content of the GG principles.

This paper questions the substantive aspects of these GG principles and their level of urgency in forming dynamic government decisions. The focus is on the regulation of GG principles in the Netherlands, international organisations and under various pieces of legislation that apply in Indonesia. The comparison is made to assess substantive suitability in order to determine the urgency of each of the GG principles on government decisions in Indonesia.

The first part of this paper explains the concepts of 'government', 'governance' and 'GG' in the theoretical as well as the practical sense. The second part deals with the principles of GG by substantive materials in the Netherlands, international organisations, and the various prevailing regulations in Indonesia. Subsequently, each of these substantive contents is determined as to their urgency with regard to government decisions. The conclusion at the end of this paper contains explanations and a grouping of substance of good governance principles relating to the subset of principles on the procedures for making government decisions (A1) and the subset of principles on the correctness of facts in government decisions (A2).

1.1 Government and governance

According to Rahman (1971), the state, as a giant organisation, has to meet special requirements for its establishment that separates it from other kinds of organisation. The elements of this are people, government, territory and sovereignty. Similarly, Oppenheim and Lauterpacht (1967) view the components of a state as the following: it must have people, occupy an area, have a ruling government, have the capacity to engage in relations with other countries, be able to hold its public officials accountable and be independent.

Government, as one of these elements, performs an important function in formulating, expressing and realising the people's aspirations, which constitute a manifestation of the power to control and coordinate the various aspects of life in society. Thompson (1997) explains that Beloff and Peele break down the government's functions into: a) defence, law and order; b) taxation; c) provision of welfare services; d) protection of individuals; e) regulating the economy; f) provision of certain economic services; and g) development of human and physical resources (Thompson, 1997, p.353). In essence, this concept of government is made up of the authority and responsibility to manage the affairs of the state and perform the will of the people, which is more directed towards the executive branch of government.

Unlike the concept of government, 'governance' is a concept of authority and responsibility directed towards the role of civil society and private entities, taken in relation to the role of the government itself. There are a number of views that explain the meaning of governance. The first is the definition of governance given by the United Nations Development Programme (UNDP & Eurostat, 2007):

> 'governance is the system of values, policies and institutions by which a society manages its economic, political and social affairs through interactions within and among the state, civil society, and private sector…Governance, including its social, political and economic dimensions, operates at every level of human enterprise'.

The second is the one formulated by the World Bank:

> 'the rule of the rulers, typically within a given set of rules…by which authority is conferred on rulers, by which they make the rules, and by which those rules are enforced and modified… the various processes by which they are selected, defined, and linked together and with the society generally…… power is exercised through a country's economic, political, and social institutions' (World Bank, 2014).

The third definition is provided the State Administrative Institution (LAN): 'administration, management, direction, fostering, delivery of governmental services, and governance… governance stands on three pillars: economy, politics, and administration…the institutions of governance have authority over three domains: the government, private sector, and civil society' (Jubaedah et al., 2007).

Fourthly, the United Nations Economic and Social Commission for Asia and the Pacific (UN-ESCAP) defines governance as 'the process of decision-making and the process by which decision are implemented (or not implemented)…All actors other than government and the military are grouped together as part of the "civil society" (UN-ESCAP, 2016). Based on such definitions, it can be concluded that these concepts of governance share some similar characteristics: firstly, the government and society are described as actors involved in governance; secondly, the focus of governance is government decisions; and thirdly, governance possesses legal, social, political and economic dimensions. Conversely, governance contains attributes that sets it apart from other organisations. The view of two of these organisations is that not only do the government and civil society constitute actors, but that the private sector is also an actor. Also, with regards to the dimensions of governance, aside from law, social matters, politics and the economy, governance also touches upon other aspects, such as the administrative aspect.

2 CONCEPT OF GOOD GOVERNANCE

As one of the main pillars of a modern state, alongside the rule of law and democracy (Addink et al., 2010), GG arises from the history of a country's growth that tended to occur unchecked and adversely affected the people. Therefore, according to Addink et al. (2010), GG as a concept will always be in line with the rule of law and the democratic state. In short, GG is a source of the norms of government and the right of the citizens that are conditioned under a special formulation, such as legislation at the internasional, regional and national level.

According to UN-ESCAP (UN-ESCAP, 2016), Asian Development Bank (ADB, 1999), United Nations Development Programme (IFAD, 1999), and the World Bank (World Bank, 2016), a governance can fall under the definition of GG if it displays the characteristics of participation, the rule of law, transparency, responsiveness, equity, effectiveness and efficiency, accountability, strategic vision and predictability, and is consensus oriented.

According to Table 1, it can be seen that participation, transparency and accountability are cited by four international organisations in describing GG. The principles of the rule of law, responsiveness, equality and inclusiveness, and effectiveness and efficiency are included by three of the international organisations. The principle of being consensus oriented is adopted by two international organisations, while the principles of strategic vision and predictability are each used by one international organisation.

Ryngaert and Wouters (2005) state that 'GG is, logically, appropriate governance' (p. 73). Furthermore, Botchway (2001) postulates that '…GG need norms and idea to concretize and enrich it. It is argued subsequently that subjects such as democracy, rule of law, decentralization and discretion perform the filling function' (p.189). This outlook focuses GG on legal norms and the framework of the rule of law, since GG as a principle cannot be implemented unless it is created within a set of legal norms under a legal system.

In countries where state administrative law is defined as a set of laws that regulate the government, the prevailing dynamics is where there is an affirmation of the nation's commitment to create good governance (Nugraha, 2005). Nugraha (2005) places the rule of law as the primary principle, as state administrative law, according to Nugraha, is a legal instrument enabling the people in power to be more actively involved with the public. On the other hand,

Table 1. Comparison of the principles of good governance as postulated by international organisations.

Principles	UNDP	World Bank	ADB	UN-ESCAP
Participation	√	√	√	√
Rule of law	√	√	–	√
Transparency	√	√	√	√
Responsiveness	√	√	–	√
Equality and inclusiveness	√	√	–	√
Effectiveness and efficiency	√	√	–	√
Accountability	√	√	√	√
Strategic vision	√	–	–	–
Consensus oriented	–	√	–	√
Predictability	–	–	√	–

Source: (UN-ESCAP, 2016), (ADB, 1999), (IFAD, 1999) and (World Bank, 2016).

state administrative law is a law that allows the people to influence the government in relation to the provision of protection and the fulfilment of human rights.

3 SUBSTANCE OF THE PRINCIPLES OF GOOD GOVERNANCE

The exercise of governmental power using a specific method, according to Addink (Curtin D.M. and Ramses R.A. (Ed), 2005), is reflected by the various elements that represent the principles embodied in GG, such as being democratic, transparent, protecting human rights, being accountable, effective and efficient, and having proper administration. The principles laid out by Addink et al. (2010) are based on the development of GG in the European Union (EU). These principles serve as parameters, in the sense that: firstly, whether the administrative actions and consequences of any government decisions are in line with or in contravention of such principles; and secondly, assessment of government decisions, in that at every step of the process an administrative decision is bound by its own principles. Such steps are preparation for the procedure of decision-making, the determination of the decision's content and the publication of the decision (Curtin D.M. and Ramses R.A. (Ed), 2005).

The principles of GG that function as the parameters of administrative law in the Netherlands are: firstly, proper administration, classified into prohibition of misuse of power, prohibition of arbitrariness, legal certainty, legitimate expectations, equality, proportionality, due care and justification; secondly, public participation in administration, divided into the principles of public participation related to persons, moment, object; thirdly, transparent administration, consisting of decision and order, meetings and information; and fourthly, human rights administration, consisting of classical human rights and social human rights (Curtin D.M. and Ramses R.A. (Ed), 2005).

Addink (Curtin D.M. and Ramses R.A. (Ed), 2005) differentiate each of these principles based on their substantive form, namely fundamental, micro and macro. Human rights administration falls under the definition of fundamental principles, while proper administration is under the definition of micro principles. Public participation and transparent, accountable, effective and efficient administration constitute macro principles. The specialised formulation of the principles of proper administration is under the micro principle group in GG, which relates to the concept of the division of power, namely the executive branch of the government (Addink et al., 2010).

4 CONCEPT OF GOOD GOVERNANCE IN INDONESIA

GG is not a new concept in Indonesia. The issuance of the Decree of the Indonesia's People Consultative Assembly (MPR) No. XI/MPR/1998 on State Governance that is Free from Corruption, which stipulates various legal principles, particularly under Article 1 on the proper procedure of governance, Article 2 paragraph (1) on accountability, Article 3 paragraph (1) on transparency and Article 4 on the principle of legal security, constitutes the cornerstone in the establishment of the values of GG in Indonesia. These principles are also present in other pieces of legislation, such as Law No. 28 of 1999 on State Governance that is Clean and Free from Corruption, Collusion and Nepotism, specifically Article 3; Law No. 30 of 2002 on the Corruption Eradication Commission, specifically Article 3; Elucidation of Article 53 paragraph 2 subparagraph b of Law No. 9 of 2004 on Amendment to Law No. 5 of 1986 on State Administrative Courts; Law No. 23 of 2014 on Regional Governance; and Law No. 30 of 2014 on Government Administration, specifically Article 5 and Article 10 paragraph 1.

According to Table 2, upon examination of the aforesaid pieces of legislation, there are a number of principles that are used as the basis for the enactment of laws and regulations, namely legal security, proportionality, accountability, transparency, public interest, proper governance, professionalism, utilisation, effectiveness and efficiency, externality, diligence, non-abuse of power, good public service and human rights.

Table 2. Legal principles embodied in Indonesian regulatory provisions.

Legal principles	TAP. MPR No. XI/MPR/ 1998	Law No. 28 of 1999	Law No. 30 of 2002	Elucidation of Law No. 9 of 2004	Law No. 23 of 2014	Law No. 30 of 2014
Legal security	√	√	√	√	√	√
Proportionality	√	√	√	√	√	√
Accountability	√	√	√	√	√	–
Transparency	–	√	√	√	√	√
Public interest	–	√	√	–	√	√
Proper governance	–	√	–	√	√	–
Professionalism	–	√	–	√	–	–
Utilisation	–	–	–	–	–	√
Effectiveness and efficiency	–	–	–	–	√	–
Externality	–	–	–	–	√	–
Carefulness	–	–	–	–	–	√
Non-abuse of power	–	–	–	–	–	√
Good public service	–	–	–	–	–	√
Human rights	–	–	–	–	–	√

Such legal principles, according to Paulus (as cited in Hadjon et al., 2010), are essentially the result of a review of court rulings in the Netherlands by R. Crince Le Roy. The study resulted in a number of principles, namely legal security, proportionality, equality, carefulness, motivation, non-misuse of competence, fair play, reasonableness of prohibition of arbitrariness, meeting raised expectations, undoing the consequences of an annulled decision and the principle of protecting the personal way of life. In Indonesian literature, these principles are more familiarly known as Asas-Asas Umum Pemerintahan yang Baik (AAUPB).

According to Erliyanna (2004), AAUPB was established in 1929 in the court of civil servants, when such principles gained acceptance as grounds for legal actions. These call upon the court to examine, firstly, whether the ruling is in conflict with the law or applicable public regulations; secondly, whether the authority exercised in making such a ruling is in line with the intended purpose for which such authority was given; and thirdly, whether there is a balance between the sanction given and the offence committed.

According to Prajudi (1994), AAUPB can be classified into two categories: first, the set of principles relating to procedures; and second, the set of principles that deal with the correctness of the facts. Prajudi (1994) further incorporates the principle of vested interest, the principle to respond to a decision and the principle of motivation into the definition of procedural principles. Principles relating to the facts are non-arbitrary action, exclusion of *detournement de pouvoir*, legal security, non-discrimination and nullification due to the neglect of a public official (Prajudi, 1994).

A different method of AAUPB classification is offered by Anna (as cited in Nugraha, 2005): first, formal principles with respect to the process of the preparation of decisions, such as the principles of carefulness, fair play, reasonableness or non-arbitrariness; second, formal principles relating to the motive for the issuance of the decision, such as the principle of motivation; and third, material principles that relate to the content of the decision, such as the principles of legal security, equality, proportionality, non-misuse of competence, fair play, meeting raised expectations and reasonableness or non-arbitrariness.

The AAUPB principles classified by Prajudi and Anna are, in fact, principles that relate to decision-making, with a focus on the government, specifically state administrative officials. According to the above view, it can be summarised that, in the context of the decision-making process undertaken by a state administrative official, not all AAUPB principles can be applied. Each subset of principles has different characteristics that reflect the nature of each stage through which a government's decision is made.

Table 3. AAUP by classification of principles.

Classification of principles		
Prajudi	Anna	AAUP
Regarding procedure	Formal procedural	Carefulness
		Fair play
		Reasonableness or non-arbitrariness
	Formal motive	Motive
On correctness of facts	Material on content	Legal security
		Equality
		Proportionality
		Non-abuse of authority
		Undoing the consequences of annulled decisions
		Meeting raised expectations
		Protecting the personal way of life

Table 3 exhibits the links between the classifications, as formulated by Prajudi, with regards to the group of principles relating to procedures and the group of principles relating to the correctness of facts. In addition, there is the formal group of principles relating to procedures, the formal group of principles relating to motive and the group of material principles relating to content, as postulated by Anna. The principles of carefulness, fair play and reasonableness or non-arbitrariness are grouped into the subset of formal principles relating to procedure, as formulated by Anna, and/or the group of principles relating to procedure, as put forward by Prajudi. Furthermore, the principles of legal security, equality, proportionality, non-abuse of power, undoing the consequences of annulled decisions, meeting raised expectations and protecting the personal way of life are included in the group of material principles and/or those on the correctness of facts.

Paulus (as cited in Hadjon et al., 2010) puts forward an argument on GG and AAUPB: '…it can be concluded there are indeed a difference and a similarity between good governance and AAUPB principles in administrative law, although it cannot be denied that there is a correlation and linkage between them' (p.47). Further, the differences and similarities between GG and AAUPB are as follows: the similarity is that both GG and AAPUB embody the philosophy, knowledge and idealism that are expressed through doctrines and theories to achieve appropriate governance; the difference is that, while GG falls under the state administrative science discipline and is a genus (general), as it can cover law, politics, the economy, social affairs and administration, AAUPB is a discipline of state administrative law that is a species (specific) that only focuses on the field of administrative law (as cited in Hadjon et al., 2010).

5 CONCLUSION

Referring to Table 4, the substantive aspects of the GG principles, based on the subset of principles relating to the procedures for making government decisions (A1) and the subset of principles on the correctness of facts in government decisions (A2), are: first, GG at the fundamental level falls under subset A2. The breakdown of the A2 fundamental level of GG is as follows: human rights administration, rule of law, protecting the personal way of life and human rights. Respectively, the characteristics of GG are adopted by the Netherlands, international organisations, AAUPB and the law.

Table 4. Characteristics of good governance as adopted by the Netherlands, international organisations, AAUPB, and regulatory instruments by grouping of principles and content.

The Netherlands		International organisations	AAUPB	Law	Principle group	Content
Principles	Sub-principles					
Human rights administration	Classical human rights	Rule of law	Protecting the personal way of life	Human rights	A2	Fundamental
	Social human rights					
Proper administration	Prohibition on misuse of power	–	Non-misuse of competence	Professionalism	A2	Micro
	Prohibition of arbitrariness	–	Prohibition of arbitrariness	Non-abuse of power	A1	
	Legal certainly		Legal security	Legal assurance	A2	
	Legitimate expectations	Predictability	Undoing consequences of annulled decisions	–	A2	
			Meeting raised expectations	Proper procedure of governance	A2	
	Equality	Equality and inclusiveness	Fair play	–	A1	
			Equality		A2	
	Proportionality	–	Proportionality	Proportionality	A2	
	Due care	–	Carefulness	Carefulness	A1	
	Justification for decision	–	Motivation	Public interest	A1	
Public Participations	Related to person	Participation	–	–	–	Macro
	Related to moment	Responsiveness	–	–	–	
	Related to object	Consensus	–	–	–	
Transparent administration	Decision and order	Transparency	–	Transparency	–	
	Meeting					
	Information					
Accountable administration	–	Transparency	–	Transparency	–	Macro
	–	Accountable	–	Accountability	–	
Effective and efficient administration	–	Effectiveness and efficiency	–	Effectiveness and efficiency	–	
	–	Strategic vision	–	Externality	–	

Source: Compiled by author.

Second, micro level GG consists of both A1 and A2. The breakdown of the A1 micro level of GG is as follows: proper administration in the GG characteristics in the Netherlands; equality and inclusiveness in the GG characteristics adopted by international organisations; non-arbitrariness, fair play, carefulness and motivation of GG characteristics under AAUPB; non-abuse of power and public interest in the GG characteristics of the law. Further elaboration of the A2 micro level of GG includes: proper administration in the characteristics of GG as adopted by the Netherlands; predictability, equality and inclusiveness in the characteristics of GG as adopted by international organisations; non-misuse of competence, legal security, undoing the consequences of annulled decisions, meeting raised expectations, equality and proportionality in the characteristics of GG under AAUPB; professionalism, legal assurance, proper procedures of governance and proportionality in the GG characteristics under the law. Third, the macro level of GG does not fall under either A1 or A2.

Thinking along these lines, it is urgent that A1 and A2 be differentiated based on substance or content into fundamental level and micro level GG. Principles under the macro level of GG that do not fall under A1 and A2, based on the substance of GG, should also be urgently given parameters for assessing government decisions. The principles of the macro level of GG, in the opinion of Addink (Curtin D.M. and Ramses R.A. (Ed), 2005), should be given parameters for assessing government decisions for the publication, in terms of: first, whether the administrative actions and consequences of such actions are in line or in conflict with the principles; and second, the review of government decisions, so that every step in the administrative decision-making process is bound by their own principles.

REFERENCES

Addink, G.H., Anthony, G., Buyse, A.C. & Flinterman, C. (2010). *Sourcebook human rights and good governance.* Utrech: SIM.

Asian Development Bank [ADB]. (1999). *Governance: Sound development management.* Manila: ADB.

Botchway, F.N. (2001). Good governance: The old, the new, the principle, and the elements. *Florida Journal of International Law, 13*(2), 159–210.

Deidre, M. Curtin and Ramses, A. Wessel, eds. (2005). *Good Governance and the European Union; Reflection on Concept, Institution, and Substance.* New York: Bureu JA Vormgevers.

Erliyanna, A. (2004). Analysis of republic Indonesia presidential decree, period *1987–1998:* prohibition of arbitrariness principle review (Ph.D. Thesis). Universitas Indonesia, Jakarta.

Hadjon, P.M., Lotulung, P.E., Marzuki, H.M.L., Djatmiati, T.S. & Wairocana, I.G.N. (2010). *Administrative law and Good Governance.* Jakarta, Universitas Trisakti.

Indonesia. (1999). *Undang-undang No. 28 Tahun 1999 tentang Penyelenggaraan Negara yang Bersih dan Bebas dari Korupsi, Kolusi dan Nepotisme* [*The Law Number 28 of 1999 regarding the State Organizer who is Clean and Free from Corruption, Collusion and Nepotism*].

Indonesia. (2002). *Undang-undang No. 30 Tahun 2002 tentang Komisi Pemberantasan Tindak Pidana Korupsi* [*The Law Number 30 of 2002 regarding the Commision for Eradication of Criminal Acts of Corruption*].

Indonesia. (2004). *Undang-undang No. 9 Tahun 2004 tentang Perubahan atas Undang-undang Nomor 5 Tahun 1986 tentang Peradilan Tata Usaha Negara* [*The Law Number 9 of 2004 regarding Amendment to Law No. 5 of 1986 on State Administrative Courts*].

Indonesia. (2014). *Undang-undang No. 23 Tahun 2014 tentang Pemerintahan Daerah* [*The Law Number 23 of 2014 regarding Regional Governance*].

Indonesia. (2014). *Undang-undang No. 30 Tahun 2014 tentang Administrasi Pemerintahan* [*The Law Number 30 of 2014 regarding Government Administration*].

International Fund for Agricultural Development [IFAD]. (1999). *Good governance: An overview.* Retrieved from http://www.ipa.government.bg/sites/default/files/pregled-dobro_upravlenie.pdf

Jubaedah, E., Dawud J., Mulyadi, D., Nugraha, Faozan, H. & Wulandari, P. (2007). *Implementation of Good governance measurement models in district/city local government* Bandung: Pusat Kajian Pendidikan dan Pelatihan Aparatur I LAN.

Nugraha, S., Erliyana, A., Mamudji, S., Hayati, T., Nursadi, H., Sunarti, E.S. & Simatupang, D.P.N. (2005). *Administrative law.* Depok: CLGS Fakultas Hukum Universitas Indonesia.

Nugraha, S. (2006). Administrative law and good governance. In: *Pidato pada Upacara Pengukuhan Sebagai Guru Besar Tetap pada Fakultas Hukum Universitas Indonesia, 13 September 2006*. Depok: Badan Penerbit FHUI.

Oppenheim, L. & Lauterpacht, H. (1967). *International law: A treatise Volume 1*. London: Longmans Green and Co.

Osborne, D. & Gaebler, T. (1993). *Reinventing government: How the entrepreneurial spirit is transforming the public sector*. New York: Penguin Group.

Prajudi. (1994). *Administrative law*. Jakarta: Ghalia Indonesia.

Rahman, H.H. (1971). *Political science and government*. Dhaka: Ideal Publications.

Ryngaert, C. & Wouters, J. (2005). *Good governance: Lessons from international organizations*. New York: Wessel.

Thompson, B. (1997). *Textbook on constitutional and administrative law*. London: Blackstone Press Limited.

United Nations Development Program [UNDP] & Eurostat. (2007). *Governance indicators: A users' guide*. New York: UNDP and Eurostat.

United Nation Economic and Social Commision for Asia and the Pacific [UN-ESCAP]. (2016). *What is good governance?* Retrieved from http://www.gdrc.org/u-gov/escap-governance.htm.

World Bank. (2014). *What is governance?* Retrieved from http://web.worldbank.org/WBSITE/EXTERNAL/COUTRIES/MENAEXT/EXTMNAREGTOPGOVERNANCE/0.contentMDK:20513159~pagePK:34004173~piPK:34003707~theSitePK:497024,00.html.

World Bank. (2016). *Municipal development partnership for Eastern and Southern Africa. Module 1: City management strategy*. Harare: Municipal Development Partnership for Eastern and Southern Africa [MDP ESA].

Does sanctity of contract exist in oil and gas contracts in Indonesia?

F.H. Ridwan, R. Agustina & J. Rizal
Faculty of Law, Universitas Indonesia, Depok, Indonesia

ABSTRACT: The Indonesian oil and gas industry is one of the largest revenue streams for the country. The management of oil and gas in Indonesia is governed in details under contracts entered into by the Indonesian government and domestic and international contractors and investors. The contractual arrangement underwent a number of phases from prior to the introduction of the Law No. 44 of 1960 to the enactment of the Law No. 22 of 2001 on Natural Oil and Gas. The most common and preferred form of contract to date is the production sharing contract. The background leading to the choosing of this issue is to what extent these PSCs executed by the government and investors or contractors have met the principle of sanctity of contract, which constitutes a requirement for the validity of an agreement. This study explains the importance of upholding of sanctity of contract since from the drafting of the agreement up to its conclusion. The reason for this study originates from the inequality of position of the Indonesian government and the investors in terms of the production sharing and the legal certainty of the contract. Another aim is to gain a deeper understanding of the latest issues relating to sanctity of contract by comparing Indonesia's oil and gas contracts with those executed by other countries. At the end of the study, it is concluded that Indonesia is in need of a production sharing contract that is more equitable, provides greater legal certainty and upholds sanctity of contract.

1 INTRODUCTION

The term production sharing contract can be found in Article 12 paragraph 2 of the Law No. 8 of 1971 on Pertamina and its amending law, the Law Number 10 of 1974. Under these laws Pertamina (Indonesia's state-owned gas and oil company) holds the mining authority over all mining areas throughout the country, insofar as such operation relates to the extraction of natural oil and gas. In performing such function, Pertamina faces limited capital and technology and is allowed to collaborate with other entities to explore and exploit natural oil and gas under a production sharing contract (Article 12 of the Pertamina Law) [Mining Info].

Meanwhile, in article 1 sub-article 19 of the Law No. 22 of 2001 on Natural Oil and Gas, the term used is cooperation contract. A cooperation contract can take the form of a production sharing contract or any other arrangement.

A production sharing contract is a model developed from a production sharing concept already adopted by traditional laws in Indonesia. The traditional concept of production sharing agreement has been subsequently codified into the Law No. 2 of 1960 on the Production.

Sharing Agreement. Under this law, the definition of production sharing agreement is an agreement that can take any name entered into between an owner as one party and another person or legal entity, referred to as the 'cultivator'. Under the agreement, the cultivator is permitted by the owner to run agricultural activities on the owner's land, and the contract provides that the product from such work to be shared. This concept further evolved to become production sharing contracts employed by the oil and gas industry [the Law No. 2 of 1960, Article 1 sub-article c].

Cooperation contracts according to the Law No. 22 of 2001 on Natural Oil and Gas encompass production sharing contracts and other forms of contract governing exploitation of resources that may better serve the interests of the government and any product resulting therefrom must be used for the greatest prosperity of the people [the Law No. 22 of 2011 on Natural Oil and Gas]. The provision of this clause does not specifically explain the definition of production sharing contract, but it is more focused on the concept of cooperation as adopted in the oil and gas sector. The Law No. 22 of 2001 does not provide a definition of production sharing contract. Instead, the definition can be found in Article 1 sub-article 1 of the Government Regulation Number 35 of 1994 regarding conditions and guidelines for production sharing contracts in the oil and gas sector. Under this regulation, a production sharing contract is "an arrangement between Pertamina and a contractor to undertake exploration and exploitation of oil and gas pursuant to the principles of production sharing" [the Government Regulation No. 35 of 1994].

Production sharing contracts in the mining of oil and gas are designed in such a way so as to get around Pertamina's lack of capital, technology and human resources for conducting exploration and exploitation of oil and gas. Under Article 6 paragraph 1 of the Law No. 22 of 2001, oil and gas production activities, particularly for upstream operations, must be implemented and controlled through cooperation contracts.

2 RESEARCH QUESTION

Based on the description above, there seems to be underlying problems in the execution of the production sharing contract in the oil and gas industry in Indonesia. There is a gap in implementing the content of the agreement and different interpretation still exists as the "Sanctity of the Contract" is breached. Therefore, in this paper, there is a search for answer to the following question: Does the oil and gas contract in Indonesia comply with the general principles of the contract law in order to reach the sanctity of the contract?

3 RESEARCH METHODS

To address the problem and to achieve the objectives of this research, the juridicial normative method or the legal doctrine research is used. Normative research is applied to reveal how norms or principles of an agreement can be applied in the oil and gas contract (das sollen), whilst an empirical research is applied to find out how those principles are applied by the parties, in this case the government and the investor (das sein). The juridicial normative method known as the doctrinal research is research that refers to a legal analysis; law as it is written in the book and law as it is decided by judge through judicial process (Dworkin, 1973).

4 RESULTS AND ANALYSIS

There are principles of production sharing contracts:

1. First Generation (1964–1977).
These contracts were the predecessors of production sharing contracts. In the years 1973–1974, oil prices spiked, prompting the government to establish that from 1974 contracts must provide additional payments to the government under the first general production sharing contracts.
2. Second Generation (1978–1987).
The principle embodied by the second generation of production sharing contracts (1978–1987) is that no operational cost recovery ceiling shall be taken into account by a contractor.
3. Third Generation (1988–2002).
In 1984 the Indonesian government introduced a new tax regulatory scheme for production sharing contracts that set the rate at 48%. However, the regulation was only enforceable on

contracts that were entered into after 1988. Contractors still preferred to apply the previous tax scheme.

4. Fourth Generation (2002 – present).

The momentum leading up to the advent of the fourth generation of production sharing contracts started with the enactment of the Law No. 22 of 2001 on Natural Oil and Gas. The structure and governing principles of the production sharing scheme under this law differs from those under the previous laws. Under the previous regulatory framework, the parties to the contract are Pertamina and contractors, while under the Law No. 22 of 2001 the contracting parties are an implementing agency [*badan pelaksana*] and a commercial enterprise [*badan usaha*] and/or a permanent commercial enterprise [*badan usaha tetap*].

The Law No. 22 of 2001 does not specifically provide the production sharing portion for the implementing agency and the commercial enterprise and/or permanent commercial enterprise. Their respective shares are elaborated in a lower ranking regulation and in the production sharing contract itself. Referring to Article 66 paragraph (2) of the Law No. 22, it is clearly stipulated in that clause that any implementing regulation of the Law No. 44 of 1960 on the Mining of Oil and Gas and the Law No. 8 of 1971 on Pertamina will continue to be effective insofar as none of their provisions are in conflict with or have been superseded by newer provisions under this law. In Article 16 of the Government Regulation Number 35 of 1994 on the Terms and Guiding Principles for Oil and Gas Production Sharing Contracts it is stipulated that the determination of product sharing portion shall be made by the Minister of Mining and Energy.

The Law also provides for the apportioning of the entitlement of the commercial enterprise or permanent commercial enterprise to meet the domestic market need at a maximum of 25% (Article 22 of the Law No. 22 of 2001 on Natural Oil and Gas). Each generation of production sharing contracts establishes different portions of production sharing for Pertamina and the contractor.

5 RIGHTS AND OBLIGATIONS OF PARTIES UNDER A PRODUCTION SHARING CONTRACT

The rights and obligations of a commercial enterprise and/or permanent business entity undertaking upstream operations under a production sharing contract are governed by Article 31 of the Law Number 22 of 2001 on Natural Oil and Gas. There are 2 types of obligations imposed on a commercial enterprise and business entity, namely payment of tax state revenues and payment of non-tax state revenues.

Tax state revenues consist of:

1. taxes;
2. custom duty and other taxes on imports and customs;
3. local taxes and retributions.

Non-tax state revenues consist of:

1. State's share, i.e., a portion of the production relinquished by the commercial enterprise or permanent business entity to the state as owner of the oil and gas resources;
2. Land rent [*iuran tetap*], i.e., the amount that must be paid by the commercial enterprise and/or permanent business entity to the state as the owner of the oil and gas resources in proportion to the concession area and as compensation for the provision of opportunity to conduct exploration and exploitation work;
3. Exploration and exploitation fee, i.e., a fee paid by the commercial enterprise or permanent business entity to the state as compensation for the extraction of non-renewable resources;
4. Bonuses, i.e., bonuses from income or signing bonus, data compensation bonus, production bonus, and bonuses in any other forms received by the implementing agency under a production sharing contract.

6 WHAT IS SANCTITY OF THE CONTRACT?

The sanctity of the contract means (a) a valid contract is binding upon the parties. It can only be modified or terminated by consent of the parties or if provided for by the law. The parties to a contract must, unless legally excused from performance, perform their respective duties under the contract [*pacta sunt servanda*] (b) a valid unilateral promise or undertaking is binding on the party giving it if that promise or undertaking is intended to be legally binding without acceptance (Tras-Lex Law Research, n.d.).

7 PROBLEMS ASSOCIATED WITH PRODUCTION SHARING CONTRACTS IN INDONESIA

A problem with production sharing contracts in the oil and gas sector in Indonesia is the varying interpretations regarding the positions of the government and the investor as contracting parties. As such, to overcome this problem the principles of law of contracts must be applied, which are binding on the parties. If these principles are complied with, it can be deemed that the contract has achieved sanctity of the contract. The principle of proportionality is one that is always raised and demanded in civil law, as it is an important part of an agreement. To be familiar with this principle, one would have to examine the substantive content of the agreement drafted and agreed upon by the parties. In any agreement the most important aspect is the performance and enjoyment of the obligations and rights in a proportional manner, as otherwise it can potentially lead to a conflict (Hernoko, 2013).

8 DOES THE SANCTITY OF THE CONTRACT EXIST IN THE PRODUCTION SHARING CONTRACTS IN INDONESIA?

As a country adopting the civil law system, the classical theory applies in Indonesia. An agreement is a legal act involving two [*een tweezijdige overeenkomst*] based upon on a consensus that carries a legal effect. The term "legal act involving two" refers to a legal act consisting of an offer from one party and an acceptance of such offer by another party. This classical outlook, however, is not accurate in this case. As there is an offer from one party and an acceptance from the other, there are two legal acts that each involves one. Therefore, the agreement is less of a legal act than a legal relationship between two persons who have agreed to create a legal consequence. Under the Indonesia Civil Code, a definition is given to agreement (Article 1313), but the wording is overly general and ambiguous, as it only describes it as "act", which is wide open to interpretation, as it can denote a legal act or factual act, and it is also unclear (Mertokusumo, 2002).

Steven L. Emanuel states that "A contract is an agreement that the law will enforce in some way. A contract must contain at least one promise i.e. a commitment to do something in the future" (Syaifuddin, 2012). The law on contract adopts the school of individualism encased in its characteristics, both under the BW 1838 (previous) and under Nieuw BW of 1992, under the themes of contractual freedom, equality, and bondage [*vrijheid, gelijkheid, en contractuele gebondenheid*]. Of the numerous legal principles embodied in a contract, only three constitute the most fundamental [*grondbeginselen*], namely:

a. Consensus, meaning that an agreement arises from a consensus reached by the contracting parties. An agreement is entered into freely and is stipulated formally.
b. Binding effect [*verbindende kracht der overeenkomst*], in that the parties must subject themselves to and abide by what has been agreed upon.
c. Freedom of contract, whereby any person is free to enter into any agreement and be bound to any person he desires.

Sudikno's take on the legal principles governing a contract as described above is different from that of Ridwan Khairandy. The difference is that Ridwan Khairandy postulates that contract law contains four interrelated principles, namely: (1) consensuality; (2) legal binding effect of contract; (3) freedom of contract; (4) good faith (Khairandy, 2013).

Meanwhile, Nieuwenhuis prescribes that the law on contracts recognizes three underlying principles that differ from the frameworks declared by Sudikno and Ridwan Khairandy. The three underlying principles are: (1) autonomy, or the authority of the parties to enter into a legal relationship (freedom of contract); (2) trust, or the trust created by the contract that needs to be protected (principle of protection for party acting in good faith); (3) cause, or a codependency for an agreement to be subject to a certain legal regime, notwithstanding the principle of freedom of contract (Khairandy, 2013).

According to Sudikno Mertokusumo, freedom of contract constitutes the third principle for contracting. Basically every person is free to enter into and determine the content of an agreement. While the principle of consensuality relates to the creation of an agreement, and the principle of legal bondage relates to the consequence of an agreement, the principle of freedom of contract relates to the substantive content of the agreement (Mertokusumo, 2002).

Two principles that can be found in a contract are the principles of equality and proportionality that are often closely interwoven and difficult to set apart due to their minor differences. In order to differentiate and understand them, we would have to trace back the lexical background of the terms. Many experts disagree on these two principles; some view them as one and the same, while others see them as separate principles.

In Great Dictionary of the Indonesian Language [*Kamus Besar Bahasa Indonesia*], the official Indonesian dictionary, equality [*keseimbangan*] means the condition of being of equal terms, similar, balanced. Meanwhile, the term '*proporsionalitas*" means proportionate, similar, balanced. W. van Hoeve translates the Dutch word '*evenredig*' as proportionate, of equal footing. '*Evenredigheid*' means proportionality or the state of being in equal footing. In this regard *evenredigheid* is taken to be the same as *evenwicht*, which means balance, of equal weight. Meanwhile, '*proportioneel*' is also taken as to mean being in balance, proportionate. It appears that the interpretation provided by W. van Hoeve is similar to that which is provided in the Great Dictionary of Indonesian Language, giving the same meaning to both terms (Hernoko, 2013).

In the Indonesian legal system, the Indonesian Civil Code [*KUHPerdata*] is the sole legal instrument that governs contracts in Indonesia. An agreement or contract is governed by the set of contract law codified in Volume III of *KUHPerdata*. The three principles that are interrelated are consensuality, binding effect of contract, and freedom of contract.

The principle of consensuality determines that an agreement is established due to the convergence of interests or intention of the parties. An agreement can essentially be in any form without having to go through any formal process, but merely through the attainment of a consensus. The binding effect of contract and the freedom of contract have their legal foundation in *KUHPerdata*, specifically Article 1338 paragraph (1), which reads as follows:

"All contracts lawfully entered into shall serve as law for the contracting parties." (Subekti, 2002).

The article states that an agreement gives rise to legal obligations on the parties who are bound to perform the contractual agreements and that any undertaking must be honored. Life in society can only continue harmoniously if a person is able to trust the words of another person; as such, compliance with an agreement is one of the conditions demanded by logic and conscience. This contractual binding power works under the assumption of freedom for society to participate in all legal interactions and imply freedom of contract. Freedom of contract is a key principle in civil law, particularly contract law that lays the foundation for all agreements. Freedom of contract that originated from the philosophical teaching of individualism of ancient Greece and the rapid development during the renaissance era determine that every person is free to receive what he desires (Atiyah, 1989). This school of thought is familiar both in the civil law system and the common law system. Freedom of contract is also the philosophical foundation on which international entities based their effort to create a unification of contract laws (Budiono, 2006, p. 101 citing Asser-Hartkamp 4-II, *De trouw aan het gegeven* word is *derhalve een eis der natuurlijke rede*, Deventer, 1997, nr. 37. 44 Id, p. 104), where it is stated that the principle of freedom of contract stems from two concepts:

(1) freedom of contract indicates that the contract is based upon a reciprocating agreement; (2) such freedom emphasizes that the establishment of the contract is the representative of the free will of the parties that is free from any external influence, such as interference from the government or the legislature).

"The parties are free to enter into a contract and to determine its content". The freedom of contract principle was born from the philosophy of natural law founded by Hugo de Groot (Grotius). According to Grotius, this principle represents the moral power of an undertaking by stating that "a promise is binding" [*pacta sunt servanda*] and that "we need to honor our promises" (*promissorum implendorum obligati*) (Friedman, 1960).

The adage *pacta sunt servanda* has carried a major significance since the sixteenth century, not only in civil law but also in state administrative law and international law. If it is conceived by an oath, the nature of the adage changes to become the binding effect of an agreement. The obligation to honor and perform the provisions of a contract is mandatory, as the agreement in essence is binding upon the parties. The principle stating that we are bound by contractual covenants and undertakings determines that they are not merely moral obligations but they also constitute legal obligations. They must be considered as independent obligations and thus should not be questioned further. This also implies the obligation to perform the agreements even though the outcome may not be desirable or the performance thereof may not be pleasant or convenient (Dirdjosisworo, 2003).

Freedom of contract does not entail absolute freedom, as such freedom is still curbed by various applicable restrictions, among others, those imposed by the national public legislation that govern public order and morality. These restrictions are known as the legal principle of '*pacta private juri publico derogare non possunt*'. This restriction on freedom of contract is also manifested in the conferral of larger significance on values of reasonableness and fairness (*redelijkheid en bijlijkheid*). Moreover, curtailing freedom of contract can also arise in formal contracts, where the parties cannot independently determine their rights and obligations. In this case, the terms of the agreement are determined by a party with a higher position of power.

Under Indonesia's Civil Code, provisions that restrict a person's freedom of contract include those stipulated in Article 1320 that lays down four legal conditions to be met to ensure validity of contracts, and Article 1338 paragraph (3) that stipulates that an agreement must be entered into in good faith. According to Subekti, good faith is the cornerstone of contract law. Subekti further postulates that the law itself always strives to achieve two objectives, namely to provide legal assurance (order) and meet the demand for justice. If Article 1338 paragraph (1) of the Civil Code can be viewed as a prerequisite for and demand for legal assurance (binding effect of contracts), then Article 1338 paragraph (3) of the Civil Code must be seen as a demand for justice (Subekti, 2002).

9 CONCLUSION

Basically a production sharing contract is an alternative form of cooperation in the exploration and extraction of resources that better serve the interests of the state and the proceeds to be applied for the prosperity of the people. The momentum leading up to the advent of the fourth generation of production sharing contract started with the enactment of the Law No. 22 of 2001 on Natural Oil and Gas. The structure of the production sharing scheme under this law differs from those under the previous laws. Under the previous regulatory framework, the parties to the contract are Pertamina and contractors, while under the Law No. 22 of 2001 the contracting parties are the implementing agency (Badan Pelaksana) and the commercial enterprise (Badan Usaha) and/or permanent commercial enterprise (Badan Usaha Tetap). The existing production sharing contracts should include general legal principles that govern contracts in order to ensure the sanctity of contract, which means a valid contract is binding upon the parties. It can only be modified or terminated by consent of the parties or if provided for by the law. The parties to a contract must, unless legally excused from performance, perform their respective duties under the contract ("pacta sunt servanda") and the parties should strictly abide by the agreed upon terms.

10 RECOMMENDATIONS

The share received by the regions under production sharing schemes should be increased, given that their share is extremely small due to the high cost expended for the exploration and extraction of oil and gas and the sophisticated technology required to undertake such operations. Therefore, to facilitate exploration and extraction of these natural resources, production sharing contracts are entered into with domestic and foreign based companies. Since foreign companies possess significant assets and skills, they are given a proportionately large portion of the share, and the state ought to intensify efforts to enhance local human resources and their skills as well as increasing capital to allow for a more profitable outcome from the exploration and exploitation of natural resources, both for the government and domestic companies. Therefore, the suggestion for the parties, both the government and the investor, is they should make a production sharing contract by taking into account four inter-related principles in the contract law: (1) consensuality; (2) legal binding effect of contract; (3) freedom of contract; and (4) good faith, and if the principles are applied in the contract, we can say the sanctity of the contract has been reached.

REFERENCES

Atiyah, P.S. (1989) *An Introduction to the Law of Contract*. Oxford, Clarendon Press.
Budiono, H. (2006) *Asas Keseimbangan bagi Hukum Perjanjian Indonesia*. Bandung, Citra Aditya Bhakti.
Dirdjosisworo, S. (2003) *Kontrak Bisnis: Menurut Sistem Civil Law, Common Law dan Praktek Dagang Internasional*. Bandung, Mandar Maju.
Dworkin, R. (1973) *Legal Research*. Daedlus, Spring.
Friedman, W. (1960) *Legal Theory*. London, Stevens & Sons.
Hernoko, A.Y. (2013) *Hukum Perjanjian Asas Proporsionalitas dalam Kontrak Komersil*. Jakarta, Kencana Prenada Media Group.
Indonesia (1945) *Undang-undang Dasar 1945* [*The 1945 Constitution of the Republic of Indonesia*].
Indonesia (1960) *Undang-undang No. 2 Tahun 1960 tentang Perjanjian Bagi Hasil* [*The Law Number 2 of 1960 regarding Sharing Contract*].
Indonesia (1960) *Undang-undang No. 44 Tahun 1960 tentang Pertambangan Minyak dan Gas Bumi* [*The Law Number 44 of 1960 on the Mining of Oil and Gas*].
Indonesia (1971) *Undang-undang No. 8 Tahun 1971 tentang Perusahaan Pertambangan Minyak dan Gas Bumi Negara* [*The Law Number 8 of 1971 regarding State Oil and Gas Mining Enterprise*].
Indonesia (1974) *Undang-undang No. 10 Tahun 1974 tentang Perubahan Pasal 19 Ayat 1 Undang-undang No.8 Tahun 1971* [*The Law Number 10 of 1974 regarding Amended of Article 19 (1) Law Number 8 of 1971*].
Indonesia (2001) *Undang-undang No. 22 Tahun 2001 tentang Minyak dan Gas Bumi* [*The Law Number 22 of 2001 regarding Natural Oil and Gas*].
Indonesia (1994) *Peraturan Pemerintah No. 35 Tahun 1994 tentang Syarat-syarat dan Pedoman Kerjasama Kontrak Bagi Hasil Minyak dan Gas Bumi* [*The Government Regulation Number 35 of 1994 regarding Requirements and Guidelines for Oil and Natural Gas Production Sharing Contract*].
Khairandy, R. (2013) *Hukum Kontrak Indonesia dalam Perspektif Perbandingan*. Yogyakarta Fakultas Hukum Universitas Islam Indonesia (FH UII) Press.
Mertokusumo, S. (2002) *Mengenal Hukum: Suatu Pengantar*. Yogyakarta, Liberty.
Subekti, R. (2002) *Hukum Perjanjian*. Jakarta, Intermasa.

Protection of personal information: The state's obligation to guarantee the right to privacy in Indonesia

N. Rianarizkiwati & J. Asshiddiqie
Faculty of Law, Universitas Indonesia, Depok, Indonesia

ABSTRACT: The Second Amendment to the Indonesian Constitution brings an acknowledgement of the right to protection of individuals as one of the fundamental human rights. The right to protection of individuals, which is more popularly known as the right to privacy, provides a guaranteed security for all people in their activities, in terms of either physical or non-physical protection, one function of which is providing security of personal information for every person. As a state operating under a rule of law that acknowledges human rights, the Indonesian government has the responsibility to provide protection concerning the personal information that is collected, through various methods, by the public sector (government agencies) and the private sector (companies or other private entities). This paper reviews the concept of privacy as it applies in Indonesia and the importance of regulating the protection of private information as mandated by the Constitution. The recommendations arising from this review provide options for regulations to ensure the security and convenience of every person with regards to the collection of personal information.

1 INTRODUCTION

From the beginning of its independence, Indonesia has established itself as a state that operates based on the rule of law. In its original Constitution of 1945, in the section of Elucidation of the State Government System, Indonesia states that the country is based on law (*rechtsstaat*), not only based on power (*machtsstaat*) (Kusuma, 2009). This statement is strengthened by the Third Amendment to the 1945 Constitution in Article 1 section (3), which states that Indonesia is a country based on law (Asshiddiqie, 2007). According to the rule of law theory, a country must guarantee the protection of human rights that is written in the Constitution of a modern state law (Davidson, 1994). The Constitution, as the highest source of law in Indonesia, includes human rights provisions in its Second Amendment.

The right to protection of individuals, as a new human right mentioned in the Amended Constitution of 1945, is still ambiguous. There is no clear definition or scope of the protection of individuals; whether it covers physical protection only, the protection of property, or also covers non-physical protection, such as the confidentiality of personal information. Furthermore, the right to protection of individuals has, to some extent, a similar meaning to the right to privacy, which is the right to be left alone (Warren & Brandeis, 1890). This means that everyone has the right to be protected from disturbance by others. The importance of the protection of personal information is to guarantee that information given by one person to another person/body/agency is used in accordance with its purpose of collection. This is to protect everyone and to avoid the misuse of personal information.

The aim of this paper is to give a better understanding of the right to protection of individuals in general, as is stated in the Amended 1945 Constitution, and the right to the confidentiality of personal information in particular. The first part of the paper explains the definition of the right to protection of individuals, more popularly known as the right to privacy. The second part of the paper explains the obligation of the government to protect the human right to privacy with regards to personal information. The last part of the paper explains a case study through several regulations about the implementation of collecting and processing personal information in both public and private sectors. The paper closes with a conclusion regarding the protection of personal information.

2 COMPARABLE FUNCTION: IS THE RIGHT TO PROTECTION OF INDIVIDUALS THE SAME AS THE RIGHT TO PRIVACY?

The history of human rights in the Indonesian Constitution can be traced back to the original 1945 Constitution. The original 1945 Constitution did not distinguish between human rights and the rights of citizens; however, there were a few arrangements to guarantee the rights of citizens. This condition can be understood as there was no Universal Declaration of Human Rights (UDHR) at the time of Indonesian independence in 1945; therefore, there was no guidance to arrange human rights in the Constitution. Citizens rights in the original 1945 Constitution consisted of 6 (six) provisions (articles and sections), namely regarding equality before the law, freedom of speech, the right to religion and the right to education (Kusuma, 2009).

As Indonesia changed its governmental political system from a unitary to a federal state in 1949, the Constitution was changed into the Federal Constitution of 1949. The spirit of the UDHR was adopted in the Federal Constitution of 1949 and, later on, in the Temporary Constitution of 1950, as they respectively consisted of 44 (forty-four) and 48 (forty-eight) provisions for human rights (Constitutional Court, 2010). Regrettably, as Indonesia's political will was to return to the original 1945 Constitution in 1959, this means that human and citizen rights were limited, and it was the decline of the spirit of human rights' implementation in Indonesia (Marzuki, 2014).

Under the New Order Regime, the original Constitution, which had not been aware of the importance of human rights, was used for almost 40 (forty) years. Finally, in 2000, there was the Second Amendment, which included a new chapter regarding human rights. In the Amended 1945 Constitution, a relatively new right was introduced: the right to protection of individuals. The word 'relatively' here means that actually the right to protection of individuals was stated in other previous Indonesian Constitutions, namely in the Federal Constitution of 1949 and the Temporary Constitution of 1950. In both Constitutions, there were arrangements for human rights with relation to the protection of individuals (Constitutional Court, 2010).

The year 2000 gave new hope to everyone in Indonesia regarding the right to protection of individuals. It was high time that the right was included in the Amended 1945 Constitution and this has become the highest basic regulation for everyone to implement their rights. After the inclusion of the right in the Constitution, some issues regarding the right to protection of individuals have emerged. The questions include what 'the right of the protection of individuals' means, what its scope is, and how the government is involved in protecting this right. These questions arose because there was no elucidation section in the Constitution, which enabled people to make an interpretation based on the text of the Constitution.

To answer those questions about the right to protection of individuals, there are 2 (two) previous legal documents that can support the inclusion of the right to protection of individuals in the Constitution. First, there is the TAP MPR XVII/1998 regarding Human Rights. In this legislation, Indonesia recognises that human rights are God-given. By recognising human rights, everyone can develop himself/herself and his/her role in order to achieve prosperity for other people (Constitutional Court, 2010). The sources of human rights derive from religious thoughts, universal moral values and traditional cultural values, which are based on Pancasila and the 1945 Constitution. In this regulation, the right to protection of individuals is placed under the chapter regarding 'rights to security'. Article 23 states that 'everyone shall have the right of the protection of individuals, his family, honor, dignity, and his/her property'.

Second, Law number 39 of 1999 regarding Human Rights is an implementation of the regulation, which gives a more detailed recognition of human rights than in the TAP MPR

XVII/1998, and which includes women's rights and children's rights. Similar to TAP MPR XVII/1998, in Law number 39 of 1999, the arrangement of the right to protection of individuals is under the chapter regarding 'rights to security', namely in Article 29 section (1) (Law on Human Rights, 1999). Moreover, in Article 31 section (1) there is a statement that 'no one may disturb the residence of any person'. In the elucidation of this article, the phrase 'no one may disturb' means anything regarding 'privacy' in their property (Law on Human Rights, 1999). By reading the chapter regarding rights to security, it can be concluded that the right to protection of individuals is in accordance with the right to privacy, in which everyone must respect other people's rights and should not be allowed to disturb their property.

During the process of making the amendment to the Constitution, around 1999–2000, there were discussions concerning the need to adopt the TAP MPR XVII/1998 and Law number 39 of 1999 into a chapter in the Constitution, with some adjustments. The chapter concerning human rights in the Constitution must contain the main idea of the human rights' arrangement. The discussions were mostly about the need to recognise human rights, in a similar way to the early discussions by our founding fathers before independence that Indonesia should apply a communalism idea rather than an individualism idea (Kusuma, 2009), in which one side argued that human rights are attached to every person and do not need to be formally recognised, while the other side argued that the formalisation of human rights is important to guarantee every person's and citizen's rights. Human rights that are recognised in the Constitution include the right to live, the right to establish a family, the right to education, equality before the law, the right to religion, the right to information and the right to security.

The right to security, especially the right to protection of individuals in Article 28G section (1), gives a guarantee for everyone in the territory of Indonesia to enjoy and practice that right. As the Amended 1945 Constitution considered the TAP MPR XVII/1998 and Law number 39 of 1999, the consequence is that Article 28G section (1) has the same meaning as the previous regulations. Although the right to privacy is not stated clearly in the Constitution, the right to protection of individuals manifests the right to privacy, and this means the recognition of the right to be left alone (Warren & Brandeis, 1890). The scope of the right to privacy is very broad; however, in the narrow sense, the right to privacy includes the protection of data, the protection of personal information or correspondence, and requiring the state to regulate who may lawfully access such information and to protect individuals from unauthorised access (McBeth et al., 2011). Based on the scope of the right to privacy, the government must therefore prepare legal instruments to implement the right to privacy of personal information.

3 OBLIGATION TO PROTECT THE HUMAN RIGHT TO PRIVACY OF PERSONAL INFORMATION

Information is important. Futurologists, such as John Naisbitt, explain that everyone has possession of information; therefore, s/he has everything (Asshiddiqie, 2007). Information has become the new form of colonialism; for those who do not have any access to take control of information, s/he will be left behind and will be controlled by others who have information (Asshiddiqie, 2007). To avoid the misuse of information, the state must compose a regulation to accommodate and balance the needs of society, between the need to gather information and the need for the confidentiality of personal information. After this regulation has been created, the most significant problem is: who takes the responsibility for or who has the obligation to control and implement the protection of human rights in general and the right to privacy of personal information in particular.

Regarding the question about who takes responsibility for controlling and implementing human rights in general, everyone must have a basic conception of how individuals and the government should exercise human rights. The performance of the government and of individuals is based on the available resources in the country (Nickel, 1996). The government and individuals are the elements that are in charge of the implementation of human rights.

Besides them, there are international institutions that also have the responsibility to implement human rights (Nickel, 1996).

The government constitutes the main element in the implementation of human rights. The role of the government is to control excessive and inappropriate power. The Universal Declaration of Human Rights (UDHR) in its Preamble states that '... Member States have pledged themselves to achieve, in cooperation with the United Nations, the promotion of universal respect for and observance of human rights and fundamental freedoms' (Office of United Nation High Commissioner for Human Rights [OHCHR], 2014). Therefore, as part of its function, the state has responsibilities to review, protect and implement human rights in its territory. These responsibilities include both negative and positive responsibilities. Negative responsibility means the responsibility not to violate human rights, while positive responsibility means the responsibility to implement human rights (Nickel, 1996).

Individuals are also an element in the implementation of human rights. The Preamble of the UDHR gives a detailed explanation that '... every individual and every organ of society... shall strive by teaching and education to promote respect for these rights and freedoms and by progressive measures, national and international, to secure their universal and effective recognition and observance, both among the peoples of Member States themselves and among the peoples of territories under their jurisdiction' (OHCHR, 2014). This means that not only the government has the responsibility for the implementation of human rights, but also individuals and organs of society. Individuals have the responsibility to achieve their own rights, and to remind the government to implement human rights under the principle of democracy (Nickel, 1996).

International institutions, which can be formed as 'every organ of society', are selected as another element that takes responsibility in the implementation of human rights, as is stated in the Preamble of the UDHR. This situation occurs as a result of governmental failure in achieving human rights, so that the government searches for help from international institutions. However, international institutions do not have any authority to force any country to use their best practice in the implementation of human rights (Nickel, 1996).

In the case of Indonesia, under the Amended 1945 Constitution, there are 2 (two) categories of responsibility regarding human rights implementation. Article 28I section (4) states that the government has the responsibility to protect, promote, enforce and fulfil human rights. However, Article 28 J section (1) states that individuals have the responsibility to respect the human rights of others in the orderly life of the community, nation and state (Indonesian Constitution, 2000). The government has a vertical responsibility to everyone in the territory of Indonesia in promoting human rights, while individuals have a horizontal responsibility to other citizens and to everyone in Indonesia to respect their rights. These responsibilities must be performed in accordance with the Preamble of the 1945 Constitution as the requirement to protect everyone and to improve public welfare (Indonesian Constitution, 2000).

The responsibility of the government to protect, promote, enforce and fulfil human rights can be conducted by creating regulations regarding human rights. The government can include arrangements about how to deal with the right to privacy in general and the confidentiality of personal information in particular in a regulation relating to the control of personal information. The arrangement should include phases of collecting personal information in both public and private sectors. The same concept is needed in both the public and private sectors to guarantee legal certainty and enforcement.

4 COLLECTING AND PROCESSING PERSONAL INFORMATION BY THE PUBLIC AND PRIVATE SECTORS

There are several methods used for collecting personal information. Census, survey, compilation of administration products and other methods in accordance with the improvement of knowledge and technology are mentioned in Law number 16 of 1997 regarding Statistics as methods of data collection (Law on Statistics, 1997). These methods can be used as a reference for everyone who has the intention to collect information about others.

The difference between these methods can be understood by viewing the definition of each method. Census is data collection by distributing questionnaires to all members of the population in the entire territory of the State in order to determine the characteristics of a population at a given time. Survey is a data collection method using a sample census in order to estimate the characteristics of a population at a given time. The compilation of administration products is a data collection method using the collecting, preparing, presenting and analysing of data from administrative records available from the government and/or the community (Law on Statistics, 1997). While other methods in accordance with the improvement of knowledge and technology are, for example, data collection by using computer technology through online communication (Makarim, 2005).

As seen by the definitions of the methods of data collection, there are also some similarities between these methods. The similarities can be observed from several perspectives. From the perspective of data collectors, they must have a specific purpose in collecting information. From the perspective of the subject of the data collected (known as the 'respondent' or 'customer'), they do not give the information voluntarily. They are being asked about something that probably includes their personal information and they must answer the questions. Moreover, respondents or customers are aware that their information is being gathered and used for a specific purpose.

Article 28F of the Amended 1945 Constitution states that 'every person shall have the right to communicate and to obtain information for the purpose of the development of his/her self and social environment, and shall have the right to seek, obtain, possess, store, process, and convey information by employing all available types of channels (Indonesian Constitution, 2000). This article guarantees the freedom of everyone to collect information, which can include personal information. 'Everyone' here may refer to people individually, or to people who work in the public or private sector collecting information for specific purposes. The implementation of this article can be scrutinised through several regulations. In this paper, there will be an explanation regarding the confidentiality of personal information on one side and the transparency of information on the other side.

As an example of the personal information collection process in the public sector, Law number 23 of 2006 regarding Civil Administration explains that every citizen shall inform the civil administration office of their important moments, which may consist of personal information. The officer has the responsibility to maintain and protect the confidentiality of that information. Personal information that has to be protected and registered is the number of family card; the identity card number of the holder, his/her father and his/her mother; information about their date, month and year of birth; physical and mental conditions; and any notes on important moments (Law on Civil Administration, 2006).

The personal information collection process in civil administration requires citizens to actively report to the civil administration office about their current conditions in order that this can be regularly updated by the officer. At the same time, the civil administration office has the responsibility to maintain and process the information confidentially, so that only authorised persons can access the personal information (Law on Civil Administration, 2006). This reciprocal relationship is important in order to guarantee the trust of citizens towards civil administration officers and to gather personal information about citizens that can be used as a consideration when making public policies.

Law number 11 of 2008 regarding Information and Electronic Transaction is an example regarding the collection of personal information in the private sector. This law regulates cyber law and the usage of information technology and communication. Electronic transaction is defined as 'legal transactions which use computer, network computer, and/or other electronics means' (Law on Information and Electronic Transaction, 2008), especially commercial transactions. This law protects the confidentiality of personal information that is collected from transactions using electronic media. To protect the right to privacy, the usage of personal information in electronic media must be subject to an agreement between the owner of the personal information and the data collector, namely the trader.

Law number 11 of 2008 explains that the right to privacy indicates the right to enjoy personal life free from any kind of disturbance or interruption. To enjoy the right to privacy when undertaking a commercial transaction, everyone who wants to make a transaction in cyberspace must submit his/her personal information to the trader. At the same time, the trader must maintain the confidentiality of the personal information of his/her client and must use the information according to its purpose.

The challenge in the arrangement of confidentiality of personal information is the transparency of information. By enacting Law number 14 of 2008 regarding Transparency of Public Information, the government stipulates that public information is collected, processed, sent and/or received by a public agency (executive, legislative and judicative). This information is related to government services and public interests (Law on Transparency of Public Information, 2008). Basically, public information is open and accessible to everyone, unless the Law stipulates otherwise. The Law forbids the disclosure of public information which, for example, could harm the state, is related to the protection against unfair commerce, is related to private rights and is related to official secrets. The Law on Transparency of Public Information balances the transparency of information and the protection of personal information.

5 CONCLUSION

A new concept of rights is introduced in the Amended 1945 Constitution, namely the 'right of the protection of individuals'. Even though there is no clear arrangement concerning the right to privacy in the Amended Constitution, the right to protection of individuals reflects the right to privacy and recognises the right to be left alone. The right to privacy in the narrow sense includes the protection of data and personal information or correspondence. The state has the responsibility to protect, promote, enforce and fulfil human rights. This responsibility constitutes the role of the government in making regulations and then controlling their implementation. With regards to the right to privacy of personal information, the government must guarantee the confidentiality of personal information gathered by both the public and private sectors in order to achieve legal certainty and legal enforcement.

From several regulations under the Constitution regarding the arrangement of personal information, there are some lessons to be learnt about the collecting and processing of personal information. Firstly, as the process of collecting personal information is not voluntary, data collectors must disclose the information only to the authorised person/institution. Secondly, there must be a mutual understanding between the owners of the personal information and the data collectors to use the information only for a specific purpose. In addition to the arrangements regarding the confidentiality of personal information, there is also an arrangement about transparency of information for the public, in which the dissemination of personal information must be in line with the concept of confidentiality.

In the future, the state, through its government, must construct laws on personal information in accordance with the spirit of the Amended 1945 Constitution. The construction of legislation must envisage a guarantee by the state to protect the privacy of personal information for everyone. Furthermore, legislation must also include the duties and rights of the owners of personal information and the data collectors. The responsibility of data collectors is vital in maintaining and using the personal information in accordance with the purpose of its collection.

REFERENCES

Asshiddiqie, J. (2006). *Hukum Tata Negara dan Pilar-Pilar Demokrasi: Serpihan Pemikiran Hukum, Media, dan HAM* [*Constitutional law and democratic pillars: Deep thinking of law, media, and human rights*]. Jakarta: Konstitusi Press.

Asshiddiqie, J. (2007). *Pokok-pokok Hukum Tata Negara Indonesia Pasca Reformasi* (*Indonesia's constitutional law after the reform era*). Jakarta: PT. Bhuana Ilmu Populer.

Davidson, S. (1994). *Human Rights*. Translated by Pudjaatmaka, A.H. [Indonesian]. Jakarta: Pustaka Utama Grafiti.
Indonesia. (1997). *Undang-undang No. 16 Tahun 1997 tentang Statistik* [*Law number 16 of 1997 regarding Statistics*].
Indonesia. (1999). *Undang-undang No. 39 Tahun 1999 tentang Hak Asasi Manusia* [*Law number 39 of 1999 regarding Human Rights*].
Indonesia. (2000). *Undang-Undang Dasar 1945 Amandemen II* [*1945 Constitution of the Republic of Indonesia Second Amendment*].
Indonesia. (2006). *Undang-undang No. 23 Tahun 2006 tentang Statistik* [Law Number 23 of 2006 regarding Civil Administration].
Indonesia. (2008). *Undang-undang No. 11 Tahun 2008 tentang Informasi dan Transaksi Elektronik* [*Law number 11 of 2008 regarding Information and Electronic Transaction*].
Indonesia. (2008). *Undang-undang No. 14 Tahun 2008 tentang Keterbukaan Informasi Publik* [*Law number 14 of 2008 regarding Transparency of Public Information*].
Kusuma, R.M.A.B. (2009). *Lahirnya Undang-Undang Dasar 1945: Memuat Salinan Dokumen Otentik Badan Oentoek Menyelidiki Oesaha2 Persiapan Kemerdekaan* [*The birth of the 1945 Constitution: Original document of investigation for independence preparation agency*]. Jakarta: Badan Penerbit Fakultas Hukum Universitas Indonesia.
Makarim, E. (2005). *Pengantar Hukum Telematika: Suatu Kompilasi Kajian* [*An introduction to telematics law*]. Jakarta: Raja Grafindo Persada.
Mahkamah Konstitusi [Constitutional Court] Republik Indonesia. (2010). Buku VIII Warga Negara dan Penduduk, Hak Asasi Manusia, dan Agama, Edisi Revisi [Comprehensive paper on the Amended 1945 Constitution: Book VIII concerning citizens, human rights, and religion, Revised Edition]. In: *Naskah Komprehensif Perubahan Undang-Undang Dasar Negara Republik Indonesia Tahun 1945: Latar Belakang, Proses, dan Hasil Pembahasan 1999–2002*. Jakarta: Sekretariat Jenderal dan Kepaniteraan Mahkamah Konstitusi.
Marzuki, S. (2014). *Politik Hukum Hak Asasi Manusia* [*Legal political of human rights*]. Jakarta: Erlangga.
McBeth, A., Nolan, J. & Rice, S. (2011). *The international law of human rights*. Melbourne: Oxford University Press Australia.
Nickel, J.W. (1996). *Hak Asasi Manusia: Refleksi Filosofis atas Deklarasi Universal Hak Asasi Manusia*. Translated by Arini, T.E. Jakarta: Gramedia Pustaka Utama.
Office of the United Nations High Commissioner for Human Rights [OHCHR]. (2014). *The new core international human rights treaties*. New York & Geneva: United Nations Publication.
Warren, S.D. & Brandeis, L. (1890). The right to privacy. *Harvard Law Review*, 4(5), 193–220.

The enforcement of payment of restitution in criminal proceedings as the base for filing of tort charges

S.L. Anindita & R. Agustina
Faculty of Law, Universitas Indonesia, Depok, Indonesia

ABSTRACT: The recovery of losses suffered by the state is an objective of prosecution of corruption cases. Aside from creating a deterrent effect on the offender in particular, it also serves as a legal education for the public in general. Such measure is incorporated in a side ruling to a permanent judgment, which requires the defendant to pay restitution in a criminal case pursuant to Law No. 3 of 1971. The filing of a tort charge by state prosecutors (under Article 1365 of the Indonesian Civil Code) as a legal basis to claim for the recovery of state assets from the offender will be the topic discussed in this article. The conclusion of this research, carried out using a normative method by applying the theories of justice, legal certainty, and benefits, is that the state prosecutor is not required to file a tort charge to recover losses suffered by the state as a consequence of the non-payment of restitution imposed on the offender through a side ruling. The mechanism that needs to be initiated by the state attorney to recover such losses is to file a civil writ of execution.

1 INTRODUCTION

The writing of this paper, titled "The Enforcement of Restitution in Criminal Proceedings as the Base for Filing of Tort Charges", was driven by a court ruling that awarded a tort charge pursuant to Article 1365 of the Indonesian Civil Code [KUHPerdata] filed by a state prosecutor. The charge was filed with respect to the non-performance of a court ruling having permanent legal force relating to corruption as defined under the Law Number 3 of 1971 on the Eradication of the Crimes of Corruption. The defendant named in the tort charge filed by the state prosecutor is a person convicted of corruption under the said corruption law who has failed to pay restitution set in a sentence supplemental to the corruption verdict passed by the court in the original hearing.

The state prosecutor, aiming to enforce the recovery of state assets against the convicted criminal, filed a tort charge pursuant to Article 1365 of the Civil Code with a district court. In this regard, the author has taken three district court rulings on tort cases filed by state prosecutors as the object of the research, namely: (1) Ruling of the Ungaran District Court Number: 22/Pdt.G/2010/PN. UNG; (2) Ruling of the Praya District Court Number: 24/Pdt.G/2013/PN. PRA; and (3) Ruling of the District Court of Mataram Number: 116/Pdt.G/2013/PN. MTR.

In each of the three cases, the grounds for the filing of the tort charges by the state prosecutors was the non-performance of a supplemental ruling compelling the defendant (originally the convicted offender) to pay restitution. The supplemental ruling was made in conjunction with the principal verdict of corruption having permanent legal force under the Law Number 3 of 1971. The tort charges were sought to recover losses suffered by the state from the convicted offender.

Although recovery of financial losses was sought by the state prosecutors by means of tort charges only as recently as the period of 2010 to 2013, given that during the period in which the corrupt act was committed the Law Number 3 of 1971 was still in effect, the state prosecutors as the executors of court judgment pursuant to Article 270 of the Indonesian Code of Criminal Procedure (KUHAP) were unable to invoke Article 18 paragraph (3) of the Law Number 31 of 1999 on the Eradication of the Crimes of Corruption.

In the key items of the charge, the prosecutor argued that the defendant's action of deliberately avoiding the performance of payment of restitution as affirmed in the court ruling constitutes a violation under Article 1365 of the Civil Code, as it is against the defendant's legal obligation.

In the court judgment reviewed by this research, in response to the charge initiated by the prosecutor the court declared that the defendant's action of accepting the verdict of being guilty of corruption yet refusing to perform such ruling constitutes a tort, as it conflicts with the defendant's legal obligation. The court in its ruling stated among others that the defendant has been found to have committed a tort and is ordered to pay restitution to the state through the state prosecutor, in the amount equal to the restitution cited in the original verdict of the corruption case.

The question as to whether the prosecutor's decision to file a tort charge against a person who has been convicted in a corruption case under the Law Number 3 of 1971 but has not carried out his obligation to make restitution payment as mandated by verdict of the criminal court to recover financial losses suffered by the state as a result of the corruption will be further discussed by the author using a legal normative legal method of research (Ibrahim, 2006). This article will use secondary data obtained from literature searches related to corruption, restitution and implementation. It will also examine court decisions that have permanent legal force, and the data collection is done through literature research.

2 TORT

The principle of tort [*onrechtmatige daad* in Dutch] was originally narrowly construed as being any action of a person that contravenes the right of another person as conferred under the law [*wettelijk recht*] and, as such, such action would be in contravention to the person's legal obligation [*wettelijke plicht*] (Djojodirdjo, 1982). As time progresses, the definition of tort has expanded to include any act or omission that violates the right of another person or the moral values or principles of propriety that must be observed within society with respect to a person or property. Tort is taken more as an "injury" rather than a breach of contract. Moreover, tort claims are usually not based on any contractual relationship (Agustina, 2003).

A violation of a criminal provision does not only constitute a violation of the law [*wederrechtelijk*], but in certain circumstances it can be an unlawful act [*onrechtmatige daad*]. Any binding regulatory provision is a "legal obligation." If a person causes an injury to another person through a violation of a statutory provision (whether in formal or material terms), then such person has committed a violation of the law, namely acting contrary to his legal obligations.

Further conditions that must be met for a person to be considered as acting contrary to his legal obligation under Dutch jurisprudence are as follows: (1) the claimant's interest has been affected or threatened by such contravention of the law; (2) the claimant's interest is protected by the provision that has been violated; (3) such interest is within the scope of matters protected by Article 1401 of BW (Article 1365 of the Civil Code); (4) the violation of regulatory provision is against propriety of conduct towards the claimant, among others in terms of the behavior and attitude of the claimant himself (Setiawan, 1992). Article 1365 of the Indonesian Civil Code only provides how a person can claim for remedy with a court for any injury suffered, but it does not provide any explanation as to what constitutes a tort.

Unlawful act is not only recognized in the realm of civil law but also in criminal law. In the Indonesian Penal Code [Kitab Undang-Undang Hukum Pidana or KUHP], the elements that make up tort are defined differently: *zonder verlof, zonder daartoe gerechtigd te zijn, met over schrijding van zijn bevoegheid dan zonder inachtneming*, whereas in court judgments 'unlawful' is often taken as 'contrary to the law', 'contrary to provided rights', or 'unrightfully' (Loqman, 2000). One can note certain similarities and differences between tort [*onrechtmatige daad*] and violation of the penal code [*wederrechtelijk*]. One of the shared elements is that they both contravene a ban or obligation. The scope of tort is broader than that of *wederrechtelijk* as violation of a criminal nature [*wederrechtelijk*] is subject to the principle that a deed shall not be punishable except as provided under a penal code already in force prior to the commission of the offense [*nulla puna sie praevia lege poenali*].

Corruption is a remarkable criminal offense [extraordinary crime] (Hertanto, 2014) and internationally recognized as one of the types of transnational organized crime. Constitutional Court Ruling Number: 012–016–019-PUU-IV/2006 dated 19 December 2006 regarding Constitutionality Review of the Law Number 30 of 2002 on the Corruption Eradication Commission to the 1945 Constitution of Indonesia stipulates that corruption is an extraordinary crime and is the enemy of society and the Indonesian people in general. The protection of fundamental rights intended to be achieved is considered to be of a smaller scale in priority when viewed against the necessary protection of the basic economic and social rights of the larger population that can be harmed by corruption and which would ultimately undermine the legitimacy and credibility of the state in the eye of the public.

As with the definition of *onrechtmatige daad*, the definition of corruption underwent some alterations. Government Regulation in Lieu of the Law Number 24 of 1960 on the Investigation, Prosecution and Hearing of Corruption Cases changed the definition of corruption which previously contained the terms "crime" or "offense" to contain "unlawfully" and "or known or ought to have been expected by such person". In the Law Number 3 of 1971 the phrase "unlawfully .... directly or indirectly" was changed to "may". The effect of change was on the unlawful nature of the offense, which previously means a violation of the formal letter of the law (*formele wederrechtelijkheid*), and now refers to the subject matter or spirit of the law (*materiele wederrechtelijkeheid*) in the sense that the crime occurred not only when written statutory provisions were violated but also when it can be established that the unwritten spirit of the law was violated, thus making it easier to collect evidence proving that the crime was committed. Finally, pursuant to the Law Number 31 of 1999, read in conjunction with the Law Number 20 of 2001 on the Amendment to the Law Number 31 of 1999, another change was made to the definition of the "unlawful" nature of the criminal offense of corruption to become a pure formal offense [*formele wederrechtelijkheid*], so the question of whether corruption has occurred only needs to be proven by establishing the elements set in the law, rather than also having to prove the consequence of such action.

Taking into account the elaboration above, it is the author's view that the state prosecutor does not need to re-establish the elements of the crime in the civil proceedings. This is due to the fact that establishing the unlawful nature of the crime of corruption under the Law Number 3 of 1971 only needs to be done in the formal sense [*formale wederrechtelijkheid*] as outlined above.

3 RESTITUTION OF FINANCIAL LOSSES SUFFERED BY THE STATE

Under the Indonesian penal law there are two types of sentence that can be imposed on a defendant by the court, including in corruption trials, namely the principal punishment in the form of death sentence, imprisonment, or fine, and the supplemental punishment in the form of asset forfeiture and, particularly to corruption related cases, payment of restitution in the maximum amount of the value of proceeds gained through the corrupt act.

The passing of judgment requiring the defendant to pay restitution in a corruption case and what legal measures are available in the event the restitution is not paid as required are governed by the following statutory provisions:

1. Article 16 paragraph (3) of Government Regulation in Lieu of the Law Number 24 of 1960 on the Investigation, Prosecution and Hearing of Corruption Cases, which stipulates that: "The convicted offender may also be ordered to pay restitution in an amount equal to the value of the proceeds gained from the corrupt act."

2. Article 34 paragraph c of the Law Number 3 of 1971 on the Eradication of the Crimes of Corruption, which stipulates that: "Payment of restitution in the maximum amount of the value of assets gained from the corrupt act." Furthermore, according to the elucidation section of the law, the provision on payment of the fine must be enforced when restitution cannot be met by the defendant.
3. Article 18 paragraph (1) sub-paragraph b of the Law Number 31 of 1999 on the Eradication of the Crimes of Corruption, which in essence contains the same stipulations as Article 34 paragraph c of the Law Number 3 of 1971 above. Paragraphs (2) and (3) of the same article provide that in the event the defendant fails to pay restitution within one month following the passing of a court judgment with permanent legal force, the defendant's assets can be seized and put up for auction to cover the required restitution. If the proceeds from the auction are not sufficient to cover the obligation, prison sentence can be ordered for a period not exceeding the maximum allowable sentence for the principal offense and such period must already be set out in the court ruling. This law does not confer authority upon the state prosecutor to file a civil suit to enforce the court order to pay restitution.

Corruption is considered as an extraordinary crime that harms the state's finances and economy, in addition to impeding growth and sustainable national development, and thus efforts to combat the crime also require extraordinary measures. Punishment imposed in corruption cases should ideally recover the losses suffered by the state in addition to providing a deterrent effect for the offender in particular and impart legal education for the population in general.

The Law Number 3 of 1971 does not allow the prosecutor to file a civil suit for the purpose of recovering the state's financial losses, as currently allowed under Articles 32, 33, and 34 of the Law Number 31 of 1999 and the Law Number 20 of 2001 on the Amendment to the Law Number 31 of 1999 on the Eradication of the Crimes of Corruption. In the elucidation of Article 34 in the Law Number 31 of 1971 it is explained that in principle the recovery of the state's financial losses is still expected to be pursued to the fullest extent, and therefore the supplemental sentence to pay restitution in corruption cases constitutes an expansion of what is governed under Article 10 of the Indonesian Penal Code.

To address the uncertainty as regards what legal measures are available to recover the state's financial losses, including to enforce the obligation to pay restitution pursuant to Article 34 sub-article c of the Law Number 3 of 1971, the Indonesian Supreme Court had issued Supreme Court Circular [SEMA] Number 4 of 1988. The Circular stipulates that the enforcement of an order to pay restitution by the state prosecutor does not require the involvement of the court as it merely requires the execution of the court's order. However, if in fact the assets of the defendant are no longer sufficient and the shortfall are to be sought by the prosecutor in another forum, the claim must be done through a civil suit. The Circular was then made as the legal basis by prosecutors to file a civil action to compel the convicted offender to pay the restitution as ordered by the court.

Payment of restitution as a supplemental sentence passed by the court pursuant to the Law Number 3 of 1971 can philosophically be taken as the enforcement of restitution as defined in Article 274 of KUHAP, wherein the prosecutor during the criminal hearing of the corruption case can also be deemed discharging his duty as a state attorney incorporating an argument for a lawsuit to seek restitution into the indictment pursuant to Article 99 of KUHAP, and thus enforcement of the order to pay restitution can be pursued through a civil action.

4 TORT CHARGE

Civil action constitutes a petition submitted to the chairperson of a competent district court, containing a claim made against another person which must be heard by the court according to a specific procedure, following which a judgment is made on such claim (Prist, 1992). Basically a civil claim is a petition made by a plaintiff – a person or legal entity – who believes that their rights have been infringed upon by another person (Mulyadi, 1998) with the intention of having the court grant legal protection to prevent citizens from taking matters into their own hands to resolve the problem (*eigenrichting*) (Mertokusumo, 2002).

The claim put forward by the prosecutor, as repeated in the court judgment and which is the object of this research, is not based on a dispute or a rights claim that must be heard and resolved, since all of the foregoing has been performed by the panel of judges presiding over the corruption case based upon the indictment filed by the prosecutor. The view that there is in fact no dispute involved in the claim submitted by the state prosecutor can be reinforced by the provisions contained in sub-paragraph 3 of SEMA Number 4 of 1988: "Only when in the present instance of enforcement the assets of the offender is no longer sufficient, then the shortfall, if the same is to be claimed by the Prosecutor's Office in another instance, must be sought through a civil action before the court." From the phrase "the shortfall to be claimed" a conclusion can be drawn that the cause for action is an incomplete performance of an order, rather than to seek the resolution of (another) dispute. As such, in the view of the author, the filing of a tort charge by the state prosecutor to enforce performance of the supplemental order to pay restitution contained in the court judgment pursuant to the Law Number 3 of 1971 is an erroneous choice of avenue. The prosecutor's petition for the civil court to rule that the defendant has committed a tort by not performing the restitution as ordered by the criminal court is inefficient, as it prolongs the process by which the state's assets are to be returned and adds to the financial burden of the state as the filing of a civil suit incurs cost.

5 ENFORCEMENT THROUGH THE CIVIL PROCEEDINGS AS A SOLUTION

The filing of a tort charge by the state prosecutor before a district court on the grounds that the defendant (convicted offender) fails to pay restitution pursuant to the Law Number 3 of 1971 is not appropriate because:

1. no dispute has occurred, and the grounds upon which the claim is initiated are merely the failure to make restitution payment (supplemental order to a final verdict);
2. even if the claim is accepted, heard and ruled by the court, all elements of a tort required to be present pursuant to the Law Number 3 of 1971 (unlawful act, wrongdoing, causal effect and loss) have already been established during the original corruption case proceedings;
3. the granting of the claim would not automatically entail the defendant voluntarily performs the ruling of the civil court. If the defendant persists in avoiding discharging his obligation or does not voluntarily perform the court's judgment, the state prosecutor would ultimately still have to require the assistance of the chairperson of the district court to enforce the judgment.

In this regard the author is of the view that the enforcement through civil execution is an effective and efficient means to obtain payment of restitution in order to recover the state's financial losses. A petition for restitution may be filed by the state prosecutor according to Article 274 of KUHAP, as previously discussed in the section on restitution. Enforcement through civil proceedings can be sought when the principles of execution have been met, namely (1) the judgment is final and has legal binding force, (2) the judgment is condemnatory in nature, (3) the judgment is not being performed voluntarily, and (4) enforcement is based upon the order under the auspice of the chairperson of the district court.

The execution must be followed through by the prosecutor by filing a petition of reprimand (*aanmaning*) in order to obtain a writ of reprimand. Once the reprimand has been issued, but the defendant continues to avoid voluntarily providing restitution, the prosecutor would then initiate a petition for enforcement of asset forfeiture containing a description of the defendant's assets to be placed under the executor attachment. Based upon the petition, the chairperson of the district court will issue a decision as the basis for the executor attachment. A third party who believes that its assets are included in the attached assets can undertake legal measures [*derdenverzet*] as provided under Article 196 paragraph (5) of the *Herziene Indonesisch Reglement* (HIR).

The direct implementation of civil execution to address a defendant's failure to perform a court's supplemental order of restitution payment as described by the author above can be applied to corruption-related cases under the Law Number 3 of 1971, as there is no legal avenue to file a civil suit to recover losses suffered by the state. In addition, it does not prescribe a clear mechanism that must be followed by the prosecutor when a defendant fails to pay a part or the entire amount of restitution. Additionally, the law expressly defines the rights of the state to initiate civil action against the assets of the defendant that are concealed or being concealed and were only disclosed following a final court ruling in order to recover the state's losses. The provisions of Articles 32, 33, and 34, read in conjunction with Article 38 paragraph (5) of the Law Number 31 of 1999, read in conjunction with the Law Number 20 of 2001, allows the filing of a civil tort suit with respect to payment of restitution. The filing can be made by the state prosecutor or another aggrieved party, in the latter case when the original criminal proceedings cannot be carried out effectively due to the defendant being deceased during the court hearings or when the court dismisses a corruption case or when the case cannot be proven.

6 CONCLUSION

The practice of filing a civil tort suit by state prosecutors under the provisions of Article 1365 of the Civil Code with the intention of recovering losses suffered by the state relating to corruption cases under the Law Number 3 of 1971 is erroneous. Article 274 of KUHAP can be invoked to enforce the court's supplemental order of restitution in order to recover the state's losses through civil execution.

REFERENCES

Agustina, R. (2003) *Perbuatan Melawan Hukum*. Jakarta, Program Pascasarjana Fakultas Hukum Universitas Indonesia.
Djojodirdjo, M.A.M. (1982) *Perbuatan Melawan Hukum*. 2nd edition. Jakarta, PT Pradnya Paramita.
Hertanto, H. (2014) Evaluasi Pengadilan Tindak Pidana Korupsi di Indonesia (Studi atas kebutuhan, peran dan kualitas putusan hakim di Pengadilan Tindak Pidana Korupsi Jakarta dan Bandung pasca pemberlakuan Undang-Undang No. 46 Tahun 2009 tentang Pengadilan Tindak Pidana Korupsi. *Jurnal Hukum dan Pembangunan*, 44 (1), 1–47.
Ibrahim, J. (2006) *Teori dan Metodologi Penelitian Hukum Normatif*. Malang, Bayumedia Publishing.
Indonesia (1960) *Peraturan Pemerintah Pengganti Undang-undang Nomor 24 Tahun 1960 tentang Pengusutan, Penuntutan dan Pemeriksaan Tindak Pidana Korupsi* [*Government Regulation in Lieu of Law Number 24 of 1960 regarding Investigation, Prosecution and Hearing of Corruption Cases*].
Indonesia (1971) *Undang-undang No. 3 Tahun 1971 tentang Pemberantasan Tindak Pidana Korupsi. Lembaran Negara Republik Indonesia Tahun 1971 Nomor 19. Tambahan Lembaran Negara Nomor 2958* [*The Law No. 3 of 1971 regarding Eradication of the Crime of Corruption. State Gazette Number Year 1971 Number 19. The Supplement to State Gazette Number 2958*].
Indonesia (1999) *Undang-undang No. 31 Tahun 1999 tentang Pemberantasan Tindak Pidana Korupsi. Lembaran Negara Republik Indonesia Tahun 1999 Nomor 140. Tambahan Lembaran Negara Nomor 3874*. [*The Law No. 31 of 1999 regarding Eradication of the Crime of Corruption. State Gazette Number Year 1999 Number 140. The Supplement to State Gazette Number 3874*].
Indonesia (2001) *Undang-undang No. 20 Tahun 2001 tentang* Perubahan atas Undang-undang No. 31 Tahun 1999 tentang *Pemberantasan Tindak Pidana Korupsi. Lembaran Negara Republik Indonesia Tahun 2001 Nomor 134. Tambahan Lembaran Negara Nomor 4150* [*The Law Number 20 of 2001 regarding Amendment to Law Number 31 of 1999 on the Eradication of the Crimes of Corruption. State Gazette Year 2001 Number 134. Supplement to State Gazette Number 4150*].
Loqman, L. (2000) *Masalah Tindak Pidana Korupsi di Indonesia*. Jakarta, Badan Pembinaan Hukum Nasional Departemen Kehakiman Republik Indonesia.
Mahkamah Agung [Supreme Court] Republik Indonesia (2011) *Statistik Perkara Perdata Tahun 2011*. Jakarta, Direktorat Jenderal Badan Peradilan Umum Mahkamah Agung Republik Indonesia.
Mertokusumo, S. (2002) *Hukum Acara Perdata Indonesia*. 6th edition. Yogyakarta, Liberty.

Mulyadi, L. (1998) *Hukum Acara Perdata Menurut Teori dan Praktik Peradilan Indonesia.* Jakarta, Djambatan.
Pemerintah Hindia Belanda (1941) *Reglemen Indonesia yang Diperbaharui RIB [Herziene Indonesisch Reglement RIH]*, Staatsblad Tahun 1941. No. 44.
Pengadilan Negeri [District Court] Ungaran, Putusan No. 22/Pdt.G/2010/PN. UNG Tanggal 5 August 2010. Semarang. Court Ruling.
Pengadilan Negeri [District Court] Praya. Putusan No. 24/Pdt.G/2013/PN. PRA Tanggal 4 September 2013. Praya. Court Ruling.
Pengadilan Negeri [District Court] Mataram. Putusan No. 116/Pdt.G/2013/PN. MTR Tanggal 16 October 2013. Mataram. Court Ruling.
Prinst, D. (1992) *Strategi Menyusun dan Menangani Gugatan Perdata.* Bandung, Citra Aditya Bakti.
Setiawan (1992) *Aneka Masalah Hukum dan Hukum Acara Perdata.* Bandung, Alumni.
Soetantio, R. & Oeripkartawinata, I. (1997) *Hukum Acara Perdata dalam Teori dan Praktek.* Bandung, Mandar Maju.
Subekti, R. (1989) *Hukum Acara Perdata.* Bandung, Bina Cipta.
Supomo (1986) *Hukum Acara Perdata Pengadilan Negeri.* Jakarta, Pradnya Paramita.

Repeat offence as aggravating circumstances in a criminal case: Differences in the definitions under the Criminal Code (KUHP) and the anti-corruption law

E. Elda & T. Santoso
Faculty of Law, Universitas Indonesia, Depok, Indonesia

ABSTRACT: Repeat offence is an aggravating factor of crime under the Indonesian Criminal Code (KUHP), allowing the addition of a third of the maximum possible sentence to the verdict. In corruption cases, repeat offence is a special circumstance that may lead to the imposition of the death penalty. In Case Number 114/Pid.B/2006/PN.Jak.Sel involving Dicky Iskandar Dinata, the prosecutors applied an aggravation argument under the anti-corruption law to seek the death penalty. However, the court in its consideration used the provisions of the Criminal Code regarding aggravation of punishment. The issue is why there are different definitions of repeat offence as an aggravating circumstance. This paper aimed at analysing how repeat offence is applied as an aggravating factor in a corruption case as reflected in the prosecutor's indictment or the court's consideration. The method employed is normative research by reviewing court verdicts in corruption cases. The resulting findings show that there are differences in the interpretation of repeat offence as an aggravating circumstance.

1 INTRODUCTION

Recidivism or repetition of criminal acts occurs when a person who has committed a criminal offence and previously been sentenced by a criminal judge with a legally binding verdict (*in gewijde krachtvan*), commits the criminal offence again. In recidivism as well as *concurcus realis*, a person commits criminal acts more than once. The difference is that there is already a judge's ruling regarding recidivism which is used as a yardstick, namely punishment for the previous criminal acts. Recidivism occurs when the same person has been convicted of committing one or more crimes, and then repeats the deeds or criminal acts, both of which are subject to by separate court verdict. Recidivism is a reason for the aggravation of punishment. (Santoso & Zulfa, 2014, p. 565).

The are two types of repetition of crimes, namely those referred to in Article 486, 487 and 488 of the Criminal Code as well as those that are not covered by the aforementioned articles, but subject to special criminal offences that have been particularly specified so that repetition may be considered as set out in the Criminal Code (Chazawi, 2002, p. 81). The provisions of the Criminal Code regarding criminal offences are set out in Article 216 paragraph (3), Article 489 paragraph (2), Article 495 paragraph (2), Article 501 paragraph (2), Article 512 paragraph (3), and Article 516 paragraph (2).

In addition to the set of the crimes and offences set out in the Criminal Code, recidivism is also found in certain criminal acts, one of which is analysed by the author in this paper, namely the criminal act of corruption. The criminal act of corruption in Indonesia is already widespread in all aspects of public life. (Law No. 30 of 2002 regarding Corruption Eradication Commission). Criminal acts of corruption range from those committed by officials (Taskforce Team, 2008, p. 2). This crime is on the increase year on year, particularly with regard to the number of cases occuring, and the amount of financial losses suffered by the state as well as in terms of the proceeds of the criminal acts committed increasingly systematic way (Law No. 31 of 1999 and Law No. 20 of 2001 regarding corruption eradication).

Corruption is the cause of great losses to the country every year. In recent years, there have been data for every state. In 2010, the loss amounted to 3.6 billion, in 2011 it amounted to 2.1 trillion, in 2012 the amount was 10.4 billion, and in 2013 it was 7.3 trillion (ICW Data, 2015).

In addition to the increasing amount of losses suffered by the country each year, there is evidently repetition or recidivism in the criminal offence of corruption, namely the repetition of the crime by the perpetrators who have completed their sentence as corruptors. Recidivism or repetition of criminal acts is one of the aggravating considerations for criminal punishments pursuant to the Criminal Code in the form of the addition of one-third of the maximum sentence. In the case of corruption, recidivism is a certain circumstance that may be subject to the death penalty, as set out in Article 2 paragraph (2) of Law Numbers 31 of 1999 and 20 of 2001 regarding corruption eradication.

This paper will discuss research findings with regard to the case of Dicky Iskandar Dinata, namely Case Number 114/Pid. B/2006/PN.Jak.Sel. Dicky Iskandar Dinata was convicted of corruption in 1991 and imprisoned for eight years. Although he had undergone imprisonment, Dicky Iskandar Dinata did not pay any compensation nor receive a reduction in the sentence as set out in the verdict. The verdict issued by the court in Case Number 114/Pid. B/2006/PN.Jak.Sel. was used as the basis for the aggravation of punishment, however, the status as recidivist was not taken into consideration. In this case, the prosecutors used the basis of the aggravation of punishment as set out in the law regarding the eradication of criminal acts of corruption as the specific circumstances which may be subject to death penalty. However, the judges in their considerations used the guidelines on the aggravation of punishment as set out in the Criminal Code.

The focus of the problem discussed in this paper in realtion to the criminal case of corruption is recidivism, namely Case Number 114/Pid. B/2006/PN.Jak.Sel in which Dicky Iskandar Dinata was the offender. The verdict showcases difference of views arising from the public prosecutors and interpretation of the judges with regard to recidivism. Such differences resulted in the consequences as indicated in the considerations of the indictment prepared by the prosecutors and the verdict passed by the judges.

The issue being analysed in this paper is why there were different views regarding recidivism as the basis for interpreting the aggravation of punishment in the criminal case of corruption involving Dicky Iskandar Dinata. This paper is aimed at answering the a forementioned question and analysing the way to apply recidivism as the basis for aggravation of punishment in criminal acts of corruption in the indictment of prosecutors and the considerations of the judges in Case Number 114/Pid. B/2006/PN. Jak. Sel involving Dicky Iskandar Dinata as the defendant.

The methods applied in this paper is a normative method by conducting a study of court rulings on criminal acts of corruption. The court verdict being analysed was the verdict in Case Number 114/Pid. B/2006/PN. Jaksel in which the defendant was Dicky Iskandar Dinata. In addition to the aforementioned court ruling, the sources of data also include various law-related literature, as well as the provisions of the legislation, namely the Criminal Code, Law No. 31 of 1999 and Law No. 20 of 2001 regarding corruption eradication, as well as journals.

2 AGGRAVATION OF PUNISHMENT

The fundamental question is why the state is entitled to punish a person. This matter is also considered as the basis of justification for a punishment or a legal judgement, as conveyed by Walker in his book titled: *Why Punish? Theories of Punishment Reassessed.* (N. Walker in Nitibaskara, 2009, p. 81). The objective is the imposition of criminal punishment meted out to people for committing a crime and inducing the commission of bad deeds. (Singer & La Fond, 2007, p. 20).

Aggravation of punishment refers to the consideration of aggravating punishment in criminal sanctions imposed on defendants based on court verdicts. There are two categories of punishment aggravation, namely general and specific aggravation (Huda, 2011, pp. 513–514). In the general category, the aggravation of punishment may be because of *concursus idealis, realis or voorgezette handeling*, even though the techniques used are different for each of them. In this case, the criminal punishment could be increased by one-third of the criminal sanctions that may be imposed as formulated in the law.

The special aggravation of punishment may be classified into two groups, namely criminal punishment due to repetition (recidivism) and criminal punishment due to the special status of the offender (subjective offence), for example, civil servants. Moreover, criminal punishment may also be imposed by considering the specific relationship between the defendant and the object of the offence, such as mother, father, wife or child, for which an additional punishment of one-third of the maximum punishment may be added.

A doctrine classifies the aggravation of punishment into two systems based on the existence of recidivism, namely general and special recidivism (Santoso & Zulfa, 2014, p. 566).

1. The general system of recidivism states that recidivism is the repetition of any kind of criminal offence committed at any time is a reason for the imposition of criminal punishment. It means that the type of the criminal offence or the period of time from the first commission and the subsequent commission of the criminal offence are not specified. As the length of time between the two offences is not specified, this system does not recognise any statute of limitation for recidivism;
2. The special system of recidivism states that not all repetition constitutes a reason for the aggravation of punishment. Punishment aggravation is applicable only to repeated commission of certain criminal offences within a certain period of time.

3 RECIDIVISM IN THE CRIMINAL LAW: A COMPARISON BETWEEN THE CRIMINAL CODE AND THE ANTI-CORRUPTION LAW

Recidivism or repetition of a criminal act is a condition where someone has committed several acts each of which is a criminal offence, including one act or more that has been subject to criminal sanction by a Court of Law [Act No. 12 of 1995 concerning Correctional Facilities: Article 1 point (2)]. Repetition is one of the factors that aggravates punishment.

Recidivism, by its nature can be divided into two types, namely general and special recidivism. General recidivism is a condition where a person who has committed a crime and has been subjected to criminal punishment for committing the act subsequently commits a different type of crime (Prasetyo, 2010, p. 121). Special recidivism is a condition where a person who has committed a crime and has been subjected to criminal punishment has undergone his punishment, but subsequently commits the same crime (Prasetyo, 2010, p. 123).

According to the provisions on recidivism in Chapter 486 of the Criminal Code, the criminal punishment shall be increased by one-third. Whereas the provisions are as follows:

1. Repeating the same crimes or by law is considered the same. For example, a fraud followed by fraud or a theft followed by theft is, by law, considered to be committing a similar crime including those intended by the articles referred to in Article 486 (even though they are other classes of crime, they are considered equal). This is also applicable to the articles referred to in Articles 487 and 488;
2. There is already a court decision between the time of the commission of one crime and another crime, or when there is no court decision and the crimes committed are of the same classification;
3. There must be a jail sentence, rather than confinement or fines;
4. Five years have not yet elapsed after wholly or partially serving a prison sentence which has been imposed.

The Criminal Code does not have any provision for the repetition of criminal offences in Book I regarding 'general rules'. It is specifically set out in a group of specific criminal acts and classified as 'felony' in Book II or 'misdemeanour' in Book III. In addition, the Criminal Code also requires the repetition to be within a certain period. Therefore, the Criminal Code adopts special provision on recidivism, namely that aggravation of punishment will only be levied for the repetition of certain types of crime and will only be imposed for a certain period of time (Santoso & Zulfa, 2014, p. 567).

In relation to the adoption of such special recidivism, the Criminal Code defines recidivism as the recidivism of certain crimes. In this regard, the Criminal Code distinguishes two types of recidivism, namely recidivism of certain 'similar crimes' and recidivism of certain crimes 'classified in the same type'. Recidivism of certain crimes that are 'similar' includes 11 types of crimes that may be used as a reason for the aggravation of punishment. These crimes are set out in several articles in Book II of the Criminal Code, namely: Article 137 paragraph (2), Article 144 paragraph (2), Article 155 paragraph (2), Article 157 paragraph (2), Article 161 paragraph (2), Article 163 paragraph (2), Article 208 paragraph (2), Article 216 paragraph (3), Article 321 paragraph (2), Article 393 paragraph (2) and Article 303 (bis) paragraph (2) (Santoso & Zulfa, 2014, p. 567).

According to Van Hamel, as quoted by Simons and Lamintang, there are consequences of the occurrence of repetition (Van Hamel in Simons & Lamintang, 1992, p. 493). The first consequence is that the crime must be similar to the crime previously committed and punished. The second consequence requires the existence of a similarity between the criminal acts previously committed. According to Smith, the doctrine has been developed into two forms, namely algemeine recidivism (repetition of general criminal offence) or byzondere recidivism (repetition of specific criminal offence) (Smith in Simons & Lamintang, 1992, p. 493).

4 RECIDIVISM AS AGGARAVATION OF CRIMINAL OFFENCE IN THE CORRUPTION CASE OF DICKY ISKANDAR

This paper discusses the repetition of a criminal act of corruption in Case Number 114/Pid. B/2006/PN.Jaksel involving Dicky Iskandar in the case of a fictitious Letter of Credit (LC) of BNI. The case occurred in 2003. The defendant had previously been convicted for a criminal act of corruption in 1991, namely in a case of a fictitious LC from Bank Duta.

The corruption case included a repetition of a criminal act, namely the use of a fictitious LC of Rp 1.7 trillion from BNI, known as Case Number 114/Pid. B/2006/PN.Jaksel involving Dicky Iskandar. The case occurred in 2003, inflicting losses to the government amounting to Rp 49,269,000,000 (approximately 49 trillion Rupiah) and USD 2,999,990. The case was a repetition of a criminal act of corruption committed in 1991, namely the case of the issuance of a fictitious LC by Bank Duta. In the latter case, the defendant was subject to imprisonment for eight years as well as mandatory payment of compensations.

In this case, the public prosecutor charged the defendant for the violation of Article 2 paragraph (1) jo Article 18 of Law Number 31/1999 and Law Number 20/2001, as well as Article 55 paragraph (1) point 1 and Article 64 paragraph (1) of the Criminal Code or Article 3 paragraph (1) sub-paragraphs a, b, and c of Law Number 15/2002 and Law Number 26/2003, as well as Article 55 paragraph (1) point 1 of the Criminal Code and Article 64 paragraph (1) of the Criminal Code. In the indictment, the prosecutor claimed that the defendant had been validly and convincingly proved to have committed a criminal act of corruption as referred to in Article 2 paragraph (1) and paragraph (2), as well as Article 18 of Law Number 31/1999 and Law Number 20/2001, and Article 55 paragraph (1) point 1 and Article 64 paragraph (1) of the Criminal Code, subject to the death penalty and a fine of 500 million Rupiah or alternatively six months in imprisonment.

The panel of judges of South Jakarta District Court, in this Case Number 114/Pid. B/2006/PN.Jak.Sel., passed a verdict judgement stating that the defendant Dicky Iskandar Dinata had been validly and convincingly found guilty of committing a criminal act of corruption. The punishment was 20 years of imprisonment and a fine of 500 million Rupiah or alternatively 5 months in prison. The Appellate Court passed a verdict Number 175/Pid/2006/PT. DKI, confirming the verdict of the District Court. At the cassation level, the Supreme Court in its verdict Number 181 K/Pid/2007 refused the appeal filed by the defendant. As a final attempt the defendant filed a petition for judicial review to the Supreme Court under Case Number 114 PK PK/Pid. Sus/2008 and the Panel of Justices refused the petition.

The interesting aspects of this case are the prosecutor's indictment and the judges' considerations confirming the charges made by the prosecutor. In this regard, the prosecutor demanded death penalty because the defendant had previously committed a similar criminal act of corruption. As the basis for such demand of the death penalty, the prosecutor used Article 2 (2) of the law on the eradication of criminal acts of corruption. This Article 2 paragraph (2) of the law defines the existence of the repetition of crime (recidivism) as one of the certain circumstances. However, the judges rejected the prosecutor's demand for the death penalty asserting that the prosecutors used Article 2 paragraph (1) with a maximum criminal punishment of 20 years in jail, rather than Article 2 paragraph (2) allowing the imposition of death penalty.

The panel of judges argued that the prosecutor's indictment did not include Article 2 paragraph (2), so that the defendant could not be charged with the provision of the article, even though the judge stated that the defendant had committed a repeated crime of corruption in the previous case, namely the corruption case of Bank Duta. Therefore, the judges only imposed a maximum imprisonment of 20 years as set out in Article 2 paragraph (1). The judges in their consideration stated that the defendant had never been convicted in the corruption case of Bank Duta, thus despite undergoing a criminal process he had not paid the compensation.

This case indicates in the eradication of corruption, that is, the use of Article 2 paragraph (2) of the law on the eradication of criminal acts of corruption, despite the new demands set in the prosecutor's indictment. This means that although the rule of law set forth in the legislation had been implemented in law enforcement there was a contradiction, as the judges on one hand refused to sustain the death penalty demanded by the prosecutor for the repetition of criminal acts of corruption, while on the other hand, the judges used repetition of criminal acts as an incriminating factor in their considerations.

At the cassation level, the Supreme Court Decision No. 14 K/Pid/1990 dated 25 June 1992, found that the defendant Dicky Iskandar Dinata was involved in a corruption case related to foreign exchange transactions with Bank Duta. The defendant had been sentenced to eight years in prison and had to pay a compensation of 811,342,000,000 Rupiah. However, up until the present moment, after the defendant completed his criminal sentence, the defendant has never paid the compensation. Currently, the State Attorney at the Attorney General's Office is planning to file a civil lawsuit for the payment of the compensation.

The considerations of the panel of Supreme Court Justices are as follows: 1) the defendant's crime was extremely detrimental to the economy and the state's finance; 2) the defendant's actions led to a decrease in public trust of the banking sector as a component of the state economy; 3) the defendant did not admit his act; 4) the defendant had previously been convicted in the Bank Duta corruption case, and even though he had completed his jail term, he had not paid the compensation.

Based on the aforementioned considerations, the court has also set a precedent for passing criminal punishment on defendants by paying attention to the purpose of punishment. This is because the purpose of punishment is not only to cause remorse to the defendants, but it is also intended to provide instruction for the defendants to be able to do well in the future. Moreover, punishment also has goals for society, such as ensuring that those guilty of crimes will be subject to criminal punishment. This is in line with its objective of causing a sense of fear of committing crimes in the future.

In this case, the court reasoning for refusing the prosecutor's demand for the death penalty pursuant to Article 2 paragraph (2) of the Corruption Eradication Law is that the defendant's recidivism with regard to the corruption committed previously was not founded in the charges. Even so, the court's consideration did include that the defendant had never been convicted of corruption. In this case, the court did not consider the special aggravating circumstances as recidivism as set out in the Corruption Eradication Law, but instead the court considered it as a general aggravating circumstance.

5 CONCLUSION

The results indicate that there is a difference in recidivism in the aforementioned discussion of aggravating circumstances in the criminal act of corruption in Case Number 114/Pid. B/2006/

PN.Jaksel involving Dicky Iskandar as the defendant. As a consequence, the public prosecutor's office applied recidivism as a special aggravating circumstance in the case of the crime of corruption, but the court applied the definition of recidivism as a basis for considering general aggravating circumstances as intended in the Criminal Code. In this regard, the court did not conduct an extensive study into the meaning of recidivism in the criminal act of corruption.

Based on the above descriptions, the court did not find any reason to apply the aggravation of punishment in relation to recidivism or repetition of criminal acts. Moreover, the prosecutor also did not carefully examine the records as to whether or not the defendant had previously been punished for a criminal act of corruption. Therefore, the definition of recidivism as provided in Article 2 paragraph (2) of the Law on the Eradication of Criminal Acts of Corruption was not applied.

REFERENCES

Chazawi, A. (2002). Pelajaran Hukum Pidana Bagian 2 [Lesson of Criminal Law, Part 2]. Jakarta, Raja Grafindo Persada.

Huda, C. (2011). Pola pemberatan pidana dalam Hukum Pidana Khusus [Pattern of Aggravation of Punishment in the Special Criminal Law. Jurnal Hukum, 18(4), 508–524.

Indonesia (1995). Undang-undang No. 12 Tahun 1995 tentang Pemasyarakatan. [Law No. 12 of 1995 regarding Correctional Facility].

Indonesia (1999). Undang-undang No. 31 Tahun 1999 tentang Pemberantasan Tindak Pidana Korupsi. [Law No. 31 of 1999 regarding Eradication of the Crime of Corruption].

Indonesia (2001). Undang-undang No. 20 Tahun 2001 tentang Perubahan atas Undang-undang No. 31 Tahun 1999 tentang Pemberantasan Tindak Pidana Korupsi [Law Number 20 of 2001 regarding Amendment to Law Number 31 of 1999 on the Eradication of the Crimes of Corruption].

Indonesia (2002). Undang-undang No. 15 Tahun 2002 tentang Tindak Pidana Pencucian Uang [Law No. 15 of 2002 regarding Crime of Money Laundering].

Indonesia (2002). Undang-undang No. 30 Tahun 2002 tentang Komisi Pemberantasan Tindak Pidana Korupsi [Law No. 30 of 2002 regarding Corruption Eradication Commission].

Indonesia (2003). Undang-undang No. 26 Tahun 2003 tentang Anggaran Pendapatan dan Belanja Negara Tahun Anggaran 2003 [Law No. 26 of 2003 regarding State Budget Revenues and Expenditures Fiscal Year 2003].

Indonesian Corruption Watch [ICW] (2015). 4 Tahun Tren Korupsi Indonesia 2010–2013 [4 Years Indonesia Corruption Trends in 2010–2013]. Kompetisi Infografis Lawan Korupsi. Jakarta, ICW.

Mahkamah Agung [Supreme Court] Republik Indonesia (2007). Case of Corruption Number 181 K/Pid/2007. Jakarta.

Mahkamah Agung [Supreme Court] Republik Indonesia (2008). Case of Corruption Number 114 PK/Pid.Sus/2008. Jakarta.

Nitibaskara, T.R.R. (2009). Perangkap Penyimpangan dan Kejahatan: Teori Baru dalam Kriminologi [The Pitfalls of Perversion and Crime: A New Theory in Criminology]. Jakarta, Yayasan Pengembangan Kajian Ilmu Kepolisian.

Pengadilan Negeri [District Court] Jakarta Selatan (2006). Case of Corruption Number 114/Pid. B/2006/PN.Jaksel. Jakarta.

Pengadilan Tinggi [High Court] Jakarta (2006). Case of Corruption Number 175/Pid /2006 /PT. DKI. Jakarta.

Prasetyo, T. (2010). Hukum Pidana [Criminal Law]. Jakarta, Rajawali Pers.

Santoso, T. & Zulfa, E.A. (Eds.) (2014). Hukum Pidana Materiil dan Formil [Criminal Law in Materially and Formyl]. Jakarta, USAid, The Asia Foundation [TAF] & Partnership for Governance Reform [Kemitraan].

Simons, D. (1992). Kitab Pelajaran Hukum Pidana [Leerbook van Het Nederlandse Strafrecht]. Translated by Lamintang, P.A.F. Bandung, Pionir Jaya.

Singer, R.G. & La Fond, J.Q. (2007). Criminal Law Examples & Explanations (4th ed.). New York: Wolters Kluwer Law and Business.

Team Taskforce Konsorsium Reformasi Hukum Nasional [KRHN] (2008). Naskah Akademis dan Rancangan Undang-undang Pengadilan Tindak Pidana Korupsi [Academic Texts and Draft Legislation The Court of The Crime of Corruption]. Jakarta, Konsorsium Reformasi Hukum Nasional (KRHN).

A new paradigm of the justice of outsourcing in Indonesia

I. Farida, S. Arinanto & J. Rizal
Faculty of Law, Universitas Indonesia, Depok, Indonesia

ABSTRACT: Globally, businesses require outsourcing both for labor dispatching/labor supply and for job undertaking/job supply to offer flexible employment to support government efforts to lower unemployment and increase production efficiency, especially amid recession. After 2010, Indonesian labor unions have demanded the abolishment of outsourcing and temporary contract work due to the fear of employment abuse. However, this movement has created difficulties for employers and the outsourcing industry, at times harming trade and investment. These conflicting interests have raised social justice issues. In this paper, whether Indonesian outsourcing regulations promote a sense of justice in the legal system or not is analyzed, and to creative work to construct a new justice paradigm is suggested. The research employed shows that outsourcing is intensely needed in Indonesia; however, the prevailing regulation is not giving sense of justice for all players. Through the doctrinal legal research and legal comparative approach applied in a micro-level examination of how regulations and laws in the developed countries found that it is necessary for the government to establish more flexible regulations of outsourcing that can provide justice to all players.

1 INTRODUCTION

Indonesia has one of the largest populations in the world and abundant human and natural resources. Modern development in Indonesia began in 1945 when the Indonesian people declared their independence. At the same time, they adopted the 1945 Constitution, which lays down the basic law guiding the performance of the state.

With the country's recent economic development, labor and employment issues have become crucial matters to both citizens and the state of Indonesia. Article 27 paragraph (2) of the 1945 Constitution mandates that jobs have to support a proper quality of life. However, the achievement of fair development and welfare for all Indonesian laborers is still constrained by serious legal issues in the present days. One crucial legal issue is the obstacles that labor contracts pose to management by investors and to the absorption of unemployed laborers into the labor supply system.

Indonesia's Labor Law only allows three kinds of direct labor contracts: permanent (unspecified time), fixed-term, and daily worker contracts. Unfortunately, Indonesia does not have any other stipulations, such as part-timer contracts, which makes Indonesia's Labor practice rigid and inflexible. Amid the inflexibility of the labor market and the labor policies of both the government and labor unions (e.g., the flexibility of working hours is not recognized, or a core job which is allowed only for a permanent contract), which could cause massive unemployment in the case of a future economic recession. Investors present a strong demand for outsourcing. Under such circumstances, the outsourcing industry requires a redress of a systematized legal framework for the purpose of labor efficiency and the flexibility of the labor market.

This study aims to review and analyze how far the regulation of outsourcing in the legal system of Indonesian manpower has given a sense of justice as the Constitution 1945 (e.g., rights of equal treatment, welfare, and protection) for all the players of outsourcing. Using a main analytical tool of theory of justice will point out the occurred unjust situation. The prevailing regulation is not in accordance with the constitutional rights promised by the 1945 Constitution which is also in ligned with the theory of justice by Rawls that represents justice with two main principles: (i) First Principle (Equal Liberty Principle) which guarantees the right of each person to have the most extensive basic liberty compatible with the liberty of others, and (ii) Second Principle (Social Inequality Principle) which adduces social and economic inequalities that are to be arranged so that they can give a just and equal principle of opportunity (Rawls, 1999). The prevailing regulation admittedly tries to protect workers as the weak party. However, the regulation should give benefits to all parties and does not take other's liberty. A theory however elegant and economical must be rejected or revised if it is untrue; likewise laws and institutions no matter how efficient and well-arranged must be reformed or abolished if they are unjust (Rawls, 1999).

2 METHODOLOGY

The doctrinal and non-doctrinal legal research method will be employed in this study. The doctrinal legal research includes (i) study of legal norm, (ii) study of systematic law, (iii) synchronism of law, and (iv) history of law. (Soekanto, 1985). Through non-doctrinal legal research, issues concerning Indonesia's outsourcing system are expected to be found. The study is such as the implementation of Transfer of Undertaking Protection of Employment (TUPE), disharmonized of outsourcing regulations, the ambiguity of the outsourcing system and legal enforcement. To solve those issues, doctrinal legal research is employed by using legal comparative and historical approaches. For the comparison, the researchers will review legal regulations, laws, and outsourcing implementation in a micro level in four countries (Germany, the United Kingdom, the United States, and Japan).

3 LEGAL SYSTEM OF LABOR LAW IN INDONESIA

Has the Legal System of Labor in Indonesia Given a Sense of Justice for all the Players of Outsourcing in accordance with the Constitution of the Republic of Indonesia?

Outsourcing as a form of operational delegation or implementation of part of the production process to other parties apart from the company has become popular since the beginning of 2000. Besides being considered as more efficient, outsourcing can be a solution to improve the efficiency of production costs (Sutedi, 2009) and labor cost (Fariana, 2012). By transferring company's non-core businesses to the third party, the user company can remain focus on their core business.

However, labor union resists outsourcing, as in their opinion outsourcing is used as a way to minimize workers' wage and social security. The employment relationship which is applied generally is in the form of fixed-term employment, while wage and social security are only paid in minimum and there is no guarantee of career development (Marjono, 2013). For that reasons, in April 2011 labor union submitted a judicial review to the Constitutional Court asking the abolishment of the fixed term employment agreement and outsourcing from Indonesia Labor Law (Constitutional Court, 2011).

In a judicial review, in 2012 the Constitutional Court issued the Verdict Number 27/PUU-IX/2011 which states that outsourcing complies to the 1945 Constitution. However, the fixed term employment applied in outsourcing is conditionally unconstitutional, unless the TUPE principal is applied (Constitutional Court, 2012).

The decision of the Constitutional Court No. 27/2011 has directly enacted the Labor Law to change the law. The change was directly defined in the implementation regulation through the Regulation of the Minister of Manpower and Transmigration No. 19/2012, which started to effectively prevail in November 2012. This Ministerial Regulation is a technical regulation intended to protect laborers' rights, particularly regarding an employment guarantee for laborers in fixed-term employment contracts or other labor contracts. Finally, the question arises as to whether the Ministerial Regulation is sufficient to ensure the guarantee, protection, and certainty of justice in law and equal treatment under the law for all parties, as amended in the 1945 Constitution.

The said Ministerial Regulation has also caused another problem for players due to its stipulation that when one outsourcing company finishes its contract with the user company and the user company wants to initiate a new contract with another outsourcing company, then previous outsourcing laborers must be employed by the new outsourcing company so laborers can still work in the user company (TUPE principal). The question is: How do laborers from the previous outsourcing company automatically transition to laborers in the new outsourcing company when labor contracts and employment relations exist between the laborers and their previous outsourcing company, not with the new outsourcing company (waiving the freedom of contract)? A labor contract cannot be changed or revoked unless both parties approve (Simanjuntak, 2011).

The goal of waiving the freedom of contract maybe means that there is certainty and protection to work continuously for fixed-term contract laborers. Unfortunately, that regulation is waiving the sanctity of the labor contract between the laborers and their previous outsourcing company. The obligation to absorb the fixed-term workers by the new outsourcing company as regulated in the Ministerial Regulation, the old outsourcing company as the party in the employment contract is likely not given space to declare its will and forced to end the employment relation. Both outsourcing companies have been damaged due to this stipulation.

If TUPE which does not set aside an outsourcing company is understood as the provision of the government to protect the weak party, in the context that workers have a bargaining position which is weaker than the outsourcing company, the theory of justice by John Rawls can answer it. John Rawls states that justice as fairness (Rawls, 1999). Also, justice is the main unsure of every law and social institutions (Rawls, 1999). "Justice is the first virtue of social institutions, as truth is of systems of thought. A theory however elegant and economical must be rejected or revised if it is untrue; likewise laws and institutions no matter how efficient and well-arranged must be reformed or abolished if they are unjust." Consequently, if the Ministerial Regulation No. 19/2012 has shown unjust values that it could be rejected or revised by the new form and content of the regulation as referred to in advance.

The main thing that has to be underlined concerning justice is to be equal and non-aligned. The same with social justice which is adhered by Indonesian people in Pancasila which is how to build social justice for all Indonesian community. Thus, if the government wants to intervene a civil relation between the citizens for the purpose of conducting protection for the weak, the intervention should not violate constitutional rights of the players of outsourcing (users and outsourcing companies) which is also guaranteed by the Indonesia Constitution. Although company is not human in reality, in legal science the legal subject consists of two kinds, which are: (i) natural person; and (ii) artificial person. Justice Buckley L describes the artificial person is a corporate with no physical existence. Existence is only in legal contemplation, and corporate does not have mind, but it comes from the mind of its management (Harris & Anggoro, 2010). Hence, management as the source of mind and company's interest at the same time have the position as citizens whose interest of basic human's right is protected by Constitutions as mentioned in Article 28 D paragraph (1) Constitution 1945.

The limitation to 5 kinds of jobs for labor supply in the Ministerial Regulation No. 19/2012 is also a regret for user companies and outsourcing companies. This limitation by regulations even becomes more confusing to be implemented when actually other ministries have also issued regulations regarding the kind of job to be outsourced; for example, Financial Service Authority (OJK) issued regulation of OJK No.9/POJK.03/2016 concerning the Principles of Safety for Private Banks Handling the Transferring Part of a Job to Another Party which is regulated in a wider kind of job than that stipulated in the Ministerial Regulation No. 19/2012. Besides that, the Ministry of Energy and Mineral Resources issued No. 27/2008 concerning Business Activity for Oil and Gas, which has their own category for the kinds of jobs for labor supply that prevails and binds all business perpetrators in oil and gas. This causes disharmonization of regulations and confusion among perpetrators.

4 OUTSOURCING SYSTEMS PROVIDING A SENSE OF JUSTICE FOR PARTIES

The experiences of some countries, including those used in the comparative analysis, have proven that outsourcing can help the economy of nations and the world. With abundant human and natural resources, Indonesia has the potential to improve the welfare of its society. By learning about and adopting appropriate regulations concerning outsourcing in the comparative analysis, it is expected that this research will identify an ideal regulation aligned with the 1945 Constitution.

Outsourcing in Germany is known by the term "labor leasing" (Germany Employee Leasing Act 1972). According to Article 1 of Germany's Labor Leasing Act, labor leasing is defined as a condition where the agency company, which in this case is the outsourcing company, rents out its employees to the user company. Kirchner, Kremp, and Magotsch (2010) explain that the first law concerning outsourcing in Germany was created in 1972. Since then, Germany has made some improvements in its laws. It does not have a jurisprudence system, but the jurisprudence stipulations are very important and influential in the court system (similar to Indonesia). There are two types of outsourcing in Germany: *Werktrag* (undertaking), which is the same as job supply in Indonesia; and *Arbetachmer Isberlassung*, the same as labor supply in Indonesia. Outsourcing in Germany can be conducted in every sector other than construction trade, and it can only be done for a maximum of 18 months (Kremp, 2016). This differs from what happens in Indonesia, where outsourcing can only be done for 5 types of jobs (cleaning services, catering, security, drivers, and supporting jobs in mining services). There are also laws to protect laborers' basic rights, such as those dealing with social guarantee, wages, anti-discrimination, opportunities to develop, and so on. There are some specific laws regulating the basic protection of workers, such as the General Equal Treatment Act, the Protection Against Unfair Dismissal Act, the Maternity Protection Act, the Federal Educational Allowance Act, even the Part-Time and Fixed Term Employment Act. Germany thus has regulations in the form of laws that are more detailed, specific, and flexible for workers and employers.

The United Kingdom as a country adhering to the Common Law system does not regulate outsourcing in particular. Outsourcing in general is regulated in the Statutory Instruments 2010 No. 93 concerning the Terms and Conditions of Employment—The Agency Worker Regulations 2010 ['The Agency Worker Regulations 2010'] and Transfer of Undertaking, which is regulated in the Employment Relation Act 1999. The regulation of outsourcing in the United Kingdom is done via the implementation of European Communities Act 1972, which is based on the Treaty concerning the Accession of the United Kingdom to the European Economic Community, the European Coal and Steal Community, and the European Atomic Energy Community signed on January 22, 1927 in Brussels (The Agency Worker Regulations 2010). Practically, the contract has to follow the principles of arrangement, rights and obligations, as well as transfer. The third party can act as the right or obligation receipt in its role as an outsourcing agent or job undertaking (European Communities Act 1972). Outsourcing laborers have protection when they are fired in an unfair manner by the outsourcing company. Outsourcing laborers are protected in terms of their pension insurance; if they do not receive this insurance, they can claim their employer has breached their contract or dismissed them (the Agency Worker Regulations 2010). Unlike the United Kingdom, Indonesia only stipulates that in cases in which the rights of laborers are breached, so the laborers may file a lawsuit to the labor court.

The United States is a federal state, consisting of 50 states under federal law (Bogdan & Taylor, 1975). Russel *et al.* (2013) states the Labor Law is dominated by three federal statutes: the National Labor Relation Act (NLRA), Title VII of the Civil Rights Act, and the Fair Labor Standards Acts (FLSA). Skonberg and Mankes (2001) explain the primary purpose of the Labor Law in the United States is to achieve equality in terms of the bargaining power between laborers and employers. There are two kinds of outsourcing contractors in the United States: employees and independent contractors. Both are referred to as outsourcing laborers, but independent contractors can work in more than one company, have flexible working hours, and use own authority to make job decisions. These privileges are not for employee contractors. On the other hand, employee contracts offer health insurance, and all costs are covered by the company (Skonberg & Mankes, 2001). However, the FLSA federal law protects both contracts. It shows that outsourcing in the United States is also flexible for both laborers and employers. In addition, it offers students an opportunity to work and obtain wages per hour.

The fourth comparative country in this study is Japan. The Constitution of Japan is the basis for the country's Labor Law. The law governing labor is Law No. 59/1947 concerning Employment. The outsourcing laborers are known as dispatching laborers. The law consists of two categories on the basis of employment relations (labor contracts): standard employment (permanent laborers), and non-standard employment (non-permanent laborers). Furthermore, non-standard employment is divided into three categories: temporary or contractual labor (specified working time), part-time labor and indirect employment, similar to the dispatching laborers. The kinds of jobs that can be outsourced in Japan are broader than in Indonesia; previously, there were 26 kinds of jobs, but since the Working Dispatching Act was amended in 2012, the types are unlimited. The four countries above all appear to have more specific and flexible regulations related to outsourcing, which can be a guideline for Indonesia to create more flexible regulations for its outsourcing system, including flexible labor contracts.

5 CONCLUSION

An outsourcing system is needed in the business world due its flexibility, which should allow employers to concentrate on improving their core businesses and assign supporting jobs to an outsourcing company. There are solutions offered by the Constitutional Court to guarantee legal certainty for laborers, protecting the right of every person to work and to receive rewards and fair treatment (Constitutional Court Decision 2011). Laborers must ensure the protection of their rights by implementing the TUPE for companies that outsource jobs, which are regulated in the Ministerial Regulation No. 19/2012. Another form of protection for laborers in the Ministerial Regulation is that in the end, if the user company no longer requires any outsourced job/labor, the outsourced laborers have the right to receive compensation for their work calculated from the time they begin to work in the user company. Although it does not mention in the Ministerial Regulation which party should be liable to pay for this, the stipulation does place an additional burden on both the outsourcing and user companies.

Consequently, by the decision of the Constitutional Court No. 27/PUU-IX/2011 the protection towards laborers' rights related to working continuation is legally protected by implementing the TUPE principle. However, the TUPE system and protection of working period payment, as well as the limitation of the kinds of jobs that can be outsourced, are considered to contradict the flexibility concept, which becomes the main concept of outsourcing. Many oppose the decision of the Constitutional Court by declaring that it is contrary to the principles of Indonesian justice written in Pancasila point 5 (social justice for Indonesian society), constitutional rights stipulated in the 1945 Constitution, and the freedom of contract and the Civil Code.

Labor market flexibility will generate employment opportunities. Laborers are free to choose the company that suits their decent needs, which are determined by how much the company can meet their financial needs. Laborers are not tied to a company in a long term, but can move from one job to another job to obtain better income. The ease of moving jobs can open up additional employment opportunities for more job seekers.

This study reviews the concept of legal protections to ensure a fair outsourcing system in the future by approaching law as a system with the framework of norms and to create a new paradigm of outsourcing. The analytical tool used in this study is Rawls's theory of justice (1999), which treats justice as a condition that is realized by laws through the general framework of the political order (a democracy like in Indonesia), which provides a foundation for actualizing it. Therefore, a law is a political statement made by a government in the forms of a constitution and written regulations. Both the law and political system should have the same goal and the manifestation of legal ideas.

The treatment for laborers' rights still has to emphasize the principle that the violation or erosion of the basic rights of outsourced laborers should be prohibited. Existing social and economic differences are related to the social, job, and career positions that can be obtained fairly by all people who have the ability and willingness to use their skills. Social and economic inequality can be supported if they can give advantages to less fortunate communities. Differences in social and economic conditions can occur at any time, in accordance principle of social and economic inequality (Rawls, 1999).

The result from this study shows that the outsourcing regulations existed today is not enough and event tendecious giving unjust sense to all players. The limitations of jobs and implementation of TUPE principle does not give solution on the contrary new burden to user company and outsourcing company. In practice, due to limitation of kinds of jobs in labor supply, the user company makes a change in the method into job undertaking, while it is actually a labor supply, but in a formal document it is stated as job undertaking.

The Ministerial Regulation No. 19/2012 also lacks of important regulations, such as sanction for those who commit the violation, or the party that is obligated to pay the labor's rights at the end of employment period, and so on. Besides, the understanding of TUPE still confuses not only to outsourcing players but also the institutions. The unjust implementation comes from different interpretations, and multi-interpretations happen between the Constitutional Court judges with the Ministry of Manpower and all players. That differentiation will cause legal uncertainty and a chance of making illegal outsourcing practice.

The reality that must be confronted is that the Indonesian economy must continue to grow to provide employment for the future 2 million laborers newly born every year. The government must enact policies to invite foreign investors to provide the funds and technology necessary for industrialization. Indonesia urgently needs to establish such an environment to ensure economic growth. Regardless of the desirability, economic growth results in the rapid growth of the labor supply businesses, which might lead to low-wage employment or human trafficking if unemployment increases in the absence of legal structures or an outsourcing system to protect laborers.

The analysis from this study indicates that Indonesian government needs to provide a new legal framework law for outsourcing, being independent and separated from today's existing Labor Law. The Ministerial Regulation has caused confusion and inconsistency; therefore, outsourcing in Indonesia needs to redress the regulation to exclusively regulate the outsourcing system.

REFERENCES

Bogdan, R. & Taylor, S.J. (1975) *Introduction to Qualitative Research Methods: A Phenomenological Approach to the Social Sciences*. New York, John Wiley & Sons.
Dewan Perwakilan Rakyat [House of Representative] Republik Indonesia (2011) DPR RI: Minutes of Constitution Court Session Number 27/PUU-IX/2011 regarding the Judicial Review of Law Number 13 Year 2003 concerning Manpower toward 1945 Constitution, Session of Examination of Appealing Revision, May 11, 2011. Jakarta.
Dewan Perwakilan Rakyat [House of Representative] Republik Indonesia (2011) DPR RI: Minutes of Constitution Court Session Number 27/PUU-IX/2011 regarding the Judicial Review of the Law Number 13 Year 2003 concerning Manpower towards 1945 Constitution, Hearing Session Government Opinion, House of Representatives, Expert Witnesses from appellant and Government DPR, July 6, 2011. Jakarta.
Dewan Perwakilan Rakyat [House of Representative] Republik Indonesia (2012) DPR RI: Minutes of Constitution Court Session Number 27/PUU-IX/2011 regarding Judicial Review of Law Number 13 Year 2003 concerning Manpower toward 1945 Constitution, January 17, 2012. Jakarta.
European Communities (1972). European Communities Act 1972.
Fariana, A. (2012) *Aspek Legal Sumber Daya Manusia Menurut Hukum Ketenagakerjaan*. Jakarta, Mitra Wacana Meida.

Harris, F. & Anggoro, T. (2010) *Hukum Perseroan Terbatas: Kewajiban Pemberitahuan oleh Direksi* [*Limited Liability Company Law, Liability Notification by the Board of Director*]. Bogor, Ghalia Indonesia.

Indonesia (1945) *Undang-undang Dasar 1945* [*The 1945 Constitution of the Republic of Indonesia*].

Indonesia (2003) *Undang-undang No. 13 Tahun 2003 tentang Tenaga Kerja. Lembaran Negara Republik Indonesia Tahun 2003 Nomor 39. Tambahan Lembaran Negara Republik Indonesia Nomor 4279.* [*The Law No. 13 of 2003 concerning Manpower. State Gazettes of Republic of Indonesia Year 2003 No. 39, Additional State Gazettes of Republic of Indonesia No. 4279*].

Indonesia (2008) *Peraturan Menteri Energi dan Sumber Daya Mineral No. 27 Tahun 2008 tentang Kegiatan Usaha Penunjang Minyak dan Gas Bumi* [*Minister of Energy and Mineral Resources Regulation Number 27 of 2008 concerning Supporting Business Activities for Oil and Gas*].

Indonesia (2012) *Peraturan Menteri Tenaga Kerja dan Transmigrasi Republik Indonesia No. 19 Tahun 2012 tentang Syarat-syarat Penyerahan Sebagian Pelaksanaan Pekerjaan kepada Perusahaan Lain* [*Regulation of the Minister of Manpower and Transmigration Republic of Indonesia Number 19 of 2012 concerning Requirements for Assignment of Part of Work to Other Companies*].

Indonesia (2013) *Surat Edaran Menteri Tenaga Kerja dan Transmigrasi Republik Indonesia No. SE.04/MEN/VIII/2013 tentang Pedoman Pelaksanaan Permenakertrans No. 19 Tahun 2012 tentang Syarat-syarat Penyerahan Sebagian Pelaksanaan Pekerjaan kepada Perusahaan Lain* [*Circular Letter of the Minister of Manpower and Transmigration Number SE.04/MEN/VIII/2013 concerning Guidelines on Regulation of the Minister of Manpower and Transmigration Number 19 of 2012 concerning Requirements for Assignment of Part of Work to Other Companies*].

Indonesia (2016) *Peraturan Otoritas Jasa Keuangan [POJK] No. 9/POJK.3/2016 tentang Prinsip Kehati-hatian Bagi Bank Umum yang Melakukan Penyerahan Sebagian Pelaksanaan Kerja Kepada Pihak Lain* [*Financial Authority Services Regulation Number 9/POJK.3/2016 regarding Prudent Principles for Commercial Bank that Handover Partially of Their Work to Other Parties*].

International Labor Organization [ILO] (2014). ILO: *Indonesia: Trend Sosial Ketenagakerjaan Agustus 2014*. Jakarta, Kantor ILO Jakarta.

Kirchner, J., Kremp, P.R. & Magotsch, M. (eds.) (2010) *Key Aspects of German Employment and Labour Law*. Heidelberg, Springer.

Kremp, P.R. (2016): Employment and employee benefits in Germany: Overview. Eagan, Practical Law-Thomson-Reuters.

Mahkamah Agung [Constitution Court] Republik Indonesia (2012) Decision Number 27/PUU-IX/2011 regarding the Judicial Review of Law Number 13 Year 2003 concerning Manpower toward the 1945 Constitution, January 17, 2012. Jakarta.

Marjono, R. (2013) Kritik terhadap Implementasi Sistem Outsourcing di Lingkungan PT PLN Distribusi Jateng-DIY. Yogyakarta, Lembaga Bantuan Hukum [LBH] Yogyakarta.

Rawls, J. (1999) *A Theory of Justice*. revised edition. Massachusetts, Harvard University Press.

Russel, W.T., Nelson, J.E., Garrote, N.E., Zottola, A.J., Gross, M.O. & Kolstad, C.K. (2013) United States. In: Lewis, M. (ed.) *Getting The Deal Through: Outsourcing 2014*. London, Law Business Research Ltd. pp. 124–129.

Simanjuntak, P.J. (2011) *Manajemen Hubungan Industrial*. Jakarta, Lembaga Penerbit Fakultas Ekonomi Universitas Indonesia.

Skonberg, J.M. & Mankes, M. (2001) International: USA. In: Edmonds, V., Feu, V.D., Gillow, E. & Hopkins, M. (eds.) *EU & International Employment Law*. Bristol, Jordan Publishing Limited.

Soekanto, S. & Mamudji, S. (1985) *Penelitian Hukum Normatif: Suatu Tinjauan Singkat*. Jakarta, Raja Grafindo Persada.

Sutedi, A. (2009) *Hukum Perburuhan*. Jakarta, Sinar Grafika.

The United Kingdom (2010) The Agency Worker Regulations 2010. In: *Statutory Instruments 2010 No. 93 concerning the Terms and Conditions of Employment*. Department of Business, Innovation and Skill UK.

Role of fiscal measures in financing responses to the financial crisis in Indonesia and their effects on fiscal sustainability

Y. Indrawati & A. Erliyana
Faculty of Law, Universitas Indonesia, Depok, Indonesia

ABSTRACT: Fiscal sustainability has become the focus in the world, particularly since the crisis that swept Asia in 1997/1998, as well as the global crisis in 2008. To respond to the crisis, many countries have relied on financing the state. In Indonesia, the management of the crisis in 1997/1998 created a burden for the state constituting 96% of the gross domestic product (GDP), and the state's budget changed from having surplus to suffering deficit at 1.7% of the GDP, with a decrease in GDP by about 13%, and inflation rate of 78%. It has created fiscal burden and limited fiscal movement, and affected fiscal vulnerability, which in turn affected fiscal sustainability. The development was unable to be resumed optimally. Such fiscal conditions have reduced the government's options to take actions and reliability to alleviate the financial crisis. Actually, the global economic situation can cause the financial turmoil occurring in a country to propagate quickly to other countries and turn into a crisis. Based on the review of legislation, the latest update shows that fiscal measures are no longer being relied upon in responding to a financial crisis. It is hoped that the government can maintain fiscal sustainability to enhance fiscal functions to achieve the objectives of the state.

1 INTRODUCTION

The state's role in financing the financial crisis will affect the level of the country's debt. Debt levels determine the fiscal capacity especially fiscal sustainability, which in turn affects the movement of the fiscal policy in achieving the objectives of fiscal. As stated in the 1945 Constitution (Article 23, paragraph 1), the State Budget (APBN) is managed for the greatest prosperity of the people. Fiscal sustainability and the movement of the fiscal policy are crucial achievements of the objectives of the budget. The resulting effects depend on the extent of the state's involvement in financing the crisis.

Based on this, the question raised in this paper is what role the fiscal measures play in financing responses to the financial crisis in Indonesia.

Understanding the role of fiscal developments and their implications on fiscal sustainability will inform us about the model of the fiscal role which is suitable to the condition in Indonesia. To get an idea of fiscal role in the financing of the financial crisis in Indonesia, the analysis was conducted using a normative research method. In terms of regulations, there are three momentums that can describe fiscal role in the financing of the financial crisis in Indonesia, namely the 1997/1998 financial crisis management policies, the policy to anticipate the crisis of 2008, and the Law No. 9 of 2016 on the Prevention and Management of the Financial System Crisis.

1.1 *The development of fiscal role in funding the financial crisis in Indonesia*

1.1.1 *The 1997/1998 financial crisis management policies*
In Indonesia, due to the absence of adequate measures in dealing with the financial crisis and the economic crisis in 1997/1998, the crisis management was carried out by issuing three policies, namely (1) Bank Indonesia Liquidity Assistance (BLBI) policy; (2) bank recapitalization

and divestment policy; (3) deposit insurance policy. BLBI policy was adopted by Bank Indonesia (BI) as the central bank according to its function as the lender of the last resort. This policy was pursued because the banking system at that time had run out of resources due to the liquidity problem of the banks. If this policy had not been pursued, the consequence would have been the destruction of the national banking system. This policy took about a quarter of the total domestic debts to finance the crisis. Bank recapitalization and divestment policy was pursued to immediately improve the banking system, so that banks can re-function to increase their economic activities. With this policy, the government had to bear some of the costs of recapitalization, and because of the collapsed banks including the state banks, the government had to spend an even larger amount at around two-thirds of the overall costs of the crisis. The blanket guarantee policy on the public deposits was pursued by the government to restore public trust in the banking system.

Before the crisis, given the resilience of fiscal in 1996, Indonesia's economic growth was about 7%, the inflation was less than 10%, and the surplus of APBN was at 1.9% of Gross Domestic Product (GDP). After the economic crisis (1997/1998), fiscal sustainability in 1998 changed, the GDP decreased to 13%, the inflation rate increased to 78%, the food prices rose more than doubled, the poverty rate increased to 28%, and the uncertain current fluctuations impact on deficit was 1.7% of GDP. In addition, the crisis increased the financing of government debts to 1226.1 trillion Rupiah, or about 96% of GDP. Of this amount, 653.8 billion Rupiah was domestic debts in bonds consisting of the costs of bank recapitalization at 425.5 trillion Rupiah (282.9 billion Rupiah for the recapitalization of four state-owned banks), 144.5 trillion Rupiah for BLBI, 73.8 trillion Rupiah for the blanket guarantee program, and 10 trillion Rupiah to handle the credit program (Kusumaningtuti, 2008). The debt accounted for about 60% of GDP. With the bond interest expense of approximately 60 trillion Rupiah per year (approximately 40% of the revenue) to be paid by APBN (Ratnawati, 2009). The impact of the crisis was not only suspected by the banking industry, but also by the community where poverty and unemployment directly increased (until the end of 2015, the number of the unemployed was 6.18% (Badan Pusat Statistik [BPS], 2015) and the poverty rate was 11.13% of the total population of Indonesia (Badan Pusat Statistik [BPS], 2016).

The government was forced to adopt those policies in handling the 1997/1998 economic crisis, even though it increased the government debts, to save the banks and the national economy, to create public trust in the banking sector and the government, and to prevent the economic breakdown from getting increasingly widespread and deep. Although at the time the fiscal capacity was at a minimum, the state had to take over (bail-out) the full responsibility of handling the crisis through bonds. This is accordance with the state's duty to guard and protect the interests of its people.

The amount of debts, the sharply increasing deficit, and the social safety net programs for the vulnerable, as the impact of those policies, lead to the increase in the state's expenditure. On the other hand, due to the decline of the economy, the domestic revenues decreased. This provides a fiscal pressure in which the fiscal policy was concentrated on fiscal sustainability in order to prevent the state from getting bankrupt due to the fiscal weakness as a result of keeping (retaining) the balance of the government's ability to meet its obligations (either explicitly or implicitly). Obviously the flexibility of APBN was reduced. Fiscal flexibility is important for APBN as one of the instruments by providing a stimulus to the domestic economy through the expansion of government's spending and investment, and also in carrying out the government's role in creating conducive external conditions for the private sector to encourage the national economy (Subianto, 2009). It changed the role of fiscal policy from fiscal stimulus to fiscal sustainability. Thus, the fiscal policy set back the realization of the objectives of the state.

1.2 *The policy to anticipate the crisis in 2008*

Learning from the experience, some changes were made to improve the resilience of the system for the financial sector, among others, by amending the Banking Law and the Bank Indonesia (BI) Law, and forming the Indonesian Deposit Insurance Corporation (Lembaga

Penjamin Simpanan [LPS]) and the Financial Services Authority (Otoritas Jasa Keuangan [OJK]).

The partial regulation and absence of a grand design for crisis prevention and resolution in the financial system, in its development, will undoubtedly be able to manage the financial crisis. Since 2005, the financial sector safety net was designed to give emphasis to the preventive action. Conceptually, preventive measures are a safety step in the financial system involving the competent authorities, such as the institution which regulates and supervises banks; the institution which acts as the lender of the last resort; and the institution which guarantees customers' deposits and handles failed banks. Preventive efforts are expected to maintain their soundness and improve the resilience of the banking system. Hence, if a crisis happens, the damage to the banking system will not be too severe and the costs can be lowered. The government is the last resort to the financing of the crisis. This concept is used in Government Regulation (PP) No. 4 of 2008 on the Financial System Safety Net (JPSK) to anticipate the impact of the global financial crisis in 2008.

According to Articles 20 and 27 of the Government Regulation on JPSK, the government is responsible for providing funding to prevent and counter the crisis. The source of the funding comes from APBN. The prevention and crisis management costs borne by the government include measures to address the following problems: (a) banks facing liquidity difficulties with a systemic impact; (b) banks facing solvency problems or defaults under the Emergency Financing Facility (FPD) with a systemic impact; and (c) non-bank financial institutions (LKBB) facing liquidity and solvency problems with a systemic impact. The handling of failed banks acquiesced to LPS. The government can provide loans when LPS funds are not enough.

The funding model in managing the crisis as set forth in the Government Regulation on JPSK and the FPD in the BI Law shows the application of the bail-out funding model in which the cost of the crisis is borne by the parties outside the banks (the state), except the funding of deposit guarantee of customers of failed banks by LPS which comes from the banking contribution.

The Government Regulation on JPSK was used as the legal basis to rescue Bank Century, which was established by the Financial System Stability Committee (KSSK) as a failed bank with a systemic impact and acquiesced to LPS. LPS achieved temporary capital participating to the Bank Century using LPS funds. Thus, it did not use the funds from APBN.

1.3 *Law No. 9 of 2016 on the prevention and management of the financial system crisis (PPKSK law)*

Following the disapproval of the Government Regulation on JPSK into Law by the House of Representatives (DPR), the regulation of the management of the financial system crisis returned to the partial regulation with the lack of the law on the crisis management protocol including fiscal role in the crisis management (except FPD borne by the state).

The bail-out funding model has some weaknesses, namely (1) it spends the state's budget; (2) it makes a bank not independent; and (3) according to Kaufman, it creates vulnerability to the government intervention and the politicians' pressure, especially in determining the eligible banks to be assisted (Wijoyo, 2015); (4) it creates moral hazard, because it provides an opportunity for banks not to apply the prudential principle. These inspired many countries to use the bail-in financing model in which the settlement of the banking problem is borne by the bank itself (its shareholders and creditors). This model is followed by Indonesia in the PPKSK Law.

The bail-in funding model was introduced by Credit Suisse in which the owners and creditors are responsible for resolving the financial problems (Wijoyo, 2015). Funding in dealing with the financial crisis is not borne by taxpayers (APBN). This model has been chosen because, in practice, the bail-out funding model triggers problems both in terms of economics, politics and law (Wijoyo, 2015).

In the PPKSK Law, there is no distinction in the management of financial problems (banking and non-banking financial institutions) in normal and crisis conditions. The crisis condition is decided by the President based on the KKSK's proposal. The concept of managing financial issues in the PPKSK Law is to retain the financial safety net system. The characteristic of the management of crisis is holistic, meaning that all the components are involved including businessmen and the relevant authorities that synergize and consolidate to maintain the stability of the financial system in accordance with their respective roles. Funding to the handling of a bank' financial problems will use the resources of the bank itself (a business approach) without using APBN. If this effort cannot resolve the problem, the liquidity problem will be managed by BI and the solvency problem will be handled by LPS.

The funds for handling distressed banks' solvency are derived from LPS, but it does not rule out the possibility of being funded by shareholders or other parties. If there is a deficiency in funds, LPS will perform (1) by selling its securities to the market, BI (decided by KKSK), or other parties; and (2) by issuing the LPS bond.

In a crisis condition, the President will establish the Banking Restructuring Program by LPS to handle the collapse banks. The funding of the Banking Restructuring Program is derived from (a) bank shareholders or other parties as an additional capital and/or to transform certain debt to equity; (b) the management of assets and liabilities (the handled banks); (c) the contribution of the banking industry; and/or (d) the loans taken by LPS from the other parties.

The funds for handling distressed banks' solvency and the Banking Restructuring Program are from the LPS' finance as a legal entity. In case there is a deficit, it will be a cost to be borne by LPS – it is not a state's financial loss. On the other hand, if there is any surplus it will be the LPS' property. This regulation approves the concept of the finance legal status based on the theory of legal entities. Based on the legal entities theory, each legal entity has separate assets, even from its owner. Although LPS was established by the state and the initial property of LPS was from the separated state property, according to the theory of legal entity, the LPS finance is separate from the state's finance—the LPS' finance is not the state's finance because LPS is a separate legal entity.

There are some important specifications in the PPKSK Law related to the fiscal role, namely (1) the BI's support for banks that are experiencing liquidity difficulties under crisis conditions changed from FPD (which according the BI Law and the Government Regulation on JPSK it is borne by the government) to short-term liquidity loans which are usually applied in non-crisis conditions (normal conditions). There is no FPD anymore. Hence, there is no APBN spending to handle banks' liquidity problems in crisis conditions; (2) The funds for the banking restructuring program (conducted by LPS) are derived from (a) the bank itself (its shareholders, investors and bondholders); (b) the management of assets and liabilities of the bank; (c) the contribution of the banking industry; and (d) the loans that are obtained by LPS of the other parties. The funds of the banking restructuring program will be managed separately from the guarantee funds for bank customers (who also come from the banking industry).

Based on the PPKSK Law, the state should not finance the excessive crisis management fund to bear the financial system, both to handle the problem of liquidity, solvency, deposit guarantee, and bank restructuring. Thus, there is no burden on APBN.

However, if handling the financial system crisis reduces the BI' and LPS' capital, the state is obliged to replace to the minimum capital prescribed by the BI Law and the LPS Law. This obligation is a fiscal risk as detailed in the APBN Law (since 2008).

Some people have claimed that the funding bail-in model has some weaknesses:(1) the creditor does not have legal standing to be a shareholder if his credit is convertible into shares; (2) the mechanism of the bail-in is not sufficiently responsive to handle the financial crisis that is rapidly contagious and requires a fast, responsive, and precise objective treatment; (3) the source of the financial crisis can arise from the exchange rate crisis, the practice of shadow banking, or capital outflow, even though the bail-in is only in terms of liquidity and solvency (Tjokronegoro, 2016).

1.3.1 *Banking role in the national economy and financial sector stability*
The role of the banking sector in Indonesia is still dominated by all activities in the financial sector (about 80%). Especially in the globalization era, the development of the financial sector

has experienced the cycle of movement (Sudirman, 2011) in the economy which becomes more dynamic and tends to be volatile (Prasetyantoko, 2008). The experience has shown that the greater financial system, the higher the risk of turmoil and crisis. That can happen because the financial sector serves as a collector and distributor of public funds that directly affect the real sector. Fluctuation in the financial sector will have an impact on the availability of capital that contributes to economic performance.

On the other hand, the dynamics of the financial sector is spreading to the exchange rate crisis and causes damage to the economic system in other sectors. The experience has shown that any turmoil in the financial sector that has an impact on the exchange rate always potentially affects the level of inflation that will increase the prices and, in turn, will weaken the purchasing power of the people and finally the quality of life itself. In a broader scope, the instability of the financial sector allude to the targets of the macro economy, such as economic growth, the fiscal condition to subsidize, ensuring the satisfaction of public needs (such as infrastructure), and prompting the alleviation of unemployment and poverty, which will have an impact at the level of prosperity of the nation and people. In other words, the prosperity and welfare of people is largely determined by the stability of the financial sector (Prasetyantoko, 2008). It shows that the financial sector has a large potential to create instability that disrupts public interests. It is clear that the turmoil in the financial sector has implications on the economy and community life. The degrees and the effects are different. Thus, the instability and inefficiency of the financial sector potentially creates vulnerability to the economy (Prasetyantoko, 2008).

In terms of funding, the experience has shown that the unstable financial system, especially as a result of the crisis, requires a very high cost and takes a long time to revive public trust in the financial system.

The above description proves that financial instability is a "public good", because the impact is not only detrimental to financial market players and major economic actors, but more than that, it is detrimental to all the people (Schinasi, 2005, Prasetyantoko, 2008).

1.3.2 *Implications of role of the state budget (APBN) in the chosen funding model*

Pancasila is the philosophical foundation of the state, especially its fifth principle which states social justice for all Indonesian people, and it gives a special characteristic to the purpose of Indonesia as a state, namely the welfare and social justice or welfare for all Indonesia people. It should be the basis and supports the achievement of the welfare state of Indonesia, which is realized through an economic system and social welfare. Furthermore, the state is confirmed as the single authority to ensure the achievement of the basic objective and a common life (Strong, 2015). This describes that the existence of the state is for the society (Strong, 2015, Simatupang, 2011). Thus, the state is obliged to pursue the state's achievement.

To realize the state objectives, the government is mandated to make a budget planning and a program through the State Budget (APBN). Article 23 paragraph (1) of the 1945 Constitution states "APBN as a form of state financial management is determined annually by the law and is managed in an open and responsible manner for the overall prosperity of the people." That regulation gives constitutional limits (Asshiddiqie, 2012) about the functioning of APBN, to achieve the purpose of the state, which is the people's welfare. Moreover, the purpose of the state is the imperative goal or obligation that must be implemented by the government in budgeting (the government's obligatory budget) (Simatupang, 2011).

This is accordance with the function of budgeting as a tool to achieve certain goals. Lee and Johnson (1977) state as follows: "Public budgeting involves the selection of ends and the selection of means to reach those ends. Public budgeting systems are systems for making choices about ends and means (Lee & Johnson, 1977)." For that purpose, the government's budget and work plan (in APBN) should reflect the efforts and concrete steps in realizing the state's objectives. The constitutional norm is the legal principle for the government's revenue, expenditure and financing.

Administratively, government spending is divided into governance expenditures and development expenditures. The development expenditures are aimed to realize fair and prosperous society. The fiscal constraints require the government to make priority programs according to APBN's objectives. The selection of priority is carried out as the fiscal policy needs to be balanced in APBN, between the (routine) governance expenditures, the development expenditures (to promote economic growth) and the social welfare expenditure. The chosen fiscal policy can be tested using the public choice theory. James M. Buchanan Jr., states that by using the public choice theory, the constitutional norm can be used as a measuring tool to assess an economic policy which is influenced by the personal interests of politicians and the non-economic force (how politicians' self-interest and non-economic forces affect the government policy) (Buchanan Jr. 1986). The constitutional norm is the highest legal norm in a state where the norm of the constitution, according to Frank I. Michelman, is a political moral principle (Asshiddiqie, 2012). Thus, fiscal policy as an operational policy is morally bound by the constitutional norms. Fiscal policy should refer to the purpose of the state (the constitutional norms).

The parameter of the chosen funding model is that the model brings a minimum burden to the budget, and results in minimum negativity to the national economy for both the short term and the long term. That burden is inevitable. Hence, the deficit on the development expenditures must be prevented to avoid the national economic downturn a bigger burden to the budget in the future.

Based on the description above, the fiscal role in managing the financial crisis poses a dilemma, namely (1) the stability of the financial sector is a public good that must be maintained and administered by the state; (2) the state is responsible to protect the people from the excesses the economic downturn; (3) the state is responsible for the use of APBN in accordance with the constitution; (4) the state is required to make neutral fiscal policies; (5) the state is obligated to maintain fiscal sustainability; (6) the chosen funding model will influence state's expenditures, macro and micro economics, and fiscal sustainability.

2 CONCLUSION

The fiscal role to fund the financial crisis is started with full state funding (bail out), limited state funding (limited bail out), and no state funding (bail in). The chosen funding model in the PPKSK Law is already accurate due to the public choice theory, in which no funding is provided by the state to finance the crisis, the development expenditures are not harmed and the purpose of the state can be achieved properly. The PPKSK Law is already based on the constitution.

REFERENCES

Asshiddiqie, J. (2013) Memperkenalkan gagasan konstitusi ekonomi. *Jurnal Hukum Prioris*, 3(2), 1–26.
Badan Pusat Statistik (2015). Agustus 2015: Tingkat Pengangguran Terbuka (TPT) sebesar 6,18 persen. In: *Berita Resmi Statistik 5 November 2015*. Jakarta, BPS.
Badan Pusat Statistik (2016). BPS: Persentase penduduk miskin September 2015 mencapai 11,13%. In: *Berita Resmi Statistik 4 Januari 2016*. Jakarta, BPS.
Buchanan Jr., J.M. (1992) The constitution of economic policy. In: Mäler, K.-G. (ed.) *Nobel Lecture: Prize Lecture in Economic Sciences 1981–1990*. Singapore, World Scientific Publishing Co.
Indonesia (1945) *Undang-undang Dasar 1945 [The 1945 Constitution of the Republic of Indonesia]*.
Indonesia (1999) *Undang-Undang No. 23 Tahun 1999 tentang Bank Indonesia, BI [The Law Number 23 of 1999 concerning Bank Indonesia]*.
Indonesia (2004) *Undang-Undang No. 24 Tahun 2004 tentang Lembaga Penjamin Simpanan, LPS [The Law Number 24 of 2004 concerning Indonesian Deposit Insurance Corporation]*.
Indonesia (2004) *Undang-Undang No. 3 Tahun 2004 tentang Bank Indonesia, BI [The Law Number 3 of 2004 concerning Bank Indonesia]*.
Indonesia (2008) *Peraturan Pemerintah Pengganti Undang-Undang No. 4 Tahun 2008 tentang Jaring Pengaman Sistem Keuangan, JPSK [The Government Regulation to Amend the Law Number 4 of 2008 concerning Financial System Safety Net]*.

Indonesia (2011) *Undang-Undang No. 21 Tahun 2011 tentang Otoritas Jasa Keuangan, OJK* [*The Law Number 21 of 2011 concerning Financial Service Authority*].

Indonesia (2016) *Undang-Undang No. 9 Tahun 2016 tentang Pencegahan dan Penanganan Krisis Sistem Keuangan, PPKSK*. [*The Law Number 9 of 2016 on the Prevention and Handling of the Financial System Crisis*].

Kusumaningtuti, S.S. (2008) *Peranan Hukum dalam Penyelesaian Krisis Perbankan di Indonesia*. Jakarta, Rajawali Press.

Lee Jr., R.D. & Johnson, R.W. (1977) *Public Budgeting Systems*. 2nd edition. Baltimore, University Park Press.

Prasetyantoko, A. (2008) *Bencana Finansial: Stabilitas Sebagai Barang Publik*. Jakarta, Kompas.

Ratnawati, A. (2009) Reformasi sistem perencanaan penganggaran Indonesia: Mempertajam efektivitas kebijakan pengeluaran anggaran. In: Abimanyu, A. & Megantara, A. (eds.). *Era Baru Kebijakan Fiskal: Pemikiran, Konsep, dan Implementasi*. Jakarta, Kompas. pp. 347–361.

Schinasi, G.J. (2005) Preserving financial stability. In: Caminis, A. (ed.) *Economic Issues 36*. Washington DC, International Monetary Fund.

Simatupang, D.P.N. (2011) *Paradoks Rasionalitas Perluasan Ruang Lingkup Keuangan Negara dan Implikasinya terhadap Kinerja Keuangan Pemerintah*. Jakarta, Badan Penerbit FHUI.

Strong, C.F. 2015. *Konstitusi-konstitusi Politik Modern* [*Modern Political Constitutions: An Introduction to the Comparative Study of Their History and Existing Form*]. Translated by SPA Teamwork. Bandung, Nusa Media.

Subianto, B. 2009. Kebijakan Fiskal dalam Menghadapi Krisis. In: Abimanyu, A. & Megantara, A. (eds.). *Era Baru Kebijakan Fiskal: Pemikiran, Konsep, dan Implementasi*. Jakarta, Kompas. pp. 80–100.

Sudirman, I.W. (2011) *Kebijakan Fiskal dan Moneter: Teori dan Empirikal*. Jakarta, Kencana Prenada Media Group.

Tjokronegoro, P.L. (2016) *Penormaan Asas Kekhususan Sistematis yang Berbasis Efisiensi Terhadap Tindak Pidana Korupsi di Bidang Perbankan*. Jakarta, Universitas Pelita Harapan. Disertasi Doktor.

Wijoyo, N.A. (2015) *Referensi Risiko Sistemik Perbankan*. Jakarta, UI Press.

Abuse of substance, restorative justice and diversion

S. Asa & S. Fitriasih
Faculty of Law, Universitas Indonesia, Depok, Indonesia

ABSTRACT: The Indonesian National Narcotics Board (NNB) reported that it is estimated that there were 5,000,000 drug abusers in 2015, ± 4% of whom were suspects in drug cases. It was also confirmed that 81,360 of 187,701 prisoners were contributed by drug cases. Prisons are overcrowded, and today Indonesia is in a state of 'drug emergency'. A qualitative study using documentary research using a descriptive analytical approach was designed to examine the possibility of implementation of restorative justice and diversion to help overcome the problem of drug addicts contributing to prison overcrowding. The concluding points of the study are; first, rehabilitation has not been properly implemented because of the dilemma in determining a drug abuser either as a medical patient or as an offender. There are some dominant factors: unclear definition of user, abuser, addict and victim; no protocol for rehabilitation before adjudication; absence of regulations to that allow minimum hand-carry of drugs for user; and lack of engagement of community in drug rehabilitation program. Second, restorative justice and diversion should be implemented through the Criminal Justice System (CJS) to enable the education of life skills, job training, circle and community conferencing program, even after prison treatment. Third, narcotic law should be amended to: differentiate or strictly define illicit drugs as a serious crime yet drug abuse as a misdemeanour; settle restorative justice and diversion; and reorganise the roles of the Indonesian National Police (INP), NNB, related ministries: Health, Social Affair, Law & Human Rights and local government.

1 INTRODUCTION

Indonesia's reaction to drugs problems has been typified by a repressive and punitive approach. The enactment of Indonesian Law No. 22/1997 on Narcotics and Law No. 5/1997 on Psychotropics that was later revised into Law No. 35/2009 generated a number of problems, including clarity on drug use, drug dealing, traffickers, dealers and individuals found in possession of drugs and their punishment. The application of this repressive and punitive approach has failed to effectively kerb the war on drugs as marked by the increase in the number of drug users (Nasir, 2011).

This paper is intended to criticise efforts to tackle protracted drug abuse that has escalated during the last decade, resulting in Indonesia's declaration of a state of drug emergency. The discussion is on the three main questions: Why is the number of drug abusers increasing yearly? What has been done to tackle drug abuse? Can restorative justice and diversion (RJ & D) be applied to overcome drug abuse? The aim is to discuss the possibility of the implementation of RJ & D to overcome the drug abuse behaviour which constitutes a criminal act punishable by up to four years of imprisonment (with further mandatory rehabilitation when the perpetrator is addicted), in anticipation of contributing to better handling of drug problems in Indonesia.

2 METHOD

The discussion is based on a qualitative study that rests on documentary research and is presented in a descriptive analysis of the three research questions above. A number of laws, policies, studies and secondary relevant documents and literature were reviewed to answer the research questions.

3 RESULTS AND ANALYSIS

In 2014 a study conducted by the National Narcotics Board (NNB) and Universitas Indonesia's Health Research Centre estimated the number of drug abusers to be 3.8–4.1 million, or 2.1–2.25% of Indonesia's total population (NNB, 2015). The data are supported by the increase in the number of major types of drugs seized, such as marijuana, methamphetamine, ecstasy and heroin and the number of users who became suspects. The number of suspects increased significantly; there were 17,326 in 2006, 26,461 in 2010 (NNB, 2015, p. 1), and 363,737 in 2015 (INP's Directorate of Drugs & NNB, http://www.bnn.go.id). The number of first-time abusers or the number of five times a year users has continued to increase in the last decade. In some provinces first-time users also increased significantly from 481,466 in 2008 to 751,786 in 2011 and later in 2014 up to 891,490 (NNB, 2015, pp. 103–136).

The Republic of Indonesia's Supreme Court [*Mahkamah Agung Republik Indonesia*] reported an increase in narcotic cases during 2010–2014. It revealed that besides cases decided by lower courts under the Supreme Court, 512 cases (16.67% of 3,072 narcotic cases) were appealed and reviewed in 2010 (Supreme Court, 2010). This number increased to 792 cases in 2014, or 29.11% of 2,721 cases (Supreme Court, 2014).

The increase has led to overcrowded prisons. Prison occupancy is dominated by drug inmates, convicted as abusers, drug retailers, dealers or manufacturers. According to the General Directorate of Corrections (*Direktorat Pemasyarakatan*) and the Ministry of Justice and Human Rights (*Kementrian Hukum dan Hak Asasi Manusia* [*Kemenhukham*]), the majority of prison occupancy is made up of 56,326 narcotic inmates (or 92.39% of 60,961 special crime inmates). If the occupants and the capacity of occupancy are compared, the prisons will continue to experience a shortage of space, that is, 37,344 in 2006; 26,461 in 2010; 47,392 in 2012, and 53,841 in 2014. The overcrowding condition has triggered frequent violence, persecutions, rebellions, fights, drug smuggling and escapes (Kemenhukham, 2014, p. 18).

In 2011 the President issued Instruction No. 12/2011 for combatting drug abuse and illicit trafficking (*Pencegahan Pemberantasan Penyalahgunaan dan Peredaran Narkotika* [P4GN]) manifested in three national policies that aim to: provide an increase of infulnerability of 97.2% of the population from drug abuse and illicit trafficking; thoroughly crush drug syndicates; and gradually provide medical and social rehabilitation services for up to 2.8% of drug abusers (P4GN Policy, 2012). The government also issued Government Regulation No. 25/2011 for the implementation of Law No. 35/2009 on Narcotics that requires the reporting of drug addicts and the joint regulation by the Chief Justice, the Minister of Law and Human Rights, the Minister of Health, Minister of Social Affairs, the Attorney General, the Chief of the Indonesian National Police and the Head of NNB regarding the control of drug addicts and victims of drug abuse in correctional institutions through an assessment of a Tim Asesment Terpadu Integrated Assessment Team (TAT).

A press release on the development of P4GN reported that throughout 2015 the NNB revealed that of 102 cases of narcotics and money laundering by national and international syndicates, 82 cases are to be tried in court. Total assets seized are equivalent to IDR 85,109,308,337. Up to mid-2015, 55 persons convicted of narcotics cases were sentenced to death, and 14 persons were on death row waiting for executions. In 2015, the NNB, government and communities rehabilitated 38,427 abusers, addicts and narcotic abuse victims. The law defines an abuser as a person who uses narcotics without rights or against the law, while an addict is a person who uses or abuses narcotics and is in a state of dependence on drugs, both physically and psychologically; and a victim of drug abuse is a person who uses drugs not by their own will and without intent, but who has been coaxed, tricked, deceived, coerced, and/or threatened to use them (NNB, 2015).

Despite the noted progress, the implementation of various policies has not shown encouraging results. The illicit trade in narcotics is still widespread and massive. With the help of technology in transportation and information the perpetrators continue to develop new substances and renew their *modus operandi*. The illicit practice of drug trafficking has economic benefits as its main motive and is controlled by a transnationally organised crime body known as the mafia. Indonesia is now become market in which drug are sold in a high price rather than was only a transit country. Drug abuse is increasingly prevalent in all walks of life. The overlapping operational policies and the lack of coherent technical guidelines related to minimum possession versus the definition of drug abuse have been identified as the main causes.

Now efforts are needed to reduce the number of first-time abusers to prevent an increase in the number of future suspected abusers and a later subsequent increase in the number of those in need of medical and social rehabilitation. The increase is likely due to the fact that addiction is a chronic brain disease which often recurs (Leshner, 1997), and about 45% of rehabilitated addicts have failed and relapsed during the resocialisation process (United Nations Office on Drugs and Crime [UNODC] & World Health Organisation [WHO], 2009). A social rehabilitation programme is required to prevent more abusers from experiencing a relapse. Families' and local community's acceptance, participation of community leaders and religious leaders, and life skills programmes through RJ & D should be considered.

RJ emanated from the previous philosophy of convictions: retribution, deterrence, rehabilitation and restitution were unsatisfactory (Mackenzie & Stobbs, 2010, p. 42). Related to the RJ approach as the philosophy of punishment, Tony Marshall states: Restorative justice is a process whereby all the parties with a stake in a particular offense come together to resolve collectively how to deal with the aftermath of the offense and its implications for the future. (Braithwaite, 1999, p. 3). Tony Marshall statement asserts four things: the parties as stakeholders such as perpetrators, victims and society; who are specialised in specific criminal acts; understanding of their effects; and the efforts of stakeholders to jointly resolve the consequences for better future.

Can RJ be applied to drug abuse as a victimless crime or a crime that is considered to have no direct victim? (Garner, 2004, p. 339). In the article *Rethinking Drug Courts: RJ as a Response to Racial Injustice*, O'Hear (2009) attempts to discuss RJ as a method of handling drug crimes. O'Hear (2009) states that the general principle of Community Conferencing Programme (CCP) in RJ can be adapted to drug crimes. Although there is no real victim, the CCP can be implemented as in other cases where community members discuss the impact of illicit drug trafficking. As in these other cases, the endorsement and requirement in CCP of drug criminal acts can be through community service and joint praying group. Through CCP an offender is asked to follow a programme for drug treatment even if he/she does not confess to be a user. A perpetrator can be referred to a service that provides care on a regular basis and if it fails another can be applied as a second chance. After that a care programme can be stopped and the offender is then referred to a conventional prosecution of a serious criminal offence if the sequence treatment has failed. However, if agreement in CCP is reached and the treatment programmes are well completed, then the public prosecutor could reduce the charge to a misdemeanour (O'Hear, 2009).

According to O'Hear (2009), RJ as a set of normative principles contains a very noticeable difference when compared with the philosophy of retributive justice in punishment. As a philosophy of punishment, RJ does not recognise the plight of the imposition of a criminal as something that contains the goodness in itself. Although RJ embodied element of forward-looking as in deterrence, incapacitation and rehabilitation in the philosophy of utilitarianism, it seriously emphasis on the recovery of the victims and the people who suffer the consequences of a criminal act as its primary objective.

In resolving social conflicts, RJ always uses an approach that involves the victim, the perpetrator, civil society networks and CJS officers through communitarian justice, community justice, family group conferencing and community conferencing (UNODC, 2006, p. 6). RJ is practised through three approaches: restorative programme, restorative process and restorative outcome. There are three main mechanisms particular to the restorative process known as the victim-offender mediation, conferencing, and circles (Morris & Maxwell, 2011).

Two forms related to diversions are the true diversion that authorize the diversion of case handling from the formal system with no follow-up program while the new diversion program included the follow up program as part of the diversion (Hughes & Ritter, 2008, p. 5). According to Garner (1999) diversion programmes refer certain criminal defendants before trial to community programmes on job training, education and the like, successful completion of which may lead to the dismissal of the charge (p. 115).

RJ & D is governed by Law No. 11/2012 regarding the Juvenile Justice System (JJS) that applies to children aged 12–18 years who are implicated in legal cases. RJ & D is even affirmed as the paramount approach in the process of JJS which is the responsibility of investigators, prosecutors and judges, followed by prosecution if they intentionally do not make an attempt for the settlement of juvenile cases.

CJS components perform the RJ & D approach before sentencing (investigation, prosecution, and court proceedings) and outside the verdict (coaching, guidance, supervision and mentoring). One side of the goal is to grant legal protection through the CJS when the child has acted against the law, and the other side is taking into account victims' interests through a mutual agreement reached between the victim and the offender; settling disputes outside the court; preventing children from being deprived of their liberty; encouraging community participation; and instilling a sense of responsibility in the child. In order to be processed through RJ & D, the threat of imprisonment for the criminal act committed by the child should not exceed seven years and it must not be a repetition of a criminal act.

The RJ & D process is carried out through consultation led by JJS officials, involving children and parents or guardians, victims and parents or guardians, community mentors, professional social workers and, where necessary, social welfare workers and the public. The RJ & D process should take into account the interests of victims; child welfare and responsibility; avoidance of negative stigma; avoidance of retaliation; the harmony of society; and propriety, decency and public order. In fulfilling the obligation to prioritise RJ & D, investigators, prosecutors and judges must carefully consider the categories of offences; ages of the children; report of community research from the children's correctional centre, and family and community support. The results of consultation set forth in a written agreement with consent from the victims and offender and their families of their own free will, except for a criminal act of infringement; misdemeanour; a criminal offence without a victim; or a criminal act with relatively minimum casualty loss.

RJ & D agreements may contain indemnification of the victims; medical and psychosocial rehabilitation; handover to parents or guardians; participation in education or training in an educational institution or the Social Welfare Organising Agency (LPKS); or community service. An RJ & D agreement is sent to the court for endorsement and then sent to the relevant supervisor of the community, investigator, prosecutor or judge. After receiving a decision from a district court, the investigator or prosecutor issues a certificate of termination of investigation/prosecution. If the RJ & D process results in no agreement or the agreement is not implemented, then the juvenile criminal procedure will continue.

JJS adheres to the double track system, namely conviction and repression. Punishments consist of the principal punishments in the form of warning; conditional punishment to be guided outside a prison, community service or supervision; work training; coaching in the institution; or imprisonment; and additional punishments, such as deprivation of the benefits derived from the crime; or the fulfilment of customs obligation. In order to respect the dignity of the child, the punishment can be replaced with vocational activity when the criminal offence is threatened by cumulative imprisonment and a fine. Further measures imposed on children can be the return of the child to the child's parents or guardian; handover to someone; treatment in a mental hospital; treatment in LPKS; enrolment in formal education or a training institution run by government or privately; revocation of driving licence; and/or cost of repairs resulting from a criminal act.

The narcotics law does not define children nor set the age limit for so-called child drug abusers. This obscures the meaning of Article 128 paragraph (2) that an addict (abuser or user) of narcotics who is not old enough and has been reported by a parent or guardian does not need to be punished. In other words, an addict who is not old enough but not reported by a parent/guardian may be subject to criminal prosecution. If a child voluntarily uses illegal drugs, he could be qualified as a minor abuser. All of them those consisting of first-time users, regular users or addicts may be prosecuted under Article 127 paragraph (1) of the narcotic law which sets a maximum of four years imprisonment. The provision has drawn criticism since as a victimless crime, drug abuse should follow the double track system and especially for children, appropriate qualifications for different treatments should be in place.

Law makers recognise the possibility of minor addicts but do not standardise detailed qualifications to accommodate their needs. Should it be considered sufficiently accommodated by the provisions of Article 127 paragraph (1), then the question is whether it is appropriate for minor abusers to undergo a judicial process that stipulate imprisonment sentence for the offenders. A clear cut of age limit and mechanism, medical treatment, psychosocial rehabilitation, education and job training as per the RJ & D agreement should have been set in JJS for minor abusers, whether as first-time users, regular users or addicts.

One of the criteria to facilitate a child who has committed a criminal act through the RJ & D approach is that the crime committed is categorised as a misdemeanour with a penalty of fewer than seven years' imprisonment and the offence is categorised as a victimless crime. Child drug abusers who meet the criteria as the longest term of imprisonment for child or adult drug abuse is four years while drug abuse is a victimless crime. Since the enactment of narcotic law is followed by the enactment of JJS law, RJ & D should have been used in the treatment of minor abusers, but in practice, the number of cases of child drug abusers resolved through RJ & D is not proportional to the number of identified child drug abusers. Hence, the absence of RJ & D implementation guidelines policy for child drug abusers is a major constraint.

Supreme Court Regulation No. 4/2014 about RJ & D in the state court determines that the parties for deliberations should be presented before the prosecution as the official delegates of the case to process RJ & D in the district court; however, there are no mechanisms, processes, or procedures for the implementation of RJ & D related to victimless crimes, especially for minor abusers.

Furthermore, how is the application of RJ & D towards adult abusers? Narcotics law does not provide a mechanism for the possible application of RJ & D for adults. Thus, the TAT assessment mechanism of the joint regulation applied for them encounters many juridical-technical obstacles which, among others, are related to its unknown position in the hierarchy of legislation. According to the constitutional system, the Indonesian Supreme Court as a high state institution (and the only other holder of judicial power other than the Constitutional Court) should not be engaged with other executive institutions under the President in the formation of the joint regulation.

Technical constraints that hinder the joint regulation are, inter alia, the absence of firm, clear and detailed guidance and the exclusion from the provisions of the minimum amount of drugs that can be stored, possessed and taken for consumption. Briefly, an abuser must have been in contact with drugs, so if being checked, the abuser's urine and blood will positively contain drugs, and to use drugs, the abuser definitely buys, stores and possesses drugs. The regulation on the minimum drug possession allowed is crucial.

For example, principally Portugal's Law No. 30/2000 asserts a legal framework applicable to the use of illegal drugs without a prescription, along with medical care and social welfare improvement for users. The law sets out the maximum amount of drug that can be possessed, stored, bought and consumed over ten days, among others, cannabis, 25 grams; hashish, 5 grams; cocaine, 2 grams; heroin, 1 gram; d-lysergic acil diethylamide LSD or ecstasy, ten pills. If the amount of drugs possessed/carried exceeds the maximum for ten-day use, then the person is treated as a drug user (Domoslawski, 2011, p. 13).

The maximum amount of drug possession has actually been set out in the Supreme Court Circular (SEMA) No. 4/2010, replacing Supreme Court Circular No. 7/2009. The circular sets the minimum amount of drugs that can be owned, consumed, bought, and stored in one day, and not to be sold/distributed to obtain economic advantages. The circular that internally binds judges is not used as it is considered a mere recommendation.

In addition to the Supreme Court, the Attorney General has also issued the Regulation of Attorney General (PERJA) No. PER-029/A/JA/12/2015 which essentially states that the public prosecutor shall prosecute the defendant (adult)and/or child drug addict and victim of drug abuse to be admitted to medical and social rehabilitation institutions if, according to the investigation report, the laboratory examination result and TAT recommendation, the defendant acts neither as a dealer, a courier nor as a manufacturer or a drug case recidivist or a person caught red-handed but do not have narcotics with him/her or the evidence does not exceed the maximum amount for one-day use.

An interesting issue for discussion is whether the maximum amount of drugs controlled, stored, or possessed to be used should be small in amount and only for the use of a day up to ten days as the practice in Portugal, two things should be considered; firstly, if the maximum amount is smaller and the criterion is of a shorter duration of use, the number of potential people abusing is higher and may cause CJS to be overloaded, and secondly, if the maximum amount is bigger and a longer duration of use is set as a criterion, tightened supervision is required due to potential distribution by drug controllers or owners.

The implementation of the Supreme Court Circular and the Regulation of Attorney General still experience technical-juridical constraints as there are provisions which are in conflict with Article 127 of the narcotics law which merely sets a maximum of four-year imprisonment as the only penalty for abusers, including first-time users.

One of the countries that applies RJ & D to overcome drug abuse is Australia. Australia applies it to obtain three benefits, namely: (1) treatment of drug abusers, (2) providing options on how to set penalties for wrongdoers, and (3) preventing repetition of criminal acts. The programme's objectives as set out in the Alcohol and Other Drugs Council of Australia in 1996 are to: reduce adverse effects of drug use; prevent or reduce drug-related crimes in the community; promote public health and the health of individuals; define an overall approach that allows the fulfilment of services to individual needs; promote the ability to change a person's behaviour in the future through an appreciation of health and thus depart from badness; and obtain more efficient and effective results than imprisonment (O'Callaghan et al., 2004, p. 188).

RJ & D in Australia is organised in three frameworks. Firstly, the RJ & D authority is attached to each component of the JJS, namely the investigators, prosecutors, judges and prison officers. Secondly, it asserts that RJ & D can only be applied against perpetrators of certain crimes such as misdemeanours, the crimes of drugs and/or criminal offences related to drugs, such as possession of prohibited substances, drunk driving and theft to buy drugs, and it sets the age category associated with the perpetrators of the offences. Thirdly, it can be applied to children and adults. The effectiveness and efficiency of its implementation require four essential prerequisites, namely referral source; eligibility criteria; programme characteristics; and administrative details (Spooner et al., 2001, p. 2).

RJ & D in Australia is divided into two main programmes: police diversion and court diversion. Police diversion is developed via three schemes, namely, programmes targeting cannabis use or possession only, diversion for other illicit drugs and diversion for drugs or drug-related offenders. Besides this, RJ & D for drug abusers includes several levels under both the traditional judicial court and drug court judges through the role of the commissioner. Court diversion programmes are divided into two, namely, court diversion for minor drug or drugs-related offenders, and court diversion for serious drugs or drug-related offenders (Australian Government, 2011).

RJ & D for drug users through a drug court and a court diversion programme is built on a framework that focuses on increasing the number of abusers who are treated, while the strategy of 'coercive' is used for drug offenders with a history of serious relapse. The RJ & D programme is divided into five phases: pre-arrest; pre-trial; pre-sentence; post-sentence, and pre-release (Spooner et al., 2001, 1)., NSW is the only state in Australia that has a Drug Treatment Correctional Centre that provide treatment to prisoners very serious drug dependence prior to their release through a major RJ & D programme (Payne et al., 2008, p. 3).

With regard to the description of RJ & D for minor drug abusers and adults, it is time to amend the law to stipulate narcotic drug abuse as a misdemeanour that will be handled using the double track system and at the same time the law should regulate the implementation of RJ & D.

4 CONCLUSION

Efforts to combat illicit drug trafficking in the country should continue through multilateral cooperation, including mutual legal assistance due to the nature of the crime as an international common enemy masterminded by transnational organised crime. As a precautionary measure, there should be early identification and education of the younger generation to prevent them from drug abuse. The government through the police, NNB, the Ministry of Health, the Ministry of Social Affairs, the Ministry of Law and Human Rights, and local governments should make the effort collectively with civil society to conduct medical and social rehabilitation. To obtain the best results, the law needs to be amended to clearly stipulate drug abuse as a misdemeanour that needs to be addressed through the implementation of RJ & D.

REFERENCES

Australian Government—Ministerial Council on Drug Strategy. (2011). *National Drug Strategy 2010–2015: A Framework for Action on Alcohol, Tobacco and Other Drugs*. Canberra: Commonwealth of Australia.
Badan Narkotika Nasional. (2011). *BNN: Ringkasan Eksekutif Survey Nasional Penyalahgunaan Narkoba 2011: Kerugian sosial dan ekonomi*. [National Narcotic Board, *Executive Summary on National Survey regarding Substances Abuse Year 2011: Social and Economic Disadvantages*]. Available from: http://bnn.go.id/read/hasil_penelitian/17488/blog-single.html.
Badan Narkotika Nasional. (2012). *BNN: Kebijakan dan Strategi Nasional di Bidang Pencegahan dan Pemberantasan Penyalahgunaan dan Peredaran Gelap Narkoba, (P4GN)*. [National Narcotic Board, *Policy Implementation and Prevention National Strategy and Eradication Drugs Abuse and Illegal Distribution*]. Available from: http://bnn.go.id/portal/_uploads/post/2012/01/26/20120126130403-10111.pdf.
Badan Narkotika Nasional. (2015). *BNN: Laporan Akhir Survei Nasional Perkembangan Penyalahguna Narkoba Tahun Anggaran 2014*. [National Narcotic Board, *Final Report on National Survey, regarding Substances Abuse, Year 2014*]. Available from: http://bnn.go.id/portal/_uploads/post/2015/03/11/Laporan_BNN_2014_Upload_Humas_FIX.pdf.
Braithwaite, John. (1999). *Restorative Justice: Assessing Optimistic and Pessimistic Accounts*, Crime and Justice, (Vol. 1, 1999), 1–85.
Domoslawski, A. (2011). *Drug Policy in Portugal: The Benefits of Decriminalizing Drug Use*. Translated by Siemaszko, H. Warsaw: Open Society Foundations.
Garner, A.B. (Ed.) (2004). *Black's Law Dictionary* (8th ed.). St Paul: Thomson West.
Hughes, C. & Ritter, A. (2008). *A Summary of Diversion Programs for Drug and Drug-Related Offenders in Australia. DPMP Monograph Series No. 16*. Sydney: National Drug and Alcohol Research Centre.
Indonesia. (1997). *Undang-undang No. 22 Tahun 1997 tentang Narkotika* [*Law Number 22 of 1997 regarding Narcotics*].
Indonesia. (1997). *Undang-undang No. 5 Tahun 1997 tentang Psikotropika* [*Law Number 5 of 1997 regarding The Psychotropic*].
Indonesia. (2009). *Surat Edaran Mahkamah Agung No. 7 Tahun 2009 tentang Menempatkan Pemakai Narkoba ke dalam Panti Terapi dan Rehabilitasi* [*Supreme Court Circular Letter Number 7 of 2009 regarding the Internment of Narcotics Users in Rehabilitation and Therapy Institutions*].
Indonesia. (2009). *Undang-undang No. 35 Tahun 2009 tentang Narkotika* [*Law Number 35 of 2009 regarding Narcotics*].
Indonesia. (2010). *Surat Edaran Mahkamah Agung No. 4 Tahun 2010 tentang Menempatkan Pemakai Narkoba ke dalam Panti Terapi dan Rehabilitasi* [*Supreme Court Circular Letter Number 4 of 2010 regarding the Internment of Narcotics Users in Rehabilitation and Therapy Institutions*].
Indonesia. (2011). *Instruksi Presiden No. 12 Tahun 2011 tentang Pelaksanaan Kebijakan dan Strategi Nasional Pencegahan dan Pemberantasan Penyalahgunaan dan Peredaran Gelap Narkoba [P4GN] Tahun 2011–2015* [*The Presidential Instruction Number 12 of 2011 regarding Policy Implementation and Prevention National Strategy and Eradication Drugs Abuse and Illegal Distribution Year 2011–2015*].

Indonesia. (2011). *Peraturan Pemerintah No. 25 Tahun 2011 tentang Pelaksanaan Wajib Lapor Pecandu Narkoba* [*Government Regulation Number 25 of 2011 regarding Compulsory Reporting of Drug Addicts*].

Indonesia. (2012). *Undang-undang No. 11 Tahun 2012 tentang Sistem Peradilan Pidana Anak* [*Law Number 11 of 2012 Juvenile Criminal Justice System*].

Indonesia. (2015). *Peraturan Jaksa Agung Republik Indonesia No. PER-029/A/JA/12/2015 tentang Petunjuk Teknik Penanganan Pecandu Narkotika dan Korban Penyalahgunaan Narkotika ke dalam Lembaga Rehabilitasi* [*General Attorney Republic of Indonesia Regulation Number PER-029/A/JA/12/2015 regarding Technical Guideline for the Management for Rehabilitative Confinement of Narcotics Addicts and Victims of Narcotics Abuse*].

Kemenhukham. (2015). *Direktorat Pemasyarakatan: Laporan Tahunan 2014* [*Directorate of Correction: Financial Statements 2014*].

Leshner, A.I. (1997). Addiction is a brain disease, and it matters. *Science*, 278(5335), 45–47.

Mackenzie, G. & Stobbs, N. (2010). *Principles of Sentencing*. New South Wales: The Federation Press.

Mahkamah Agung [Supreme Court] Republik Indonesia. (2011) *Laporan Tahunan 2010* [*Financial Statements 2010*]. Jakarta. Available from: https://www.mahkamahagung.go.id/en/the-annual-reports-of-the-supreme-court-of-the-republic-of-indonesia.

Mahkamah Agung [Supreme Court] Republik Indonesia. (2015) Laporan Tahunan 2014 [*Financial Statements 2014*] Jakarta. Available from: https://www.mahkamahagung.go.id/en/the-annual-reports-of-the-supreme-court-of-the-republic-of-indonesia.

Nasir, S. (2011). Misguided Drug Law. (July 31, 2011). *Jakarta Post*. Available from: http://www.thejakartapost.com/news/2011/07/31/misguided-drug-law.html.

O'Callaghan, F., Sonderegger, N. & Klag, S. (2004). Drug and crime cycle: Evaluating traditional methods versus diversion strategic for drug-related offences. *Australian Psychologist*, 39(3), 188–200.

O'Hear, M.M. (2009). Rethinking drug courts: Restorative justice as a response to racial injustice. *Stanford Law & Policy Review*, 20(2), 463–499.

Payne, J., Kwiatkwoski, M. & Wundersitz, J. (2008). *Police Drug Diversion: A Study of Criminal Offending Outcomes*. Canberra: Australian Institute of Criminology.

Spooner, C., Wayne, H. & Mattick, R.P. (2001). Overview of diversion strategies for Australian drug-related offenders. *Drug and Alcohol Review*, 20(3), 281–294.

Strang, H. (2001). Justice for victims of young offenders: The centrality of emotional harm and restoration. In A. Morris, & G. Maxwell (Eds.) *Restorative justice for juveniles: Conferencing, mediation and circles*. Oxford: Hart Publishing, 183–193.

United Nations Office on Drugs and Crime & World Health Organization. (2009). *UNODC & WHO: Principles of Drug Dependence Treatment*. New York, United Nations. Discussion Paper.

United Nations Office on Drugs and Crime (2006). *UNODC: Handbook on Restorative Justice Programmes*. New York, United Nations. Criminal Justice Handbook Series.

Waqf shares to create equitable economic distribution in Islam in relation to the Law Number 41 of 2004 on *Waqf*

H.N. Lita
Law Faculty of Universitas Padjadjaran, Bandung, Indonesia
Law Faculty of Universitas Indonesia, Depok, Indonesia

U. Hasanah & Y.S. Barlinti
Law Faculty of Universitas Indonesia, Depok, Indonesia

ABSTRACT: Among the social institutions that exist in Islam, *waqf* has played a very important role in developing the support for the development of *waqf* in the country through the Waqf Law. In economic terms, *waqf* is a means to distribute assets in such a way so as to allow economic prosperity to be enjoyed by the general population. Shares as a *waqf* object, in the study of Islamic jurisprudence, are associated with *waqf* cash are included in the category of movable objects. The proposition used to develop shares as *waqf* is strongly related to the purpose of public interest. The legal arguments of permissibility are based on *Mashlahāh mursalah* and *isthishan*. On the basis of *Mashlahāh mursalah*, *waqf* of shares bring benefits for the people. The establishment of shares as a *waqf* object in Indonesia is based on the Waqf Law and the Principles of Sharia. The comparison with the situation in a number of countries has found that *waqf* share is a part of the development of monetary waqf. *Waqf* Shares can be made as an effective and sustainable model of CSR. This would be more in line with the expectation attached to the development of *waqf* in a professional and integrated manner in accordance with the corporate culture and values.

Keywords: *Waqf* Law; Sharia Shares

1 INTRODUCTION

The development of *waqf* in Indonesia has experienced significant changes since the enactment of Law Number 41 of 2004 on *Waqf*. Based on Article 16 paragraph (1) of the *Waqf* Law, it is stated that *waqf* property consists of immovable and movable goods. Furthermore, shares as movable goods can be *waqf* based on Article 16 paragraph (3) and Article 21 of the Government Regulation No. 42 of 2006 on the implementation of the *Waqf* Law.

Shares are securities that represent equity participation in a company. However, according to Islamic principles, equity investment must be made in companies that do not violate the principles of sharia, so an investment must not be made in fields such as gambling, usury, or in the production of prohibited goods, such as beer.

This article will discuss whether Islamic shares as an object of *waqf* can contribute to equitable economic distribution, the application of sharia law, and the applicable law in Indonesia.

The purpose of this article is to analyse and explain in what way *waqf* (endowment) can realise fair economic distribution, and the application of shares as an object of *waqf* under Islamic law and the positive law in Indonesia. To discuss these issues are used a normative method (legal research). Soetandyo Wignjosoebroto explains that normative legal research works to find the correct answers by validating applicable law or religious texts, following the teachings or doctrines that underlie them (Soetandyo Wignjosoebroto: 2011).

Associated with Soetandyo Wignjosoebroto's view on normative legal research, this article seeks to find the correct answers with the proof of prescriptive law concerning *waqf* shares, in relation to the rules on *waqf* themselves and shares as an object of endowment through various legal searches on applicable law in Indonesia and Islamic Law. Because endowments are derived from Islamic legal institutions, the development of waqf shares must remain in the corridor of the principles of Islamic law.

This research was conducted by library research. The data used are both primary and secondary data. Primary data is the data obtained directly from its sources/informants. To support this primary data are conducted field study and interviews how to implementation *waqf* shares. To analyse the data and conclusions from the results of the study, the researchers used a qualitative method and normative analysis. The normative analysis for this study starts from the analysis of the applicable law related to *waqf* shares.

The hyphothesis is that the *waqf* (endowments) is not just an application of faith in Islam, but it also has a social dimension and significant benefits which are in public's interest. Theoretically, the goal of *waqf* is to realise social justice. In the history of Islamic civilisation, endowments were widely used for charitable or public purposes. *Waqf* is also an institution of wealth distribution for economic justice, and a symbol of the desired economic system of Islam. Thus, the management of *waqf* which is in accordance with what has been governed by Islamic law and the Waqf Law will make a major contribution to the realization of social welfare in the community. Waqf shares can be equated with *waqf* cash. Imam az-Zuhri explained that *waqf* cash is permissible, by making dinar or money as capital. Shares as securities by the Indonesian Ulama Council through its Fatwa Commissian are included in the category of money and can be a *waqf* object. Meanwhile, shares can be a *waqf* object according to Indonesian positive law based on Article 16 of the *Waqf* Law and Articles 15 and 21 of the Government Regulations of the *Waqf* Law.

2 THE RESULTS

2.1 *Waqf in Islamic law and the law of Indonesia*

The word *waqf* is derived from an Arabic verb, which means causing a thing to stop and stand still, which is a procedure of call letters, from where it starts and where it stops (Mohammad Daud Ali: 2006). According to the sharia term, *waqf* is the principal detention and development. It is the detention of the property and the use of the property and its benefits in the way of Allah. In terms of jurisprudence, *waqf* is generally defined as durable property which can be used to provide benefits without being damaged or depleted, and is commonly used for charitable purposes.

Waqf is also referred to as *al-habs* (*al-ahbas*, plural); *al-habs* means *al-sijn* (prison), silent, prevention, hindrance, impediment, custody, and security; and when the word ahbasa (*al-habs*) is combined with *al-mal* (wealth) it means *waqf* (*ahbasa al-mal*) (Ahmad Rofiq: 1977).

The term and concept of *waqf* is not mentioned in the Qurán, but it is mentioned in some *Hadith* literature and the history of the companions. Therefore, *waqf* doctrinal legitimacy comes from the Hadith, although there is no doubt that the Qur'an emphasises the importance of charity on behalf of others. Scholars argue that the command of *waqf* is part of the command to carry out *al-Khayr* (literally, goodness) as mentioned in Surah al-Hajj [22]:77 which states '... and do good that you may gain the victory'.

The role of *waqf* according to Abul Hasan M. Shadeq:

> 'there are many ways to handle the poverty problem, starting from preventive to curative measures. An important way to allivate poverty is charity, which has been playing a role and can potentially be an effective means to overcome the poverty problem. Charity is again *[sic]* many forms, but a long-lasting form of charity is one that has the characteristic of perpetuity. The institution of *waqf* is such a perpetual charity in the Islamic ethical system' (Ahmad Rofiq: 2002).

The principle of *waqf* for social solidarity means to utilise the property for the benefit of the people. It is related to the view of Abdurrahman Raden Aji Haqqi in his book entitled 'The Philosohy of Islamic Law of Transactions'. He noted social solidarity in economic transactions in Islam as follows:

> *'Transactions, according to Islam, are a system for a happy life on the basis of needs of the people. Islam has laid down principles for each, thus, the rules and methods have been enunciated relating to people, state affairs, etc. Such principles provide a kind of social solidarity. There is freedom and right to work, to own property and to enjoy other property rights. Nevertheless, these principles are bound by the langer interests of the community in Islam. In fact, the right of private ownership in the sharia is owned by God, man is to enjoy its benefits, in other words his title is limited to the benefits accruing on the property, but the legal title vests in God. The exercise of this right is circumscribed by the larger interest of the community and, thus, the principle of social solidarity prohibits exploitation under which the economic system of the sharia exists'.* (Abdurrahman Raden Aji Haqqi: 1999).

Al Azhar Islamic University in Egypt is one of the successful models of university in the world that is based on the benefits of investment and it has allocated a major part of its financial sources to educational centres. *Waqf* in Western countries is known as an Islamic custodianship or piety foundation of Muslim citizens which allocates property to social welfare programmes and support of their vulnerable religious obligations to attract attention and consent of God Almighty (Salarzehi, et al., 2010).

The term and concept of *waqf* is not mentioned in the Qurán, but it is mentioned in some *Hadith* and the history of the companions. Therefore, *waqf* doctrinal legitimacy comes from the Hadith, although there is no doubt that the Qur'an emphasises the importance of charity on behalf of others. Scholars argue that the command of *waqf* is part of the command to carry out *al-Khayr* (literally, goodness) as mentioned in Surah al-Hajj [22]:77 which states '... and do good that you may gain the victory'.

Waqf is valid if the conditions have been met and are in line with the four pillars (Tholhah Hasan: 2011):

a. *Shighat* (a word or statement showing willingness to donate some property);
b. *Waqif*: People who are donating and are required to have legal capacity or *kamalul ahliyah* (legally competent) in spending their money.
c. *Mauquf 'alaih* (parties to the allotment): This means the purpose of the *waqf*. *Waqf* must be used within the limits of what is appropriate and allowed according to Islamic law. This is basically because, *waqf* is to draw closer to God. Therefore, *mauquf 'alaih* should be the virtue.
d. *Mauquf bih* (*waqf* goods); *waqf* goods must meet certain requirements in order to be valid. The requirements for property for waqf are as follows: First, the object must be *mutaqawwam*. According to the Hanafi school of thought, *mutaqawwam* is anything that can be stored and used under normal circumstances (not in an emergency). Second, to be *waqf* goods must be known to fully belong to *waqif* (the giver of *waqf*) or *ainul ma'lumun*, so it will not give rise to disputes (Wahbah az-Zuhaili: 2011).

In the general explanation of the Waqf Law, it is stated that the purposes of *waqf* is to promote general welfare. To achieve this purpose, it is necessary to explore and develop the potential of religious institutions that have economic benefits. One strategic move to promote general welfare is to increase the role of *waqf* not only to provide various means of worship and social purpose, but also to realise the potential economic power, among others, to promote general welfare, so that its is developed in accordance with Islamic principles.

As a comparison, the development of *waqf* in Malaysia, according to Mashithoh, is as follows:

> 'The practice of *waqf* is not new to the Malaysian community. The fairly high number of creations of *waqf* nowadays is in fact, an extension of the practice of Malays in the

past, which had been influenced by their religious spirit as early as the acceptance of Islam in Malaya. One may find that different classes of individuals were appointed to manage *waqf* assets, and it gradually grew to entrusted to the State Islamic Religious Council (SIRCs) for each state. The institution of *waqf* in Malaysia seems to have improved from time to time with the introduction of various systems of laws and regulations. Nevertheless, it cannot be denied that, to a certain extent, civil regulations, financial and administrative constraints and others. Based on the existing legal framework Malaysia, mention of *waqf* can be found in the federal constitution that provides fundamental position as one of the country's affairs. Under the provisions of the constitution, the rule of law both *waqf* related provisions of substantive and administrative has been set in the legislation states'. (Mahamood: 2006).

Waqf intended for general interest (*waqf khairi*) is one of the sources of the public finance in Islam. Beside that, this *waqf* is one form of the contribution of community participation in the government's efforts to improve the welfare of the community in a variety of forms, including for community economic empowerment. Given the important role of the state in Islam, including its role in managing the economy and the public sector to enable the equitable distribution of welfare which can be enjoyed by the whole society, the state may intervene to keep justice in the distribution of income and wealth. This can be done with the support and policies on the protection of *waqf* so that it can be used optimally, including efforts to prevent and impose a sanction on any fraudulent attempt to use or abuse with regards to the possession and use of objects of *waqf*. To secure property from the unauthorised use of *waqf*, including the third party's intervention against the interests of *waqf*, the Waqf Law also establishes both administrative and criminal sanctions against the misuse of *waqf*.

2.2 *Waqf of shares and its implementation in realising the distribution of economic justice*

2.2.1 *Distribution of economic justice in Islam*

Theoretically, in Islam, the purpose of the *waqf* is to realise social justice (Sukron Kamil: 2011), at least justice in economy, health and, education. Therefore, the institution of *waqf* in Islam is not only a measure of one's faith but also helpful in distribution of wealth to realise economic justice and the symbol of the desired economic system of Islam. Islam recognises property as individual rights which also have social functions (Sukron Kamil: 2011). Through *waqf* and ZIS (*zakat, infaq/shadaqah*), it is expected that the wealth is not only concentrated on a few individuals, but is rather among all the people (Sukron Kamil: 2011).

Because *waqf* has a great potential for human, it becomes one of the basic considerations of the Indonesian Government to strengthen the position of the institution of *waqf* with the issuance of the *Waqf* Law. *Waqf* is a religious institution that has the potential of economic benefits and needs to be managed effectively and efficiently for the benefit both of worship and to promote general welfare.

Based on Article 5 of the *Waqf* Law, the purpose of *waqf* is not only for worship but also to serve and realise the potential and economic benefits to promote general welfare. This provision also confirms that Article 22 of the Waqf Law is intended for:

a. facilities and religious activities;
b. facilities and activities of health and education;
c. assistance to the poor and neglected children andorphans, and scholarships;
d. advancement and economic improvement of the people; and/or
e. advancement of general welfare of others that is not contrary to sharia and legislation.

The provision reflects an effort to create balance in the realisation of society as part of the implementation of social justice in Islam.

2.2.2 *Waqf of shares in realising economic justice*

Shares are certain ownership rights over the wealth of an individual or a portion of ownership in the network of the business of a company (Paul Redmond: 2002). Shares represent the amount of money invested by investors in a company. For their investment, shareholders (*aandeelhouder*) receive the benefits from the company in the form of dividends in proportion to the money invested (Paul Redmond: 2002).

The shareholders are given a proof of share ownership, as mentioned in Article 51 of the Company Law in Indonesia. Based on Article 52 of the Company Law, the form of proof of share ownership is specified in the Articles of Association of a company. Shareholders have the rights to:

a. attend and cast a vote at the annual general meeting of shareholders;
b. receive payment of dividends and liquidation proceeds from the remaining assets;
c. exercise other rights under the law.

The rights are granted to the holder of such shares after the relevant shares are recorded under the name of the holder in the register of shareholders.

In the practice of Islamic law, shares did not exist at the time of the Prophet and the Companions. At the time only commodities or real goods were known and traded in regular markets. Recognition of the ownership of a company (*shirkah*) at that time was not represented in the form of shares as it is nowdays.

Due to the lack of texts of the Qur'an and Hadith that would provide a clear and definite law about the existence of shares, scholars and jurists have tried to find a formula of their own to reach a conclusion about shares. This effort is known as *ijtihad*, which is an earnest effort to obtain and issue an Islamic law that has not been clearly shown either in the Qur'an or *Hadith*, which are usually referred to for basic legal legitimacy.

In *muamallah* jurisprudence, classical scholars do not recognise the term shares, but the concept is included in '*shirkah 'Inan*'. In the chapter on Islamic jurisprudence, *syarikah* is mixing. It is a treasure to treasure the other so they cannot be distinguished anymore. In the definition according to jurists, *syarikah* is an agreement between two people who combine their capital and profits. Scholars later used the term to refer to a specific transaction called *syirkah* that causes mixing of property transactions.

Waqf shares integrated into a company is an appropriate pattern which ensures the development of *waqf* through modern, profesional and, transparant management. According to Syafi'i Antonio's opinion about cash *waqf*, there are three basic waqf basic philosophies (Syafi'i Antonio in Writing Nur Faizin Muhith: 2013). First, to allocate *waqf* there must be an integrated project, inseparable from each other, especially in terms of costs. Secondly, the welfare of *nazhir*, as a profesional job like a manager in a profesional company to achieve development of *waqf*. Thirdly, transparancy and accountability principles, which refer the *waqf* board and the supporting bodies to assist in the making of report every year as the responsibility to the public for the use of *waqf* cash (*audited financial report*).

The dividends of a waqf of shares will be distributed to *mauquf a'laih* or partially reinvested to gain benefits. The benefits can be used directly for charitable or religious purposes or some can be reinvested to gain more benefits. Hence, the benefits of *waqf* are long term, and the profits of the company can be distributed to social activities. This is the distribution economy in Islam as referred to in Qurán, *Surah* Al-Hasyr, in that wealth can be distributed to all human beings. Through *waqf* of shares, dividends are not just for investors but also for social activities. It is one way to create social security and help the programmes of the goverment to serve society and help the poor people.

2.2.3 *The implementation of shares waqf based on Islamic law and positive law in Indonesia*

Waqf shares could be analogous with *waqf* cash. There are divergent views related to *waqf* cash, some of which approve of it some do not. The view of Hanafiyah allows *waqf* cash. Likewise, Imam Az-Zuhri found that donating dinars (money) as working capital is similar to shares. Thus, the functions of *waqf* of money and shares are the same, that is, to be used as working capital. In shares as securities, the nominal value of shares includes the amount of money that has been set by the issuing company. This shows the limit of the rights and responsibilities of shareholders of the company, including the dividends as profits.

The provisions of *waqf* based on Islamic law in the Qur'an are not mentioned directly. Scholars have argued that the *waqf* command is part of the command to *al-Khayr* (literally, goodness) as mentioned in the Qur'an, *Surah* al-Hajj [22]: 77. In the *Hadith* it is mentioned that the endowment is called *jariah* alm (*sadaqah jariyah*). Due to the absence of texts in the Qur'an and the *Hadith* of the Prophet associated with the endowment of shares, based on the methodology of Islamic law, endowment of shares is determined through ijtihad. *Waqf* of share is based on *mashlahah mursalah* and *istihsan*.

Mashlahah as a proposition of law implies that *al-mashlahah* is the foundation and benchmark for law enforcement. *Mashlahah mursalah* also called *Masalih al-mursalah* is how to find the law of something provisions on which are not contained either in the Qur'an and in the *Hadith*, based on the consideration of the benefits for society or the interests of society to ensure equal distribution of income or the collection of the necessary funds in the maintenance of public interests, which is not mentioned in the Qur'an and as-Sunnah (Hadith) (Muhammad Daud Ali:1999). Based on *Masalih al-mursalah*, *waqf* of shares provides benefits to society and ensures equal distribution of income and collection of the necessary funds in the maintenance of public interests.

Meanwhile, *istihsan* according to the literature of '*Adda wa ash-syai'a 'Itiqaduhu Hasanan*', is regarded as a 'good' belief in something (Wahbah az-Zuhaili: 1987). According to its definition in Arabic, istihsan is to make/assume something is good (Sulaiman Abdullah:1995). Share as an object of *waqf* is also based on istihsan. It is strongly associated with the purpose of *waqf* itself for social interests, and in particular for the distribution of economic justice in society.

Waqf of shares in the new Indonesian positive law was set after the enactment of *Waqf* Law. Not all types of shares governed by the positive law in Indonesia can be a *waqf* object. *Waqf* of shares is not only recognised by the Indonesian law as the provisions of the legislation in force, but it also does not conflict with the principles of sharia. It is also mandated by the provisions of the *Waqf* Law.

The types of shares in capital market instruments that are clearly forbidden according to Islamic principles are as follows:

1. prefered shares (special shares), which are forbidden by Sharia because there are two main characteristics, namely: a) the existence of a fixed profit (pre-determined revenue), which according to scholars is categorised as usury; and b) preferred share owners have the privilege, especially in the event of liquidation, which suggests an element of unfairness.
2. Forward contracts are prohibited because any form of sale of debt (*dayn dayn bi*) is not in accordance with the principles of sharia. It is forbidden because selling debts/receivables is an element of usury, since the sale and purchase transactions are made before the due date.
3. Option right to buy and sell items that are not accompanied by underlying assets or real assets.

The implementation of *waqf* of shares in Indonesia, for example, has been executed by Tabung Wakaf Indonesia (TWI). TWI has earned the trust of capital market securities, such as shares of a number of donors. Through this TWI, *waqf* is integrated in the existing management and used in the conduct of productive business activities.

In practice there are two models that can be executed in the mechanism of the *waqf* of shares:

1. *Waqf* cash is set out in the shares scheme as an investment in the *waqf*.
2. *Waqf* of shares of a company.

The cash *waqf* is set out in the shares scheme as an investment in the *waqf*. *Wakif* buys units of shares offered and then donates the units of shares for *waqf* purposes such as the construction of houses of worship, education, health care, or some other Islam charitable purposes. Steps taken are as follows:

1. Wakif buys units shares in a certain amount;
2. Wakif is given a certificate of *waqf*;
3. *Nazhir* accumulates funds available for the purposes stated in the certificate of *waqf*, such as for the development of facilities of education, health services, mosques, acquisition of land and, so on in accordance with the purposes of the *waqf*;

4. Indonesian *Waqf* Board (BWI), Indonesia Ulama Council (MUI), society and, *wakif* will controling or supervise the implementation of the *waqf* of shares.

Meanwhile, the mechanism of *waqf* of shares is as follows:

1. Wakif donate shares (registered shares) as *waqf*.
2. Pledge of *waqf*/transfer of ownership of shares is listed/reported in accordance with the share registry management through the company's directors and secretary on behalf of *nazhir* (after they become shares that may be transferred in accordance with the provisions of the Company Law and the Articles of Association of the company).
3. *Waqf* of shares cannot be assigned by anyone, including the company.
4. *Nazhir* will represent the *waqf* of shares, including representing it at the General Meeting of Shareholders, distribution of dividends and, so on.
5. *Nazhir* will distribute the dividends to *mauquf ala'ih*.

Based on the two models of *waqf* of shares above, the eligible waqf of shares pursuant to the Waqf Law is the one in the form of donating shares. The first form of *waqf* of shares is through buying units of shares which are then allocated for *waqf* purposes such as the construction of houses of worship, facilities of education, health care and, so on for the purposes of interests and benevolence towards Muslims. Basically, this is a type of waqf through money or cash *waqf*. The second type is *waqf* of shares of a company which are eligible for waqf of shares pursuant to the *Waqf* Law.

The *waqf* of shares of a company, cannot be separated by a transitional provision pursuant to Law No. 40 of 2007 of the Limited Liability Companies (the Company Law). The shares which have been transfered to be a *waqf* object are registered with the company's directors to be recorded in accordance with the provisions of Article 50 of the Company Law. However, pursuant to the provisions of Article 3 of the Government Regulation on the Waqf Law, despite the registration of *waqf* of shares, nazhir does not reveal the owner. Nazhir's only duty is to take care of it in accordance with the appropriate designation of *waqf*.

The supervision by the government is through monitoring the management of the *waqf* of shares. The state should be included in the socio-economic life of the community in order to realise common prosperity. Islam recognises the freedom to carry out economic activities, but it must not cause damage to individual or public interests. *Waqf* is a part of wealth distribution activities. In distribution, the government has an important role in controlling the economy. The government established Indonesian Waqf Board (BWI) which participates in promoting and developing the *waqf* in Indonesia.

Based on the Waqf Law, the *nazhir* is given the mandate to directly manage the productive *waqf*. Besides this, the nazhir has the obligation to carry out its functions as mandated by the pledge of *waqf* and is subject to legal sanctions in case of abuse. The *nazhir* must administer, manage, develop, supervise and, protect the *waqf* property. In order to perform this task, the *nazhir* is obliged to make periodic reports to the Minister and BWI about *waqf* activities. Thus, the mechanism of supervision and management of *waqf* objects by a *nazhir* should also be reported to the government, in this case through the Ministry of Religious Affairs and the Indonesian *Waqf* Board.

To ensure the development of *waqf* of shares, a n*azhir*'s professional management is needed. Effective and professional supervision is necessary to safeguard the assets of *waqf*. It needs the synergy of all parties, especially those directly related to *waqf*.

Related to the protection of shares as *waqf*, the insurance is one step that can be taken, although in practice today, the insurance of *waqf* of shares is not yet generally used. According to the researchers, a synergy in the formation of regulation on the waqf of shares is required. In connection with the principles of *waqf*, to keep the sustainability of *waqf*, then the principal value of *waqf* of shares must be guaranteed, including *waqf* shares. To maintain the sustainability and maintenance of shares as *waqf*, insurance would provide the necessary guarantees, so the shares could be continuously guaranteed, and their value and benefits enhanced. However, to date, shares insurance coverage has not been widely known as one of the insurance products in Indonesia. The shares as *waqf* could be insured in an effort to anticipate the risk of a decline in the value of the shares or the loss of value of the shares in case the company goes bankrupt.

The *waqf* of shares is one form of professional development of *waqf* which could be combined with a sustainable CSR programmes, in an integrated synergy between the business to make a profit that is managed by a professional, transparent trust management company, and the programme which contains social values for those in need of assistance. In a broader concept of democracy, it could create real economic justice in accordance with the spirit of the economy based on Pancasila as Indonesian Ideology.

The utilisation of *waqf* of shares could include the goals which are related to the objectives of economic empowerment of the people. Dividends from shares as *waqf* could be used to help to support the economy of the people, especially for the economically weak groups, so that ultimately they have the ability to become economically independent and meet the minimum standards of living in general.

The utilisation of dividends of shares as *waqf* could be used as capital for financing in accordance with the principles of sharia. One example that could be developed is to utilise the dividends as capital financing through the *qardhul hasan* system, that is a programme of capital paid in instalments without any additional effort and aim for the independence. Moreover, it could also be developed with a profit-sharing system in economic groups that benefits them, while all but the profits are aimed at empowerment. Thus, the fund could continuously be used by other economic groups. It is expected that this attempt to achieve economic justice through the distribution of dividends from the shares can be performed for the economic empowerment of the people. Ultimately it is expected that improve the social welfare and poverty reduction can be achieved.

3 CONCLUSION

Based on the discussion, analysis and, prescriptions that have been described, it can be concluded that:

1. *Waqf* is one of the methods of distribution that, according to the teachings of Islam, seeks to create social justice, especially justice in economy and, has a very broad potential to be developed, since *waqf* can be carried out by many parties, not limited by the status of a person, either rich or poor and, without having to be limited to family ties. *Waqf* for public interests (*waqf khairi*) is for the benefit of the public and can be enjoyed by society at large. It is one of the means of the implementation of public welfare which is not only for the sake of ritual worship, but also for social, cultural, educational benefits, including the improvement of social welfare in the economy.
2. The provisions on *waqf* based on Islamic law in the Qur'an are not mentioned directly. Scholars have argued that the *waqf* command is part of the command to *al-Khayr* (literally, goodness) as mentioned in the—Qur'an *Surah* al-Hajj [22]: 77. In the Hadith it is mentioned that the endowment is called *jariah* alm (*sadaqah jariyah*). Waqf of shares in the study of Islamic jurisprudence is associated with the *waqf* of money. The argument used is the developing of cash *waqf* and dinars as *waqf*. However, it is not limited to the form of dinars and dirhams only, but it could also include shares as a *waqf* object. Due to the absence of texts in the Qur'an and the *Hadith* of the Prophet about *waqf* of shares, it is determined through *ijtihad*. Thus *waqf* of share is based on *maslaha mursalah* and *istihsan*. Meanwhile, a share as a *waqf* object is based on the Waqf Law in Indonesia, recognised by Indonesia law as the provisions for the legislation in force, which does not conflict with the principles of sharia.

REFERENCES

Abdullah, S. (1995). *Sumber Hukum Islam* [The Source of Islamic Law]. Jakarta: Sinar Grafika.

Ali, M.D. (2006). *Sistem Ekonomi Islam Zakat dan Wakaf*, [System of Islamic Economy]. Jakarta: UI Press.

Az-Zuhaili, W. (2011). *Fiqih Islam Wa Adillatuhu*, [Al-fiqh Al-Islami wa Adilatuhu]. *Translated by A.H. Kattani, et al.(Eeds)*, Jakarta: Gema Insani.

Haqqi, A.R.A. (1999). *The Philosophy of Islamic Law of Transactions*. Kuala Lumpur: International Islamic University Malaysia Univision Press.

Hasan, T. (2011). Pemberdayaan Nazhir [Empowerment of Nazhir]. *Jurnal Al-Awqaf* 4(4), 1–14.

Indonesia (2004). *Undang-Undang Nomor 41 Tahun 2004 tentang Wakaf [The Law Number 41 of 2004 concerning Waqf]*.

Indonesia (2006). *Peraturan Pemeritaj Nomor 42 Tahun 2006 tentang Pelaksanaan Undang-Undang Nomor 41 Tahun 2004 tentang Wakaf [The Government Regulation Number 42 of 2006 concerning The Implementation of Law Number 41 of 2004 regarding waqf]*.

Indonesia (2007). *Undang-Undang Nomor 40 tahun 2007 tentang Perseroan Terbatas [Law Number 40 of 2007 concerning Limited Liability Companies]*.

Kamil, S. (2011). Wakaf untuk Keadilan Sosial antara Teori dan Praktik [Waqf for Social Justice, Theory and Practice]. *Jurnal Al-Awqaf* 4(4), 60–87.

Mahamood, S.M. (2006). *Waqf in Malaysia: Legal and Administrative Perspective*. Kuala Lumpur: University of Malaya Press.

Muhith, N.F. (2013). *Dahsyatnya Wakaf* [The Miracle of Waqf]. Surakarta: al-Qudwah Publisihing.

Rendmond, P. (2000). *Companies and Securities Law, Commentary and Materials* (3rd edition). New South Wales: LBC Information Services.

Rofiq. A. (1977). *Hukum Islam di Indonesia* [Islamic Law in Indonesia]. Jakarta: Raja Grafindo Persada.

Sadeq, A.M. (2002). Waqf Perpetual Charity and Property Alleviation. *International Journal of Social Economic*. 29 (1/2), 135–151.

Salarzehi, H., Armesh, H. & Nikbin, D. (2010). Waqf as a Social Entrepreneurship Model in Islam. *International Journal of Business and Management*. 5(7), 179–186.

Wignjosoebroto, S. (2011). Penelitian Hukum dan Hakikatnya sebagai Penelitian Ilmiah. In Sulistyowati, I. & Sidharta (Eeds). *Metode Penelitian Hukum; Konstelasi dan Refleksi* [Writing in Legal Research Methods: Constellation and Reflection]. Jakarta: Yayasan Obor Indonesia.

Paradox of groundwater tax collection

E.S. Sunarti & T. Hayati
Faculty of Law, University of Indonesia, Depok, Indonesia

ABSTRACT: Tax is the most important source of revenue for a state. One of the local taxes that can be collected by the local government is the groundwater tax regulated in the Law No. 28 of 2009 on local taxes. Groundwater tax is the most crucial tax as the exploitation of groundwater causes damage to the environment. This paper will discuss whether the tax collected compensates the damage caused by the exploitation of groundwater. What is the purpose of groundwater tax collection? Do the results of the groundwater tax realize a prosperous society? This paper uses a normative legal review to examine whether the meaning contained in the rules of law is conceptually well aware of the application in practice. This paper used the statute approach and the analytical approach. Data processing was done by the qualitative approach and the quantitative approach. This research was conducted in Jakarta, Bogor, and Sukabumi. Collection of the groundwater tax creates a paradox in its implementation. On the one hand, the local government requires funds to finance regional governance, and the funds derive from local taxes. On the other hand, the exploitation of groundwater damages the environment. The collection of taxes should be for the purpose of financing public interest and enhancing the prosperity of the entire community.

1 INTRODUCTION

1.1 *Paradox of groundwater tax collection*

Just like the Central Government which levies a tax to finance its activities, the Local Governments also use taxes as one of their sources to finance their activities (Sidik, 2002). Local tax is a tax levied by local tax regulations set by the Local Government to finance their internal budget as a public legal entity (Darwin, 2010). It is said in Article 23 the 1945 Constitution that "all taxes and other levies for the needs of the state of a compulsory nature shall be regulated by law." Tax collection by the Central Government is regulated in an Act, and tax collection by the Local Government is regulated in the local regulation.

In order to materialize the implementation of regional autonomy and fiscal decentralization, the Central Government gives authority to the Local Governments to determine their sources of revenue based on their own potentials. Collecting local taxes and retributions is an example how this authority can be implemented (The Law No. 23, 2014). The purpose of giving the freedom to levy local taxes and retributions to the Local Governments is to improve the ability of the Local Governments in implementing regional autonomy; therefore, this authority must be supported by public awareness to pay such taxes. Sularno (1999) defines local taxes as: local taxes or state taxes are transferred to Local Governments, collected by Local Governments in their own territory, and used to finance local expenditure on their territory in relation to duties and responsibilities to regulate and manage their own affairs. According to James and Nobes (1997), "a tax is a compulsory levy made by public authorities for which nothing is received in return." Such levy is used to finance the Local Governments to govern and develop their territory (Davey, 1988).

Revisions and improvement of the local taxation system in Indonesia are needed to implement local autonomy and fiscal decentralization by transferring the sources of income to the Local Government gradually. The Law on Local Taxes and Local Retributions divides Local

Taxes into a provincial tax and a district/city tax. One type of a tax levied by the local district/city government is the Groundwater Tax. The groundwater tax is levied on the exploitation of water from the groundwater in the respective area. The groundwater tax is a tax that is the most crucial because the exploitation of water from the ground is related to the environment. The Groundwater Tax collection has to give attention to the availability of groundwater related to environment and other ecosystems (Kompas, 2012, p. 13). Uncontrollable exploitation of groundwater could damage the environment (Asdak, 2002).

As a basic human need, water is very important for human life (Dewi, 2009). The increasing human population has resulted in the increase of the need for water every day. The increase of population, the development of sciences, politics, economics, the social aspect, and culture have influenced the number of extraction and/or utilization of groundwater, for domestic use, irrigation, worship, or for business purposes, such as hotels, offices, recreation, bottled water, factories, hospitals, or houses with a pool that becomes a trend in the present time, etc. Groundwater tax is imposed on exploitation and/or utilization of groundwater for business purposes.

The availability of water which is a natural resource is absolutely necessary anywhere and anytime. Each groundwater has a different quality for human consumption (Kompas, 2013, p. 6). The excessive exploitation of groundwater will cause damage, such as decreasing water level, the formation of groundwater basins critical in some areas, and other effects, such as land subsidence, sea water intrusion, and intrusion of pollutants as well as the scarcity of groundwater itself, rob, and flood. The impact of an uncontrollable extraction of groundwater in some big cities, such as Jakarta, Bandung, Sukabumi, and Bogor City has caused the decrease of soil surface. The research focus asks: what is the purpose of groundwater tax collection? Do the results of the groundwater tax realize a prosperous society?

The problem in this research is the paradox of the taxation of groundwater with respect to the environmental ecosystem and groundwater earmarking taxes for people in the district of Bogor, Sukabumi, and Jakarta. This research is normative research, using secondary data, which include primary legal materials, secondary law, and tertiary legal materials. To examine whether the meaning contained in the rules of law is conceptually well aware of the application in practice, this paper uses the statute approach and the analytical approach. Data processing was done by the qualitative approach and the quantitative approach. Collection of the groundwater tax creates a paradox in its implementation.

1.1.1 *Taxes for welfare*

Tax is aimed for welfare, and this is based on two arguments. First, tax is as development funding, especially for public facilities; and, second, tax functions as a fair instrument to increase social fairness in society. Rochmat (1990) states that tax is an obligation fee for state budget (from the private sector to the public sector, and it is a mandatory).

Welfare is at the core of the collection of taxes. The success of tax collection should aim for prosperity. The state that collects taxes, consequently, has to try to raise the well-being of the society (Devereux, 1996). The benefits of tax collection are based on two reasons. First, tax is a means for the state to finance the construction for the welfare of the society. The task of the state is to provide prosperity and justice in various facets of life, especially in terms of building public facilities. If the source of tax revenue is not achieved, then the various development programs will not run smoothly. Second, taxation is the only instrument fairest in providing public welfare. Only with taxation can each state create prosperity and justice for the society (Ilyas, 2011). The countries that have an ideology that is oriented to the welfare of the people will have the support of the people in the form of tax payments.

The legal basis for the authority to collect groundwater tax is the following:

1. The Law No. 23 of 2014 about local governments, Article 285 on regional finance, gives the authority to the regions to levy local taxes and retributions;
2. The Law No. 28 of 2009 on local taxes, Article 2, authorizes the district/city to collect groundwater tax;

3. The Law No. 11 of 1974 on irrigation which replaces the Law No. 7 of 2004 on water resources is based on the decision of the Constitutional Court Number 85/PUU-XII/2013 on canceling the enforceability of the overall the Act No. 7 of 2004 on water resources because it does not meet the six basic principles of water resources management restrictions, namely:

First, any water utilization should not undermine and eliminate the rights of the people because in addition to being state-controlled water is intended for the welfare of the people.
Second, the state must meet the people's right to water as a human right, which is based on Article 28 i paragraph (4) of the Constitution, which should be the responsibility of the government.
Third, water management must be given to the environment.
Fourth, as a branch of vital production for the life of many, in accordance with Article 33, paragraph 2 of the 1945 Constitution, water should be under the supervision and absolute control of the state.
Fifth, the absolute water management rights belong to the state; the top priority given to water concession lies in the hands of the state or local government.
Sixth, the Law of Water Resources is declared contrary to the 1945 Constitution in order to prevent a vacuum in the management of water resources and wait for the formation of a new Act, and then the Act can be reinstated.

1.1.2 *Welfare state*

The Republic of Indonesia in its Constitution has mandated this nation as a welfare state. Welfare is the most fundamental rights for every citizen. The Constitution states that the state should be responsible for the basic rights of every citizen. In this case the state should be fully responsible to improve the lives of its citizens so that they can live decently. The objectives of the Indonesian state are set on the fourth paragraph of The Constitution Preamble, which in essence is "to promote the general welfare and educate the nation based on the principles of social justice for all Indonesian people". Objectives contained in the opening are then poured into the articles of the Constitution which regulate various provisions related to social welfare. Provisions governing economic and welfare issues are contained in some articles: Article 26, Article 27 Paragraph (2), Article 28, Article 31, Article 32, Article 33, and Article 34.

On the other hand, Article 33 regulates economic issues and determines that the production branches, which are important for the people and the Earth and the water, and natural resources that exist on top of it, are controlled by the state. It means that the people's welfare is the responsibility of the State. Spicker explains that a welfare state is welfare which is provided comprehensively by the state to the best possible standards (Spicker, 1982).

There are five principles and characteristics of a welfare state, which therefore should be strived for countries that embrace the welfare state system in the framework of its efforts to achieve the goal of reducing economic disparities and improve the welfare of its citizens (Budiardjo, 1982).

First, the branches of production that are important concerning the livelihood of the people are controlled by the state. The purposes and objectives of the control of the production branches which are important for the lives of many people are that the needs of the people on the production of the goods can be obtained by the people at affordable prices, but not onerous and burdensome for the life of the people. Examples of the production branches which are important for the interests of the people mostly include electricity, gas and petroleum, water, and public transportation.

Second, private enterprises outside production branches concerning the livelihood of the people are allowed, but the state governs (by regulation) and supervises so that no monopoly or oligopoly would distort the market or other forms of adverse welfare. Examples include establishing schools, hospitals, shipping, transports, and trade.

Third, the state is directly involved in the efforts of well-being of its people, by providing various forms of health care and education services. A variety of services that, with the system implemented, should be accessible to all people without exception. Of course, the number of the types of services that must be provided by the state depends on the development and/or the ability of the state, but both types of these services (health and education) are a necessity that cannot be abandoned. In addition, the state also provides social security and old age insurance for every citizen. In fact, many experts consider that the state's role in providing welfare services to its citizens should include the life of the people "from the cradle to the grave."

Fourth is to develop a progressive taxation system, which puts higher tax rates in percentage (enlarged) for people who are getting richer and for greater efforts. Through the tax system and the security system developed by the state, the expected difference between the rich and the poor in the country concerned will be reduced, and the number of poor people will also decrease.

Fifth, public policy making should be done democratically. This means that the welfare state adheres to the democratic system of managing its country. The community is involved in the proses of policy making and supervises the implementation of policies, so the policies issued by the government is something that is needed by the community.

Governing the administration and the welfare of the country require funding. The state obtains funds from various sources, among others, from taxes, retributions, contributions, dues, the natural wealth of the country, and others. Most of the funds are obtained by the state from taxes levied by the state to its citizens and any other legal entities to finance the state administration and the public welfare. Taxes are paid for the purposes of the state to establish and improve social welfare (Rahman, 2012). Tax is a tool of development progress, a tool to equalize economy, and a tool for prosperity of a country.

1.1.3 *The right to water*

Water is part of natural resources that are fundamental for human life and cannot be replaced. All aspects of human life are associated with water. Water is vital for the society; therefore, water is a *res commune* (public goods) or common property (Sunarti, 2016).

In history, the right to water has been admitted since a long time ago. For instance, there are traditional and old rights in water known as Riparian Rights, and these traditional Riparian Rights existed across the Himalayan. In the beginning, the Riparian Rights aimed to divide the water and protect the sustainability of the water, and it did not regulate ownership of the water (Shiva, 2002).

Under the 1945 Constitution (UUD 1945) Article 33 states that land, water, and natural resources inside are regulated by the state and used for the people's prosperity (Gautama, 1980). The article provides a basis for recognition of the right to water as part of the right to life of spiritual and physical welfare, and the substance is a human right. There are several principles of water democracy (Suteki, 2010): (1) Water is a gift from nature; therefore, exploitation of water must not create a negative impact to ecology; (2) Each species has the right to water access; (3) Water must be accessible freely for agriculture; (4) Water is a limited resource; therefore, groundwater should not be consumed more than the ability of the ecology to restore the water; (5) no one is allowed to produce pollution in water.

The state must ensure that everyone has access to water; therefore, the state should be involved in water management or public utilities (Khusaini, 2006).

2 PARADOX IN THE GROUNDWATER TAX

The Local Government is funded by local revenue which consists of Local Revenue (Local tax, Local retribution, local SOE, etc.) and Revenue from the Central Government (Bohari, 1993). The Law No. 28 of 2009 on Local Taxes and Retributions is the law which regulates the basic provisions that provide policy guidance and directions for regions in the implementation of the collection of taxes and retributions, and establish arrangements to ensure the application of common procedures for Local Taxation and Retributions.

The Law No. 28 Year 2009 on Regional Taxes and Retributions has been set under the authority of the Provincial Government. There are five types of taxes: (1) Tax on Motor Vehicles, (2) Duty of Vehicle, (3) Tax on Motor Vehicle Fuel, (4) Surface Water Tax, (5) Tax on Cigarettes. Moreover, the local district/city government manages eleven types of taxes, consisting of: (1) Hotel Tax, (2) Restaurant Tax, (3) Amusement Tax, (4) Advertisement Tax, (5) Street Lighting Tax, (6) Mineral Tax (non-metal or rock), (7) Parking Tax, (8) Groundwater Tax, (9) Swallow's Nest Tax, (10) Land and Building Tax, (11) Tax on Acquisition of Land and Buildings.

Groundwater tax is a tax that is mostly at risk in the implementation because, on the one hand, the government wants the revenue of the groundwater tax. On the other hand, the utilization of this creates environmental degradation in its territory. Therefore, the taxation will lead to contributions of revenue, but it is not balanced with the impact. In the district of Bogor, Sukabumi, and Jakarta, the revenue of groundwater taxes is of little benefit to the region concerned. The amount of the tax revenue of groundwater cannot repair the environmental damage caused from water withdrawal territory. Groundwater tax revenue is not comparable to the cost of the damage caused by groundwater abstraction. The following will explain the large percentage of the groundwater tax revenue compared to local financial revenue.

The table shows groundwater tax revenue is very small compared to the local finance revenue. Furthermore, the environmental damage resulting from the extraction of groundwater requires rehabilitation with enormous cost. The cost to repair the environmental damage is not proportional to the amount of the revenue of the groundwater tax.

The excessive exploitation of groundwater will cause damage, such as decreasing water level, the formation of groundwater basins critical in some areas, land subsidence, sea water intrusion, and intrusion of pollutants, as well as the scarcity of groundwater, which is increasingly widespread (Konrad Adenauer Stiftung, 2006), rob, and flood. The impact of uncontrollable exploitation of groundwater has happened in some big cities, such as Jakarta, Bandung, Sukabumi, and Bogor City, which have caused the soil surface to decrease. In Jakarta, people's needs for water are taken through an underground water pump so that the groundwater level is lowered. Suction of groundwater in Jakarta is 280 million cubic meters per year, but only 22 million cubic meters per year is reported. There are still many companies that have not paid the groundwater tax. Now the decreasing water level in Jakarta has reached 11 cm per year. The groundwater is sucked, and it causes infiltration or intrusion of sea water, and it further infiltrates into urban areas, not only in the area around the beach. In Sukabumi, more than 120 companies are taking groundwater directly from water sources. There are still many companies that have not paid the groundwater tax since 2006 (Kompas, 2012, p. 22). The tax revenues are used to improve conservation programs, and the national park area is a buffer zone of water demands in Jabodetabek through DAS Cisadane and Ciliwung watershed Cimandiri (Asdak, 2002). Conflicts have occurred due to the exploitation of groundwater by bottled water companies (Surono, 2015). Occupation area and exploitation of the water are forms of perversion of the Constitutional Court judgment in the case verdict 58-59-60-63/PUU-II/2004 and case 8/PUU-III/2005, regarding judicial review of the Law No. 7 of 2004 on Water Resources (Al Afghani, 2006). The norm diversion impact on

Table 1. Groundwater tax to local financial revenue: District Bogor.

	2011	2012	2013	2014	2015
Financial revenue	697,294,749,223	1,068,548,465,318	1,258,766,010,367	1,712,852,487,027	1,674,547,617,000
Groundwater tax	21,601,213,132	27,992,374,686	40,350,069,968	51,916,322,226	68,635,365,000
(%)	3.10%	2.62%	3.21%	3.03%	4.10%

Annual report of the local government of Bogor district.

Table 2. Groundwater tax to local financial revenue: Sukabumi district.

	2011	2012	2013	2014	2015
Financial revenue	125.879,976,500	163,476,799,856	218,478,439,414	355,346,307,000	447,751,711,000
Groundwater tax	23,824,309,608	24,465,369,907	47,553,729,776	74,480,701,249	67,015,210,281
(%)	18.93%	14.97%	21.77%	20.96%	14.97%

Annual report of the local government of Sukabumi district.

Table 3. Groundwater tax to local financial revenue, DKI Jakarta.

Description	2011	2012	2013	2014	2015
Financial revenue	17,825,987,294,430.80	22,040,801,447,924.00	26,852,192,452,636.00	31,274,215,885,719.00	40,355,853,087,978.00
Groundwater tax	114,442,293,835.54	102,046,137,531.32	95,346,034,925.00	101,880,754,052.00	104,929,684,748.00
%	0.64%	0.46%	0.36%	0.33%	0.26%

Annual report of the local government of DKI Jakarta.

practice is likely to provide opportunities of privatization and commercialization of water, which are detrimental to society (Manar, 2009).

The Act No. 28 of 2009 governs the tax object of groundwater as retrieval and/or utilization of groundwater. Several things excluded from the tax object of groundwater are: (a) retrieval and/or utilization of groundwater for basic household purposes, irrigation farming and artisanal fisheries, and worship; and (b) retrieval and/or utilization of groundwater for other purposes stipulated by the regional regulation. Taxpayers of the groundwater tax is a private person or agency that does the fetching and/or utilization of groundwater. The basis of the groundwater tax is Earned Value. The groundwater tax rates are set by the regional regulation. Uncontrolled groundwater gives an impact environmental degradation (Fadjar, 2005).

The paradox occurs (1) when regional revenues from the groundwater tax represent revenue that will fill the local budget. The greater revenue from the groundwater tax, the more groundwater is exploited and more groundwater will be taken, which will have an impact on water availability and the soil, and create other environmental impacts. (2) The above explanation describes that the small percentage of local financial acceptance of the results of the groundwater tax is not worth the cost to be borne by the community and the government, due to the consequences of groundwater abstraction.

Water has functions in social life, environment, economy, organization, and harmony. Water resources are managed under the principles of sustainability, balance, public benefit, integrity and harmony, justice, independence and transparency, and accountability. Water resources are controlled by the state and utilized for the welfare of the people (Ogus, 2004). Viewed from the aspect of welfare, welfare is the goal of the establishment of the state of Indonesia, as stated in the preamble of the 1945 Constitution, the fourth paragraph. All activities and events are geared towards the welfare of the entire people of Indonesia. In a modern state, interfering hands of the state in dealing with the public interest aims to create a prosperous state. The state plays an active role in managing the affairs of the people.

Reviewing the aspects of taxes aims to improve the welfare of society. However, groundwater tax collection in several regions in Indonesia has no impact on the welfare of the community. Proceeds from the groundwater tax are not proportional to the damage it has caused from the imposition of the tax object. The above explanation describes that the groundwater taxation is not in accordance with the purpose of taxation to create prosperity for the community.

3 CONCLUSION

Welfare is the right of every human being and the right of every citizen to be realized by the state. Welfare also includes the right to clean water. For the public welfare, the government collects taxes. One type of tax levied by local governments is the tax of groundwater collection and utilization. However, problems arise in implementation. The paradox occurs in terms of tax revenue. Groundwater is required for the government income to be utilized by the Local Government, but the impact of groundwater is causing environmental damage. The groundwater tax does not help the people to be prosperous as intended by the purpose of taxes.

REFERENCES

Al Afghani, M.M. (2006) Constitutional Court's Review and the Future of Water Law in Indonesia. *Law, Environment and Development Journal*, 2 (1), 1–18.
Asdak, C. (2002) *Hidrologi dan Pengelolaan daerah Aliran Sungai.* 2nd edition. Yogyakarta, Gadjah Mada University Press.
Bohari, H. (1993) *Pengantar Hukum Pajak.* Jakarta, Raja Grafindo Persada.
Budiardjo, M. (1982) *Masalah Kenegaraan.* Jakarta, Gramedia.
Darwin (2010) *Pajak Daerah dan Retribusi Daerah.* Jakarta, Mitra Wacana Media.
Davey, K.J. (1988) Pembiayaan Pemerintah Daerah. Jakarta: UI Press.
Deveux, M.P. (1996) *The Economic of Tax Policy.* New York: Oxford University Press.
Dewi, R.I. (2009) *Pendayagunaan sumber daya air untuk kesejahteraan masyarakat lokal: Kajian mengenai pengusahaan air di Kecamatan Cidahu—Sukabumi dan Polanharjo—Klatem.* Master Tesis, Depok, Universitas Indonesia.
Fadjar, A.M. (2005) Pasal 33 UUD 1945, HAM, dan UU Sumber Daya Air. *Jurnal Konstitusi,* 2 (2), 7–10.
Gautama, S. (1980) Tafsir Undang-undang Pokok Agraria. Bandung, Citra Aditya Bakti.
Ilyas, W.B. (2011) Kontradiktif sanksi pidana dalam hukum pajak. *Jurnal Hukum,* 18 (4), 525–542.
Indonesia (1945) *Undang-undang Dasar Republic Indonesia 1945 [The 1945 Constitution of the Republic of Indonesia].*
Indonesia (1974) *Undang-undang No. 11 Tahun 1974 tentang Pengairan [The Law Number 11 of 1974 regarding Water Resources Development].*
Indonesia (2004) *Undang-undang No. 7 Tahun 2004 tentang Sumber Daya Air [The Law Number 7 of 2004 regarding Water Resources].*
Indonesia (2009) *Undang-undang No. 28 Tahun 2009 tentang Pajak Daerah dan Retribusi Daerah [The Law Number 28 of 2009 regarding Local Taxes and Retributions].*
Indonesia (2014) *Undang-undang No. 23 Tahun 2014 tentang Pemerintahan Daerah. Lembaran Negara Republik Indonesia Tahun 2014 Nomor 244. Tambahan Lembaran Negara Republik Indonesia Nomor 5587 [The Law Number 23 of 2014 regarding Local Governments. The State Gazette of Year 2014 Number 244. The Supplement to the State Gazette Number 5587].*
James, S.R. & Nobes, C. (1997) *The Economic of Taxation: Principles, Policy and Practice.* Lincoln, Anybook Ltd.
Khusaini, M. (2006) *Ekonomi Publik: Desentralisasi Fiskal dan Pembangunan Daerah.* Malang, Badan Penerbit Fakultas Ekonomi Universitas Brawijaya.
Mahkamah Konstitusi [Constitutional Court] Republik Indonesia (2013) *Putusan No. 85/PUU-XI/2013 tentang Uji Materil Undang-undang No. 7 Tahun 2004 [Decision Number 85/PUU-XII/2013 concerning Testing Act Number 7 of 2004].*
Manar, D.G. (2009) Krisis kekuasaan negara di balik privatisasi air. *Majalah Pengembangan Ilmu-ilmu Sosial FORUM,* 37 (2), 15–20.
Novalinda & Waddell, S. (2006) *Air Perkotaan dalam Pembangunan Kota yang Berkelanjutan.* Jakarta, Asosiasi DPRD Kota Seluruh Indonesia [ADEKSI], Konrad-Adenauer-Stiftung [KAS] & Program Lingkungan Hidup Indonesia Jerman Gesellschaft für Technische Zusammenarbeit [ProLH GTZ].
Nugroho, W. (ed.) (2012) Iuran air tak dibayar. *Harian Kompas,* April 26, 2012.
Nugroho, W. (ed.) (2012) Regional: Krisis pangan dan air mengkhawatirkan. *Harian Kompas,* April 21, 2012.
Nugroho, W. (ed.) (2013) Wacana: Air: Masalah yang kian akut. *Harian Kompas,* Maret 19, 2013.
Ogus, A.I. (2004) *Regulations: Legal Form and Economic Theory.* Portland, Hart Publishing.

Rahman, F. (2012) *Kesadaran Pajak untuk Kesejahteraan Sosial*. Direktorat Jenderal Pajak Kementerian Keuangan Republik Indonesia. Artikel. Available from http://www.pajak.go.id/content/fadjroel-rahman-kesadaran-pajak-untuk-kesejahteraan-sosial.
Shiva, V. (2002) *Water Wars: Privatization, Pollution, and Profit*. Cambridge, South End Press.
Sidik, M. (2002). Perimbangan Keuangan Pusat dan Daerah Sebagi Pelaksanaan Desentralisasi Fiskal. In: *Seminar Setahun Implementasi Kebijakan Otonomi Daerah di Indonesia, 13 Maret 2002, Yogyakarta, Indonesia*. Yogyakarta, Universitas Gadjah Mada.
Soemitro, R. (1979) *Dasar-dasar Hukum Pajak dan Pajak Pendapatan 1944*. 9th edition. Bandung, Eresco.
Spicker, P. (2014) *Social Policy: Theory and Practice*. 3rd edition. Chicago, University of Chicago Press.
Sularno, S. (1999) *Pajak Daerah dan Retribusi Daerah*. Jakarta, STIA-LAN Press.
Sunarti, E.S. (2016) *Pajak Air Tanah di Indonesia: Konseptualisasi Penyelenggaraan Otonomi Daerah dalam Pengaturan Pajak Air Tanah yang Berkeadilan*. Depok, Universitas Indonesia. Disertasi Doktor.
Surono (2015) Pengelolaan Air Tanah Pasca Putusan Mahkamah Konstitusi No. 85/PUU-XI/2013. In: *Seminar Nasional Implikasi Pembatalan UU 7/2004 tentang Sumber Daya Air terhadap Pengelolaan Air Tanah, 16 April 2015, Bandung, Indonesia*. Bandung, Perhimpunan Ahli Air Tanah Indonesia & Program Studi Magister Teknik Air Tanah Institut Teknologi Bandung (ITB).
Suteki (2010) *Rekonstruksi Politik Hukum Hak Atas Air Pro Rakyat*. Malang, Surya Pena Gemilang.

The concept of a regulation of collateral under the *mudharabah* financing contract according to the Law No. 21 of 2008 on sharia banking in Indonesia

M.Y. Harahap & U. Hasanah
Faculty of Law, Universitas Indonesia, Depok, Indonesia

ABSTRACT: According to the Indonesian Law No. 21 of 2008 on Islamic Banking, any financing that is based on *mudharabah* contract is technically intended to meet the interests of businesses for gaining capital or additional capital to run a productive business. If a sharia bank requires collateral under a *mudharabah* financing contract, it would violate the principles of *mudharabah* financing itself, as *mudharabah* is not a loan arrangement that requires any collateral. Therefore, if a *mudharabah* financing contract is not recognized as a mutually beneficial arrangement, it can cause a problem of injustice. The first part of this paper discusses the concept of a regulation of collateral in the *mudharabah* financing contract. The second part constitutes an analysis on the imposition and application of the collateral system in the *mudharabah* financing contract under the Law No. 21 of 2008. This research applies a legal normative method by the statute approach namely the Law No. 21 of 2008 on sharia banking. The findings of the research show that the application of the collateral system is to avoid business risks, complying with the principle of prudence and caution against the moral hazard of business actors.

1 INTRODUCTION

The Law No. 21 of 2008 concerning Islamic Banking stipulates that the *mudharabah* financing contract is a business cooperation agreement between the first party as the capital owner, or an Islamic bank that provides all the capital, and the second party as a customer who acts as a fund manager by dividing business profit, which is in accordance with the agreements set forth in the contract, while the loss is borne entirely by the Islamic bank unless the second party commits willful misconduct, neglects, or violates the agreement. The principle of financing activity that is based on the *mudharabah* contract is intended to meet the demand and interests of businesses actors to obtain capital or additional capital to conduct a productive business, between two or more parties. The data that have been collected by the authors show that the growth and the distribution of financing service in Islamic banking in Indonesia from 2010 to 2015 have been increasing significantly. It can be shown from the data in Figure 1.

As a form of cooperation, *mudharabah* is important to be understood as the basis or foundation of thinking. If the mudharabah financing contract is not well understood as a form of cooperation, then it will produce some problems of injustice. This discussion will be interesting to explore because the concept of *mudharabah* financing with encumbrances and assurance binding is not justified under Islamic law, for *mudharabah* financing is not a credit agreement of financing grants, which puts the capital owners and entrepreneurs in a position that is not equal, nor does it establish any creditor-debtor relationship. Therefore, the normative construction that is established from such a relationship should have been in the form of legal relationship between collaborating parties, instead of legal relationship between the creditor and the debtor, which may ask for collateral.

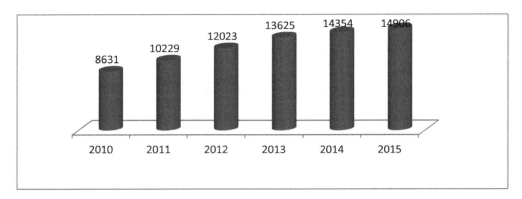

Figure 1. The growth and the distribution of financing service in Islamic banking in Indonesia from 2010 to 2015.

Based on the description of the background above, the Law No. 21 of 2008 as a legal framework in Indonesia has been used as a perspective to examine these issues, which are formulated in the research questions. First, how is the concept of a regulation of the collateral system in the *mudharabah* financing contract in Sharia Banking? Second, how are the imposition and application of the collateral system in the *mudharabah* financing contract according to the Law No. 21 of 2008 on Sharia Banking? From the description of the background issues and the formulation of the problem in this research, the paper aims to describe and analyze the concept of collateral arrangement, imposition, and application of collateral in the *mudharabah* financing contract by using al-*maslahah* theory. This study used the normative legal research method based on secondary data collection consisting of primary and secondary legal materials. The approach used is the statute approach since this study focuses on the Law No. 21 of 2008 as a perspective to analyze the issue of the collateral system in the *mudharabah* financing contract.

2 CONCEPT OF A REGULATION OF COLLATERAL IN THE *MUDHARABAH* FINANCING CONTRACT IN SHARIA BANKING

In general, any guarantee in Islamic law is known as a collateral in the form of property, which is usually called a *rahn* and warranty that is usually termed as *kafalah* (personal guarantee), either from *mudharib* or from third party. *Rahn* can be defined as having permanent and sustainable qualities, and it is also known as *al-habsu*, which means holding one of the borrower's properties as collateral for a loan received. According to Islamic law, *rahn* can be defined as holding the saleable properties that have economic value, either movable or immovable as collateral to pay off debts (Zuhaili, 2002). Some Islamic jurists argue that any legitimate saleable assets or property can be used as collateral. According to the history, the Prophet (كل ما جاز بيعه جاز رهنه) said that any goods that could be traded may also be used as a collateral. The transfer of such guarantee does not have to be real property, but it must be legal in nature for example, the form of transfer of certificates of property ownership can be considered as collateral.

The philosophical aspect of guarantee is one of the most important safety instruments to deal with the possibility of potential losses in terms of an Islamic banking financing contract. The principle of Islamic banking can take advantage of a collateral object to secure any financing that is provided for the customers (Muhammad, 2008). In terms of *rahn*, the Fatwa of the National Sharia Board defines it as a collateral in the form of assets or property, but the customers are still able to utilize such collateral and take an advantage from them. However, the certificate of ownership of the guaranteed property must be given to the capital owner. The core definition of *Rahn* is 'holding the utilizable property of a debtor that has economic value. *Rahn* must provide assurance for the creditor in order to guarantee the repayment of the debt, and the agreement of *rahn* must be conducted in the principles of *ta'awun* (mutual help) in order to allow the creditor to be able to help someone without pushing away their profit orientation, which is in accordance with the sincerity of Allah SWT (Burhanuddin, 2010).

The assets, or property as collateral in the form of *Rahn* is permitted under the following conditions: 1. *Murtahin* (consignee) has the right to hold *marhun* (goods/assets or property) until all *rahin* debts (which deliver the goods/assets or property) could be completely repaid. 2. *marhun* and its benefits will remain the property of *rahin*. In its principle, *marhun* must not be utilized by *murtahin* unless it is permitted by the *rahin*, without reducing the value of *marhun,* and its utilization is only for the replacement cost of maintenance. 3. The maintenance and the storage of *marhun* are basically a *rahin*'s liability, although it could also be conducted by *murtahin*; meanwhile, the storage and maintenance costs are still the obligation of *rahin*. 4. The amount of maintenance and storage costs should not be determined by the amount of the loan. 5. *Marhun* should be sellable. At the time of maturity, *murtahin* should warn *rahin* to immediately pay off the debts. If *rahin* is still unable to pay off the debt, then the *marhun* could be executed through an auction process that is based on sharia law. The proceeds from the sale of *marhun* should be used for repayment of debt, any unpaid maintenance and storage costs, and the cost of the sale itself. The profits from the sale belong to *rahin,* and the loss from the sale must become the liability of *rahin* (Dewan Syariah Nasional Majelis Ulama Indonesia [DSN-MUI], 2002).

Besides *rahn, kafalah* can also be used as a form of collateral. Etymologically, *kafalah* means *al-dhamanah, hamalah,* and *za'amah*. These three terms have a similar meaning, which is 'to guarantee'. *Dhamman* applies when the guarantee is associated with the property; *hamalah* applies when it is associated with *diyat* (fines in the *qishas* law), *za'amah* applies when it is associated with property (capital goods), and *kafalah* applies when the guarantee is associated with life. Meanwhile, according to the terminology, *Kafalah* is defined as: "a guarantee that is given by *kafiil* (insurer) to a third party for the liability that has to be fulfilled by the second party (the insured) (Juzairi, 2001). According to the perspective of Sabiq (1995), *kafalah* consists of *kafalah bi al-nafsi* and *kafalah bil al-maal*. *Kafalah bin nafsi* is a guarantee by the third party to present the second party if necessary. The property guarantee is an assurance by the third party for the first party that the third party is acquainted with the property held by the second party. Sabiq (1995) then divided the guarantee by property into three categories. First, *kafalah bi al-dayn* is the assurance that is guaranteed by the third party on the loan conducted by the second party. Second, *kafalah bi al-'Ayn* or *kafalah bi al-taslim* is the third party's guarantee to return the property held by the second party to the first party. For example, the guarantee provided to ensure the return of leased goods at the end of tenancy or to ensure the delivery of goods to the buyers. Third, *kafalah al-dark* is a guarantee to complete the work of an incorrect task; for example, one assures to conduct works on the goods that have been sold by the pledgor, the borrower, to other parties (Sabiq, 1995).

According to the elucidation of the Law No. 21 of 2008 on Islamic Banking, *kafalah* is an agreement of a guarantee by one party to another party or other parties, and the guarantor (*kafil*) is responsible for the repayment of debt, which is the right of the insured (*makful*). DSN-MUI's fatwa stated that *kafalah* is the guarantee given by the insurer (*kafil*) to the third party to meet the obligations of both parties or the obligation of the incurred (*makful anhu*). In principle, *kafalah* is similar to *dhaman*, which can be defined as the provision of guarantee as mentioned earlier. However, in its development, *kafalahi* is similar to personal guarantee, while *dhamman* has a similarity with a guarantee in the form of assets or property. Imposition of a guarantee in the *mudharabah* financing contract is an agreement which is determined by *mudharabah* financing agreement as the principal agreement.

According to the author, the condition of guarantee in the Islamic law that consists of *rahni* is identical with a guarantee in the form of a pledge, fiduciary, deed of mortgage, and mortgage. Therefore, we can conclude that the security rights, fiduciary, and pawn have been included in *rahn*. However, *Kafalah* can be defined as the agreement of *borgtocht* in the form of personal guarantee or corporate guarantee, which has been widely known in conventional insurance systems (Ayub, 2007). Technically, *kafalah* in Islamic banking is a service of guarantee for the customers, where the bank acts as a guarantor, while the customer acts as a guaranteed. The implementation of *Rahn* in Islamic banking can be conducted in two forms: as a complementary product and as a separate product. If *rahn* is viewed as a complementary product of *mudharabah* financing, the bank will hold the assets of the customers as a collateral to ensure that the customers can take the obligations, which arise from the contract (Usmani, 2002). However, if *rahn* is viewed as a separate product, the bank will receive a contract of *rahn* as a collateral for the customer's debt arising from the financing provided by the bank, and the value of the debt assurance is determined by the bank. The application of *kafalah* in an Islamic bank can be conducted when the customer provides the funds to obtain the facility of financing. Islamic banking will receive the funds in accordance with the principles of *wadi'ah*, and banks will obtain a reward for services rendered.

3 THE ANALYSIS OF THE LAW NO. 21 OF 2008 ON THE IMPOSITION AND APPLICATION OF COLLATERAL IN THE *MUDHARABAH* FINANCING CONTRACT

Given the fact that Islamic banks are institutions that put the highest priority on the principle of trust in its operations, Islamic banks must also fulfill other management principles, which include the precautionary principle (prudential principle). Prudential principles normatively can be understood through Article 23 of the Law No. 21 of 2008, which stipulates that the Islamic banks shall be careful in providing *mudharabah* financing services to the customers because the fund is obtained from public's money entrusted to them. It can be concluded that the precautionary principle can be defined as the risk management through the implementation of laws and regulations that apply consistently, and therefore, the application of the precautionary principle could directly maintain the trust given by customers to the bank.

The implementation of the precautionary principle in Article 23 of the Law of Islamic Banking asserts that before the Islamic banks disburse the loans to the customers, Islamic banks must be confident about the willingness and the ability of the prospective customers to responsibly repay all the obligations on time. In order to obtain the trust and confidence as intended, Islamic banks are required to conduct an in-depth assessment and an analysis of the character, capacity, capital, collateral, and business prospects of the prospective customers who will receive the financing service, and the analysis should stick with the principles of "the five C's of credit analysis". If the banks are confident about the ability of *mudharib* based on the result of the assessment of the character, ability, capital, and business prospects, then the primary guarantee or the credit-funded project can be used as a collateral (Anshori, 2010).

In the practice of Islamic banking in Indonesia, a collateral for *mudharabah* financing can be a necessity. The reason for legal argumentation is based on the fact that Islamic banks as the financial institutions must uphold the precautionary principle (prudential principle), and one of the elements for the implementation of such principles requires a guarantee for the financing services provided by Islamic banks. The prudential banking principle and the concerns of moral are the main reasons for the requirement of a certain additional guarantee to be provided by the customers in order to ensure that the customers will be able to perform their obligations, and it is an important factor that must be considered by the banks.

Given the fact that a guarantee constitutes one of the elements that should be provided by the customers to the bank, and if other elements have ensured the customers' ability to repay the debt, then the guarantee can only be provided in a form of a business project or the right to demand, which is funded by the financing. It should be understood that the guarantees in the *mudharabah* financing contract must be the last alternative source of repayment of the financing, which is provided by the bank in the event that the customer is unable to pay off the capital that has been loaned. Moreover, the function of a guarantee is provided to the creditors in order to maintain the security and the interests of creditors as the owners of capital, and therefore, there should be a binding agreement on the collateral, which serves as an additional agreement to the main agreement of *mudharabah* financing (Manan, 2012).

Mudharabah financing contracts provided by banks could impose the risk of payment failure or a bad debt, so in practice the banks must pay attention to the principles of financing based on the healthy Islamic principles. Given the fact that the source of funds comes from public's money, which is managed by the Islamic banks, then the risks will be able to affect the safety and credibility of the banks. The guarantee of *mudharabah* financing is quite important because it is one of the efforts to anticipate the risks that may occur within the grace period of debt repayment. Therefore, the role of financial guarantees is an absolute requirement in order to minimize the risk to the bank in the distribution of fund. In principle, the distribution of fund or finance does not always require any guarantee because the types and the prospects of the businesses owned by the debtors are basically a guarantee for the business itself. However, if the financing system is not secured by any guarantee, the provision of finance will possess a huge risk because if the customers' business fails to survive or experiences bankruptcy, then it will produce a significant loss for the bank because the distributed capital will be unable to be recovered. By contrast, when there is a guarantee, the bank will be able to withdraw the disbursed funds by utilizing the existing collateral.

The failure of the debtors to meet their obligations could occur due to the failure of their business (business risk). However, this failure could also be caused by other factors, such as changing market conditions, the increase of raw material and other production facilities, as well as poor management of the debtor, which include moral hazard or bad characters of the debtor. Banks also bear the risk of un-re-obtainable disbursed or distributed capital because "banking business is a business of risk" (Rahman, 1995). The risks must be well analyzed and considered, and the banks must conduct a survey and a proper analysis to the debtors in terms of their characters, capacity, capital, condition of economy, and collateral. In addition, the assessment of the debtors can be emphasized from the aspects of the debtor's business, which include the legal aspects, as well as the aspects of production, management, marketing, and finance (Dimyati, 2008).

In the practice of Islamic banking in Indonesia, a collateral on *mudharabah* financing is certainly a necessity. The reason for legal argumentation is based on the fact that Islamic banks as financial institutions must uphold the precautionary principle, and one of the elements for the implementation of such principles requires the guarantee for the financing services provided by Islamic banks. The prudential banking principle and the concerns of moral hazard are the main reason for the requirement of a certain additional collateral in addition to a principal collateral to be provided by the customer, which will assure the Islamic bank that the customers (*mudharib*) will be able to perform their obligations, and it is an important factor that must be considered by the bank (Mawdudi, 2011).

The perspective of *al-Maslahah* in the Islamic law is the most crucial factor in determining the legitimacy of the law against a case, especially in terms of the application of the guarantee in the *mudharabah* financing system. The aim of the application of assurance in the *mudharabah* financial system is to maintain and to manage other invested properties in order to ensure the realization of *maslahah* society. When the government considers the benefit and advantages of the Islamic banking system, the intervention as well as the implementation of the application of a guarantee on the *mudharabah* financing system aims to preserve and protect one's assets or property (Abidin, 2003). There are several specific conditions that allow the use of a guarantee in the *mudharabah* financing system, such as to save and protect somebody's asset or property from any irresponsible act. In Islamic banking, the source of funds for the provision of *mudharabah* financing is not solely-owned by the bank but from various customers, and therefore, it requires a guarantee or collateral to ensure that these funds are safe and protected.

The Decision of Fatwa of the National Sharia Board, Indonesian Council of Ulema (DSN-MUI, 2000) No. 7/DSN-MUI/IV/2000 concerning the *mudharabah* financing system stated in the seventh point that: In principle, there is no collateral in the *mudharabah* financing system. However, in order to prevent the malfeasance that might be committed by *mudharib*, Islamic Financial Institutions may request assurance or a collateral from *mudaharib* or any third parties. This collateral can only be disbursed if the *mudharib* has been proven to violate the agreement set forth in the contract.

There are several points in the argumentation on the permissibility of imposition of financial guarantees in the contract: (A). There are many service users of Islamic banks, and the capital owners/banks do not have the capability to assess the credibility and capability of the *mudharib*, and this condition is not similar to the practice of *mudharabah* and is also different from the situation in the era of the Prophet. (B). Commitment to the values of mutual trust as the basis for non-requirement of guarantee is no longer reliable in today's situation, and the condition of society has also changed. These are the reasons for inadaptable sharia law modification, and it is in accordance with the principle of sharia law: "the existence of law is determined by two or by the absence of *illat*" (*al-ahkamu yaduru ma'a illat wujudan wa 'adaman*). (C). A collateral is in the case of the risk of violation (*ta'addi*), negligence (*taqsir*), and violation of the agreement that has been negotiated (*mukhalafatu al-syuruth*).

4 CONCLUSION

Given the fact that the capital provided by an Islamic bank to the customers is actually owned by the third party, not wholly owned by the bank, then the requirement of a guarantee on the *mudharabah* financing system is permitted and justified. According to the perspective of *al-Maslahah* theory, the purpose of collateral imposition in the *mudharabah* financing contract system is to protect and preserve the property or the assets of the third parties. In accordance with the DSN-MUI's fatwa, the imposition of collateral aims to prevent any moral hazard that may be committed by the customers, and it also aims to avoid the risk of violations, negligence, and violation of an agreement that has been consolidated. In accordance with the fatwa from DSN-MUI on the *sharia* banking system, one of the purposes of collateral imposition is to implement the precautionary principle (prudential principle).

REFERENCES

Abidin, I. (2003) *Radd al-Muhtar ala al-Durr al-Mukhtar Juz 4*. Beirut, Dar al-kutub al-'Ilmiyah.
Al-Juzairi, A. (2001) *Fiqih Empat Madzhab*, translated by Umam, C. & Hurairah, A. Jakarta, Darul Ulum Press.
Al-Zuhaili, W. (2002) *Al-Fiqh Al-Islamy Wa Adillatuhu Juz 4*. Beirut, Dar al-Fikr.
Anshori, A.G. (2010) *Pembentukan Bank Syariah melalui Akuisisi dan Konversi: Pendekatan Hukum Positif dan Hukum Islam*. Yogyakarta, UII Press.
Ayub, M. (2007) *Understanding Islamic Finance*. London, John Wiley and Sons.
Burhanuddin, S. (2010) *Aspek Hukum Lembaga Keuangan Syariah*, Yogyakarta, Graha Ilmu.
Dewan Syariah Nasional Majelis Ulama Indonesia (2000) DSN-MUI: *Fatwa Nomor 7 Tahun 2000 tentang Pembiayaan Mudharabah [Qiradh]*. Jakarta.
Dewan Syariah Nasional Majelis Ulama Indonesia (2014) DSN-MUI: *Fatwa Nomor 25 Tahun 2002 tentang Rahn*.
Dimyati, A. (2008) *Teori Keuangan Islam: Rekonstruksi Metodologis Terhadap Konsep Keuangan Al-Ghazali*. Yogyakarta: UII Press.
Indonesia (2008) *Undang-undang No. 21 Tahun 2008 tentang Perbankan Syariah [The Law Number 21 of 2008 regarding Islamic Banking]*.
Manan, A. (2012). *Hukum Ekonomi Syariah dalam Perspektif Kewenangan Peradilan Agama*. Jakarta, Kencana Prenada Media Group.
Mawdudi, S.A.A. (2011) *First Principles of Islamic Economics*. Ahmad, K. (ed.). Translated by Hashemi, A.I.S. Markfield, The Islamic Foundation.
Muhammad (2008) *Paradigma, Metodologi dan Aplikasi Ekonomi Syariah*. Yogyakarta: Graha Ilmu.
Rahman, H. (1995) *Aspek-Aspek Hukum Pemberian Kredit Perbankan di Indonesia*. Bandung, Citra Aditya Bakti.
Sabiq S. (1995) *Fiqh al-Sunnah Juz 3*, Beirut, Dar al-Fikr.
Usmani, M.I.A. (2002) *Meezan Bank's Guide to Islamic Banking*. Karachi, Darul Ishaat.

Mediation as a means to provide *ishlah* (peace and harmony) in the resolution of *sharia* banking disputes in Indonesia

Wirdyaningsih
Civil Law Department, Faculty of Law, Universitas Indonesia, Depok, Indonesia

ABSTRACT: Growth in the application of *sharia* in economic activities in Indonesia has been marked by the establishment of many increasingly diverse economic institutions that are built upon these religious principles. However, the implementation of a *sharia*-based economy is not free from disputes. One means of resolution is through mediation. However, there are certain impediments to carrying out mediation in this area, including the applicability of varied regulatory provisions, the limited number of mediators, and a limited degree of implementation. The issue in this paper is whether mediation can create *ishlah* in resolving disputes in *sharia* banking and how it ought to be implemented in order to abide by the principle of benefit for the general public. The research discussed in this paper uses a normative method of analysis and practices in *sharia* banking. The recommendation offered herein is that dispute resolution through mediation in *sharia* banking requires regulations that incorporate procedures that are in line with the *ishlah* concept, mediators capable of carrying out such mediation who are deeply knowledgeable about *sharia* banking, and support from the government in building a *sharia* mediation institution inside as well as outside the courts.

1 INTRODUCTION

Currently the application of Islamic law in *muamalat* (civil and commercial interactions under Islamic principles) has its own place in the world, not only in Muslim countries but also in societies where Muslims are not the majority. Banks have two important functions in a country's economy, namely as a depository of public funds, and as a financier for the public or business community (Abdullah, 2006). In the intensive interactions that occur between banks and their customers, there is always the possibility of problems occurring which, if not promptly resolved, can lead to a dispute between the customer and the bank (Abdullah, 2006).

The *sharia* banking institution is a part of the fundamental framework of Islam related to the economy (Ali, 2007). One principle in Islamic economy is the prohibition of *riba* (interest), which must be supplanted by other schemes such as a profit-sharing arrangement (Q.S. al Baqarah (2):275 and Q.S. an Nisa (4):29). *Sharia* Banking is expected to create a healthy and fair investment climate by ensuring that the parties share the gains as well as the potential risks that occur, thus creating a balance of positions between the bank and customer (Lewis & Algaoud, 2001).

A recommended act in Islam to achieve the greatest good for all human beings (*maslaha*), which in turn is one of the ultimate purposes of Islam, is *ishlah* (amicable settlement) in resolving a dispute. Allah has commanded His people to make peace whenever there is a conflict. This is explained in the Qur'an, the *sunah*, and *ijma* (Sabiq, 1997). Allah recommends a peaceful settlement among others in Q.S. Al Hujurat (39):9. This command is further expanded beyond situations of war, and has come to also cover problems within the family and society.

Dispute resolution of *muamalah*, particularly commercial dealings using *sharia* banking, can be undertaken in a number of ways. A dispute can be addressed through litigation in court proceedings, or through non-litigation methods, such as negotiation, mediation and arbitration. The legal proceedings in litigation are often considered as failing to satisfy the mutual interest of the parties. Court rulings tend to create new problems, are slow in reaching a resolution, are costly and unresponsive, create an adversarial relationship between parties, and are often fraught with infringements in their implementation (Wirdyaningsih et al., 2007). This condition can bring great harm when applied to the commercial world, particularly banking.

Currently the Indonesian Supreme Court is hoping to reduce the courts' backlog of cases by optimising the use of mediation before a dispute reaches the judicial system. One measure is integrating mediation into the courts' proceedings. This adds to the responsibilities of the judges to act as mediators over every civil case that is filed in the court. Mediation through the courts is different from extra-judicial mediation. Within the court system, judges are already burdened by their primary responsibility, which is to examine and resolve cases that are submitted to them. Mediation would add to their already extensive responsibility. Moreover, judges are not specifically trained to act as mediators. Currently the number of sitting judges and the number of cases that are filed in the courts is out of proportion, causing mediation to be carried out merely as a formality to avoid the final judgement being null and void. Meanwhile, independent mediators serving outside the judicial system are not being effectively utilised by disputing parties. This condition has led to challenges being faced by the mediation process in courts (Wirhanuddin, 2016).

Another problem is that models of alternative dispute resolution often cannot be applied, despite having been provided for in various legislations. This situation can be seen when the parties fail to perform what has been agreed upon. Agreements made between parties facilitated by a mediator lack the power to force the parties to perform the accord. When mediation fails to effectively resolve disputes, the public tends to bring their cases to court (Wirhanuddin, 2016).

According to the Assistant Director of the Banking Mediation Department of Bank Indonesia, to resolve banking disputes outside the courts, Bank Indonesia uses a mediation process that does not segregate disputes involving conventional banks and *sharia* banks (Bank Indonesia, 2006). Consequently, mediators who are called upon to resolve *sharia* banking disputes are not required to possess special qualifications, such as being well versed in *sharia* principles. In fact, there are fundamental differences in the operating principles of *sharia* banks and conventional banks. This practice is not in line with the spirit of *sharia* economic dispute resolution as implemented in the religious courts and the National *Sharia* Arbitration Body (BASYARNAS), where a judge or arbiter must possess an understanding of *sharia* principles.

2 METHODOLOGY

Based upon the background of the issues, the author intends to analyse the proper process by which mediation in *sharia* banking can manifest *ishlah* and be implemented in accordance with the principles of public interest. In this study, the author limited the scope to non-litigation dispute resolution using mediation in *sharia* banking. To address the issues and achieve the objectives of this research, a legal normative approach has been used. The present research analyses the ability of *ishlah* to bring peace and order to society (*maslaha*) through mediation in *sharia* banking. Based on such analysis, governing provisions are recommended to be adopted by *sharia* banks that are in line with the principles of public interest.

3 RESULTS AND ANALYSIS

3.1 *Mediation and ishlah as a measure to reach amicable resolution*

Mediation and *ishlah* are two entirely different terms that originate from different root words and different processes, but impart a common meaning. Mediation comes from the English language, whereas *ishlah* has its origins in Arabic.

'Mediation is an informal process in which a neutral third party helps others resolve a dispute or plan a transaction but does not (and ordinarily does not have the power to) impose a solution' (Riskin & Westbrook, 1987). Unlike judges or arbiters, a mediator does not have the power to make a ruling on a dispute. The mediator merely assists the parties to work through the issues entrusted to him/her in order to allow them to achieve their objective to find a win-win solution. All disputes are resolved in an amicable manner, making the mediation outcome take the form of a consensus between the parties (Boulle, 2005).

Mediation consists of the following elements: (Margono, 2000)

- Mediation is a process by which disputes are resolved through negotiations.
- A mediator is involved in and is accepted by the disputing parties into the negotiations.
- The mediator has the duty to assist the disputing parties in reaching a resolution.
- A mediator does not have the power to make judgements throughout the process.
- The purpose of mediation is to reach or create an agreement mutually acceptable to the parties in order to end the dispute.

From a linguistic perspective, the root word of *ishlah* comes from *shalaha*, which means 'good', representing a change of form from decadence to goodness. *Ishlah* is the *mashdar* form of *lafazh aslaha yuslihu*, which means to correct, refine, and make peace (to end a conflict). In particular, *ishlah* is commonly used to settle a dispute occurring in society (Mubarok, 2003).

As an expression, *ishlah* can be taken as a noble act in terms of human behaviour. As such, in Islamic terminology, it can be interpreted as an undertaking which intends to make a change from bad to good. Meanwhile, according to *fuqaha* (Islamic scholars and experts on Islamic jurisprudence), the term *ishlah* can mean settlement, which is an agreement drawn up to conclude a dispute between people, either individuals or groups (Mubarok, 2003).

As-Shulh or settlement through amicable negotiation is one of the fundamental principles of society that must be implemented, as it is based on the principles of morality in Islam, and even the Prophet Muhammad was commanded by Allah to better the morality of mankind (Q.S. al Ahzab (33):21).

It can be concluded that *ishlah* and mediation have a common purpose, which is to reach an agreement to resolve a dispute in an amicable manner. The difference lies in the process. *Ishlah* can be performed using a number of methods, either through amicable discussion between the parties or with the help of a third party acting as a facilitator or mediator. Mediation, on the other hand, is done by bringing in an impartial third party to facilitate efforts by the parties to resolve the dispute and reach a mutually beneficial solution.

Since the birth of Islam up to the present, amicable negotiations or *musyawarah* have been well institutionalised. *Musyawarah* is often conducted to decide upon important matters, such as the economy, law, international politics, selection of a war commander, and other such matters. Achievement of settlement or *shulh* happens through amicable negotiations between parties having a title over a property or the sincere willingness of the parties to achieve peace as the ultimate purpose of Islamic law. Many experts on Islamic law view that *musyawarah* is a principle of Islam, the practice of which has been commanded by Allah upon mankind. *Musyawarah* even forms the foundation on which a state is established, and therefore Islam mandates that *musyawarah* be performed on all matters and can also be seen from the title of one of the *surah*, namely '*surah As-Syura*' (Aliyah, 2004).

As-Shulh constitutes one of the basic principles of human life that must be practised, as *musyawarah* is based on the foundation of Islam and morality as taught by Islam, and the Prophet Muhammad was commanded to improve morals (Q.S. al Ahzab (33):21). The system and form of *musyawarah* are not provided in detail in Islamic law, and as such it depends on the people of that faith to practise it. Given the varying characters and cultures that exist within society, *musyawarah* to achieve the good of mankind must be adapted to suit the current time and local culture, as long as it does not go against the prescribed *nash*.

In Indonesian society, *musyawarah* has become a national culture and is a *basic norm* adopted by the people (Sulistiyono, 2002). It was affirmed by the nation's founding fathers as the people's philosophy and incorporated in the fourth *sila* or principle of the country's five underlying principles (Pancasila): 'Democracy guided by the inner wisdom in the unanimity arising out of deliberations among representatives'. Therefore, the government is called upon to promote dispute resolution through amicable negotiations or *musyawarah*, as it fully reflects the nation's underlying philosophy. One measure that needs to be taken by the government is to formulate laws to govern the resolution of disputes by amicable means. The government should encourage parties in dispute, particularly commercial disputes, to utilise dispute resolution methods without going through the justice system.

3.2 *Settlement through mediation in sharia banking*

3.2.1 *Regulation*

The presence of mediation in Indonesia was affirmed with the issuance of the Supreme Court Regulation No. 1/2016. *Endeavours* must be made to resolve all civil cases filed *in* the court of first instance through mediation. *Sharia* banking mediation in indonesia can be *conducted* through two mechanisms, namely extra-judicial mediation and mediation within *the* court

Law No. 3/2009 on the Religious Court, specifically its Article 49, expands the powers of the Religious Court to also cover *sharia* economic disputes, including *sharia* banking matters, making its prospects very promising. As such, the Religious Court, as one of the four institutions with judicial power that can hear *sharia* economic disputes, must take into account the following: (1) Religious Court justice that possesses the necessary knowledge, skills, and integrity; (2) continual development and promotion of *sharia* economic laws, and (3) positive perception of the people on Islamic laws (Ahmad et al., 1996).

3.2.2 *Mediation in the religious court*

Dispute resolution through the court system has the advantage of being able to produce an outcome that is legally binding, albeit involving a circuitous and lengthy process that requires a considerable amount of time, money and energy to be spent by the parties (Amriani, 2011).

The benefit of mediation within the court is that the parties are able to determine the resolution process and thus potentially make it simpler, lower cost and suited to the desires of the parties. The final award, however, does not carry a strong legal binding power (Margono, 2000). The combined method of dispute resolution through court mediation is an interesting process as it tries to bring together the strong suits of each method (Usman, 2012).

The Supreme Court Regulation 1/2016 accommodates mediation prior to litigation, during litigation, at the appeal stages, cassation, and court decision review (Articles 33 and 34) as well as extra-judicial mediation (Article 36). At every stage of the litigation process, the judges are required to encourage settlement prior to the passing of judgement. Failure to observe the mediation process under the regulation is deemed to be a violation of Article 130 HIR/Article 154 RBg, causing the judgement to become null and void.

When the parties to a dispute agree to initiate mediation during litigation and to appoint one of the judges as a mediator, the chairperson of the panel of judges will give the opportunity to the parties to start mediation during the litigation process. If the parties can be reconciled by the mediator, they will have to enter into a written settlement agreement co-signed by the mediator. If the mediation process is represented by the parties' legal counsels, a written agreement must be drawn up, which can be corroborated by a deed of reconciliation (Article 17).

Based upon research conducted by Indonesia's Institute for Conflict Transformation (IICT) in 2003, it was found that efforts to implement mediation have not been optimal. There were only 6,518 cases mediated of the 20,717 filed (31.46%). In 2012, 6,811 cases out of the 20,643 cases filed were mediated (32.99%).

According to IICT, measures to popularise mediation through the court institution taken by the Indonesian Supreme Court have not provided an effective outcome. Some of the factors that impede mediation in the courts are as follows:

- Not all justices have received training to act as mediators, and thus understanding of the process is unequal among the justices.
- The number of judges in some regions is limited, thus necessitating them to concentrate on hearing disputes through litigation.
- Lack of knowledge on the part of the disputing parties regarding mediation.
- Legal counsels that do not encourage parties to choose mediation as a means to resolve their dispute.
- Some courts do not respond positively to non-judges who wish to register as mediators. Even registered non-judge mediators are seldom chosen by disputing parties as they prefer to appoint a judge as their mediator who provides service without charging the parties.

3.2.3 *Extra-judicial mediation*

Mediation outside a court of law can be conducted by independent mediators or by Bank Indonesia. According to Bank Indonesia Regulation No. 8/5/PBI/2006, banking mediation is an alternative means to resolve disputes between customers and banks who are unable to come to an agreement. The process involves a mediator to assist the disputing parties to reach a resolution in the form of a mutually accepted agreement with regard to the entire or part of the matter in dispute (Regulation 8/5/PBI/2006, Article 1 sub-articles 5 and 6).

The Deputy Commissioner of Consumer Education and Protection, explains that the Indonesian Financial Services Authority [Otoritas Jasa Keuangan [OJK], 2014] has formulated a mechanism to resolve disputes in the financial service sector:

- internal dispute resolution (resolution disputes through the financial service institution);
- external dispute resolution (resolution of disputes through the court or outside the court).

The OJK has established the Alternative Dispute Resolution Agency (LAPS) for the financial sector. A similar body was also established for the capital market, pension funds, and insurance sectors in 2014, and another for banking, financing, insurance and pawning in 2015. Currently banking mediation is still managed by Bank Indonesia, and thus dispute resolution in banking matters still refers to the relevant Bank Indonesia regulation.

With regard to resolving complaints made against a bank, OJK Regulation No.1/2013 requires every bank to have a specialised unit established specifically to handle and resolve, free of charge, any complaints made by its customers. A complaint must be based upon factual/potential financial losses suffered by the consumer as a result of the bank's misconduct or negligence.

Supreme Court Regulation No. 1 of 2016 also mentions extra-judicial mediation that can lead to an agreement. Article 36 of the regulation provides for the legal procedure to prepare a deed of resolution at the court of first instance based on an agreement resolving a dispute or based on the result of an extra-judicial mediation. The procedure starts with the filing of a claim accompanied by reconciliation agreement or document resulting from negotiations between the parties under mediation or facilitation of a certified mediator. The agreement to reconcile can be filed in the form of a claim in order to obtain a deed of reconciliation from the competent court. The claim shall be filed by the aggrieved party in the dispute.

The requirement to file a claim in order to obtain a deed setting forth a reconciliation agreement reached outside the court may be considered anomalous: a dispute that has been resolved amicably by the parties, and yet one of the parties has to file a claim against the others to the court. Given the fact that not every person who has bound himself to another person under an agreement is willing to abide by such contract, this provision actually has a distinct rationale. In Indonesia, the courts are bound by procedural regulations under the applicable legal system, where they can only act based upon a claim or suit filed with respect to a dispute or a petition for matters other than disputes.

The result of mediation is essentially an agreement reached voluntarily by the parties. If the outcome of the mediation is in the form of an agreement, it can be enforced in accordance with contract laws. In Indonesia, as is the case in certain other countries, the result of mediation must be registered with the court and enforcement can be petitioned to such court. This is provided under Article 6 paragraph (7) of Law No. 30/1999.

As such, certified mediators are needed to enhance the quality of extra-judicial mediation. Individuals who have received certification following mediation training would acquire knowledge and skills as a mediator, and therefore would not be inclined to change a mediation process, which is essentially a process leading to a consensus driven by the autonomy of the parties into a process of adjudication, such as arbitration.

Article 6, Chapter II of Law No. 30/1999 clearly states that mediation relies on the good faith of the parties, and the outcome depends on their intention. There is no punitive consequence faced by a party if it decides to set aside the agreement, other than the possibility of a suit being filed by the interested party.

Specifically with regard to mediation, however, the implementation of which is encouraged by regulators through the relevant legislations, a slight exception is made. In the case of mediation, an element of compulsion is brought upon by regulators for institutional parties, specifically in the form of an obligation to perform the agreement and the threat of (administrative) sanction in the case of failure to do so. An example of such requirement is contained in the Bank Indonesia Regulation No. 8/5/PBI/2006, Articles 13 and 16.

Additionally, if a party breaches the mediation agreement reached out of court and for which no deed of reconciliation has been sought from the court, the legal action that can be taken is to file a charge of breach of contract. The legal status of a settlement agreement without a deed of reconciliation from the court is equivalent to that of a private agreement between the parties.

4 CONCLUSION AND RECOMMENDATIONS

History has shown that *ishlah* is able to realise the good of mankind, particularly in Indonesia, which has a life philosophy of deliberation in strengthening the dispute resolution process in a way to make peace. In order for mediation in *sharia* banking to be able to achieve *ishlah* and be implemented in accordance with the principle of the good of mankind, the following needs to be realised. The government needs to formulate regulator instruments at the level of law on mediation. Indonesia's population consists predominantly of Muslims who uphold the principle of *ishlah* (peace) in dispute resolutions. Moreover, Indonesians highly regard the principle of *ishlah* in their life's philosophy. Such legislation must accommodate the needs of the disputing parties. Where a dispute requires specialised regulatory provisions, the government should have a regulatory body to formulate the necessary operating regulations. A mediation process needs an individual with mediating skills as evidenced by the relevant certification. A mediator who settles a *sharia* economic dispute has to know the principles of *sharia* banking. This is to ensure that the agreement prepared by the parties does not conflict with *sharia* principles, such as the prohibition on *riba* (interest), speculation and risk-taking. The integration of mediation into the court process could become an effective means of reducing the courts' case backlogs and enhance the function of the judicial institution in resolving disputes, complementing the judicial process which is more adjudicative in nature. The combination of the two methods of dispute resolution is expected to be able to address their respective shortcomings by bringing in their advantages.

REFERENCES

Abdullah, B. (2006). *Jalan Menuju Stabilitas: Mencapai Pembangunan Ekonomi Berkelanjutan [Pathway to Stability: Achieving Sustainable Economic Development]*. Jakarta, Pustaka LP3ES Indonesia.

Ahmad, A., Munji, M.S., Djazuli, M., Kamil, A. & Hakiem, L. (1996). *Dimensi Hukum Islam dalam Sistem Hukum Nasional: Mengenang 65 tahun Prof. Dr. H. Busthanul Arifin, S.H.; Pemikiran dan Peranannya dalam Pelembagaan Hukum Islam [Islamic Law Dimension in National Legal System: In*

Memoriam of 65 Years Prof. Dr. H. Dusthanul Arifin, S.H.,; His Thoughts and Roles in Islamic Law Institutionalization]. W. Bambang (ed.) Jakarta: Gema Insani Press.

Ali, M.D. (2007). *Hukum Islam: Pengantar Ilmu Hukum dan Tata Hukum Islam di Indonesia [Islamic Law: Introduction to Legal Studies and Islamic Legal System in Indonesia]* (6th ed.). Jakarta: Raja Grafindo Persada.

Aliyah, S. (2004). *Sistem Pemerintahan, Peradilan & Adat Dalam Islam.* Jakarta [Governance System, Court and Customary Law in Islam]: Pustaka Al Kautsar.

Amriani, N. (2011). *Mediasi: Alternatif Penyelesaian Sengketa Perdata di Pengadilan [Mediation: Alternative Civil Dispute Resolution in Court].* Jakarta: Raja Grafindo Persada.

Boulle, L. (2005). *Mediation: Principles Process Practice* (2nd ed.). Chatswood: LexisNexis Butterworths.

Indonesia (1999). *Undang-undang No. 30 Tahun 1999 tentang Arbitrase dan Alternatif Penyelesaian Sengketa* [*The Law Number 30 of 1999 regarding Arbitration and Alternative Dispute Resolution*].

Indonesia (2006). *Peraturan Bank Indonesia No. 8/5/PBI/2006 tentang Mediasi Perbankan* [*Bank Indonesia Regulation No. 8/5/PBI/2006 regarding Banking Mediation*].

Indonesia (2006). *Undang-undang No. 3 Tahun 2006 tentang Perubahan atas Undang-undang No. 7 Tahun 1989 tentang Peradilan Agama* [*The Law Number 3 of 2006 regarding Amendment to The Law Number 7 of 1989 regarding Religious Court*].

Indonesia (2013). *Peraturan Otoritas Jasa Keuangan No. 1 Tahun 2013 tentang Perlindungan Konsumen Sektor Jasa Keuangan* [*Financial Service Authority Regulation Number 1 of 2013 regarding Consumer Protection in the Financial Service Sector*].

Indonesia (2016). *Peraturan Mahkamah Agung Republik Indonesia No. 1 Tahun 2016 tentang Prosedur Mediasi di Pengadilan* [*Supreme Court Republic of Indonesia Regulation No. 1 Year 2016 regarding Mediation Procedures in Court*].

Lewis, M.K. & Algaoud, L.M. (2001). *Perbankan Syariah: Prinsip, Praktik, dan Prospek* [*Islamic Banking*], Translated by B. Wirasburata. Jakarta: Serambi Ilmu Semesta.

Margono, S. (2000). *ADR (Alternative Dispute Resolution) & Arbiterase: Proses Pelembagaan dan Aspek Hukum [Alternative Dispute Resolution and Arbitration: Institutional Process and Legal Aspect].* Jakarta: Ghalia Indonesia.

Mubarok, A. (2003). Rekonsiliasi: Perspektif Al Qur'an [Reconciliation: The Holy Quran Perspective]. In M. Iqbal (ed.). *Islam dan Perdamaian.* Jakarta: Progress.

Otoritas Jasa Keuangan No. 1/2013.

Riskin, L.L. & Westbrook, J.E. (1987). *Dispute Resolution and Lawyer.* St. Paul: West Publishing Company.

Sabiq, S. (1997). *Fiqh Sunnah Jilid 13.* Translated by K.A. Marzuki, et al. Bandung: PT. Al-Ma'arif.

Sulistiyono, A. (2002). Mengembangkan Paradigma Non-Litigasi Dalam Rangka Mendayagunakan Alternatif Penyelesaian Sengketa Bisnis dan Hak Kekayaan Intelektual di Indonesia [Developing Non Litigation Paradigm to Utilize Alternative Dispute Resolution in Business and Intellectual Property Right in Indonesia]. Disertasi Doktor. Semarang, Universitas Diponegoro.

Usman, R. (2012). *Mediasi di Pengadilan: Dalam Teori dan Praktik [Mediation in Court: Teory and Practice].* Jakarta: Sinar Grafika.

Wirdyaningsih, Perwataatmadja, K., Dewi, G. & Barlinti, Y.S. (2007). *Bank dan Asuransi Islam di Indonesia [Islamic Bank and Insurance in Indonesia].* Jakarta: Kencana Prenadamedia Group & FHUI.

Wirhanuddin. (2016). Deskripsi Tentang Mediasi di Pengadilan Tinggi Agama Makassar: Perspektif Hukum Islam [Description on Mediation in Religious High Court of Makassar: Islamic Law Perspective]. *Al-Fikr, 20*(2): 279–303.

Waqf banks under the Indonesian legal system

A.K. Munthe & F. Prihatini
Faculty of Law, Universitas Indonesia, Depok, Indonesia

ABSTRACT: This study analyses the possibility of the establishment of a *waqf* bank. Banks taking on the role of a *nazhir* in the legal system would not be possible as the *waqf* banks are merely a *nazhir* partner (namely: custodian) in the management of *waqf*. This study, with the use of normative methodology, asks what would be the rationale of potential opponents and supporters of the establishment of *waqf* banks in the Indonesian law system. This study found that Islamic banks cannot act as *nazhir*, because they have different purposes. Banks are ultimately commercial institutions, and *nazhir* is a non-profit social institution. This is stipulated in Government Regulation No. 42/2006 paragraph (3) letter 'c' and Government Regulation No. 42/2006, which also defines the role of Islamic banks merely as custodian. *Waqf* banks as banks that manage cash *waqf* can be established through a number of alternative ways, namely: the establishment of a new bank, acquisition, conversion or *waqf* window in Islamic banking. In the establishment of a *waqf* bank, cash *waqf nazhir* can act as a shareholder of the *waqf* bank, and as a continuation of the *waqif* **(the donor)** mandate provision. Therefore, the amendment to the *Waqf* Act and its implementing regulations and special laws for the establishment of a *waqf* bank are required.

1 INTRODUCTION

Creating a rule of law and economic equality is of paramount importance for achieving the welfare of society (Rokan, 2014). There are many sources of capital that can be utilised to develop the national economy, for example, funds from financial institutions including non-bank financial institutions, or funds collected from governmental organisations. Among the funds collected from the public is *waqf*. *Waqf*, when utilised optimally, can effectively reduce poverty. Historically, *waqf* has been instrumental in the development of economic, social, and cultural activities of Muslim communities (Hasanah, 2011).

Waqf is an act of handing over a property that can be exploited and utilised by the custodian to distribute the profits to those who deserve it (Al-Syarbini, 1994). In this sense, *waqf* has elements of: (1) the person who acts to give up property, (2) the use of property, (3) the static property, and (4) beneficiaries of *waqf* property. Law No. 41 of 2004 on *Waqf* (the *Waqf* Act) explains that the *waqf* is: 'Legal actions of *wakif* to separate and/or hand over part of their wealth either permanently or for a specified period in accordance with their interests for purposes of worship and/or general welfare according to *sharia*'.

Management of cash *waqf* in Indonesia is carried out by the Islamic Financial Institutions (IFI) such as Islamic banking. The problem is that the Islamic banks act merely as a custodian in the management of *waqf*. As a result, the development of cash *waqf* in Indonesia is slow. In a study conducted by Ubaid (2014), it was found that the development of cash *waqf* managed by the Indonesian *Waqf* Board [Badan Wakaf Indonesia [BWI]] is not optimal, because currently the cash *waqf* collected is invested only towards *wadhi'ah* (Ubaid, 2014).

Therefore, more progressive management of cash *waqf* is needed, for example in the form of *waqf* banks. Economically, the existence of *waqf* banks is needed to reduce the economic costs charged to the lender. The advantages of these arrangements can be used for both business and public interests. It would allow the *waqf* bank to lend to those in need of capital without burdening the borrower excessively. The issue is how to establish *waqf* banks within the laws and regulations in Indonesia?

This study uses a normative juridical research method and a conceptual approach. It aims to map the legislation related to the establishment of bank endowment in Indonesia, by looking at the liberties and constraints of the relevant legislation in Indonesia.

1.1 Problems of legislation in Indonesia related to the establishment of waqf banks

1.1.1 Legislation on waqfs

Management of *waqf* is an inseparable part of the *waqf* system. *Nazhir* acts as a fiduciary for maintaining and managing to keep the *waqf* property stable, productive and beneficial. To that end, the Act of *Waqf* Article 9 lists parties that can become *nazhir*, namely: an individual, organisation or legal entity. Independent candidates can become *nazhir* when appointed by *waqif*. The *Waqf* Act on *Nazhir* Article 4 requires the individual to be a citizen of Indonesia, Muslim, adult, *amanah*, able physically and spiritually and not hindered by illegal actions. In addition, individual *nazhir* should constitute a group of at least three people, one of whom acts as chairman in order to avoid abuse of authority and support the division of tasks.

The *nazhir* organisation can have two forms, namely profit organisations and non-profit organisations. Law No. 17 of 2013 on Community Organisation Article 1 paragraph 1 states that the definition of community organisation is 'an organisation founded and formed by people voluntarily based on common aspirations, wishes, needs, interests, activities, and goals to participate in development in order to achieve the purpose of the Unitary Republic of Indonesia based on *Pancasila*.' The organisation can become *nazhir* when complying with the requirements stipulated for the individual *nazhir*. Another requirement that must be met is that the organisation is engaged in social, educational, social and/or religious activities as set forth in the statutes and by-laws [*AD/ART*] of the organisation.

The third is *nazhir* legal entity. To be *nazhir*, a legal entity *nazhir* must meet the conditions stipulated in the terms outlined for *nazhir* individuals. Coupled with that, it must also meet the following requirements: (1) Indonesia legal entity established in accordance with the legislation in force, and (2) the legal entity is engaged in social, educational, social and/or religious activities.

There is no sufficient explanation in differentiating between an organisation and legal entities. Purwosutjipto (2007) argues such an institution can be categorised as a legal entity when recognised by the state as a legal entity. Such recognition may be in the form of ratification from the Ministry of Law and Human Rights. In addition, the material requirements are as follows:

- There is wealth (rights) with specific purpose that can be separated from the personal wealth of the founder of the agency or the allies.
- The objective of a legal entity is to gain common interest.
- There are some people acted as the trustees of the agency (Khairandy, 2009).

If the above requiremments are connected to the current conditions, there is no difference between a legal entity and an organisation. Related to *nazhir*, a cash *waqf* is different from other *waqfs*. The *Waqf* Act Article 28 governing the management of cash *waqf* states '*waqif* can submit his movable *waqf* property in the form of money through an IFI designated by the minister.'

In the *Waqf* Act, the Islamic Financial Institution [IFI] acts as the party receiving cash *waqf* and issues a certificate of cash *waqf*. As for the procedures stipulated in the law of cash *waqf*, the *Waqf* Act Article 28 jo Government Regulation [PP] No. 42 Year 2006 Article 24 paragraph (1) states that the IFI that is entitled to receive cash *waqf* is the Lembaga Keuangan Syariah [LKS] appointed by the minister on the recommendation of BWI.

When referring to IFI duty as stipulated in the act, the IFI authority only plays the role as the party receiving cash *waqf* as described in the letter "c". However, receiving the waqf is only one part of the *nazhir* duties, while the other functions are handed over to *nazhir*.

Viewed from the legislation on *waqf*s, there are four factors supporting the establishment of *waqf* banks in Indonesia. First, the receipt of cash *waqf* as one property of *waqf*. Prior to the Act of *Waqf*, cash *waqf* was debated. With the *Waqf* Law, legal certainty on the permissibility of the cash *waqf* has been stipulated by law. Second, the permissibility of a legal entity acting as *nazhir* is an opportunity to establish a *waqf* bank. Legal entities could be in the form of a limited liability company (Perusahaan Terbatas [PT]), cooperative, foundation or other forms. Third, there is law and order to carry out the management of *waqf* property productively. The institutions that are best suited to do that are financial institutions, in this case the *waqf* bank.

1.1.2 Legislation on islamic banking

Islamic banking as a bank engaged in Islamic finance is regulated in Law Number 21 of 2008 concerning *sharia* banking. The Islamic Banking Act regulates the position of banks as institutions that receive and manage cash *waqf*. The Islamic Banking Law as *lex specialist* lacks specificity compared with other business entities law, especially regarding conventional banks.

The main function of Islamic banks is as intermediary or intermediaries. This function is set in the Islamic Banking Act Article 1 paragraph 2, which explains that Islamic banks are business entities that collect funds from the public and distribute it to the public. In carrying out these functions, Article 1 Number 20 explains *akad wadi'ah* or other *aqd* which is not in conflict with *Sharia Principles* in the form of demand deposits, savings or other forms.

Therefore, Islamic banks can act as a fundraiser in the form of the cash *waqf* but not as *nazhir*. The *Sharia* Banking Act raises funds not only for commercial purposes but also for social purposes, such as *wadhi'ah*, grants, gifts and others, which are not contrary to Islamic banks. IFI who can raise cash *waqf* should have cooperation with *nazhir*, as regulated in Article 5 (1) in the Government Regulation and BWI Regulation No. 1 of 2009 on Guidelines for Management and Development of Cash *Waqf*.

Another function that is owned by the Islamic banking but not by the conventional banks is a social function. The Islamic Banking Act Article 4 paragraphs (2) and (3) describes the bank as a social institution. Both of the above verses distinguish social fund management of *waqf* with social funds other than *waqf*. Besides, the management of *waqf* fund as stated in paragraph (2) puts the Islamic bank like a treasury house (*Bayt al-Mal*). The bank in the management of *waqf* as described in Article 4 paragraph (3) is obliged to channel funds collected to *waqf nazhir*. Therefore, the position of the bank in this case is only as a party that receives deposits. Thus, in this case, the bank does not perform management, until prompted by *nazhir*.

Interested limited liability company (PT) as a legal entity in this case cannot act as *nazhir* because it has a different orientation. The orientation of the PT is to gain as much as possible. Banks in the structure of law are *lex specialist*. There are so many rules governing the establishment of more stringent banks compared with regular PT. *Nazhir's* primary purpose is for social activities, whereas PT's primary goal is profit. The purpose of *waqf* is a combination of these two objectives: to be a productive *waqf* property, while also benefiting the wider community, in accordance with the designation of the *waqf*.

1.2 Alternative establishment of waqf banks

There are four alternatives to establish *waqf* banks: establishing a new bank, acquisition, conversion and window *waqf* in Islamic banks.

1.2.1 Establishing new banks

Nazhir can establish a new *waqf* bank. The *nazhir* in this case is acting as a shareholder of PT *waqf bank*. Related to this, there are two issues, namely the ownership by *waqf nazhir* property and the issue of shares/capital in the establishment of *waqf* banks. Malikiah scholars believe that timed *waqf* can be done to properties except mosque. To support their opinion, Malikiah scholars offered two arguments: first, based on the understanding of Hadith it 'substantially resists alms as his benefit' (Al-Zuhaili, 2002), so *waqf* property still belongs to *wakif*, because of the wherewithal of his benefit. The second reason is *qiyas*. *Waqf* is considered similar to *al-Hijr* (guardianship or trusteeship) because the owner of property is incompetent law. Treasure managed by guardians may not be sold, assigned or inherited.

Ownership in Islam is restricted by moral responsibility (Kementerian Agama, 2006). Treasure being *waqf* has a legal effect, and such ownership is returned to God (Kementerian Agama, 2006). *Nazhir* is not the owner but the person entrusted by *waqif* to manage *waqf* property.

In the establishment of a *waqf* bank, the *nazhir* becomes the owner of the *waqf* bank's shares. However, for the *nazhir*'s legal entity or organisation, it must pay attention to its AD/ART. In AD/ART, a *nazhir* organisation or a legal entity of *nazhir* might include establishing financial business, such as banking. Dividends will be distributed to shareholders, and *nazhir* will distribute it to *mauquf'alaih*.

The problem is the status of capital ownership by *nazhir*. As explained above, *nazhir* is not an owner of *waqf* property, but he only has a mandate to manage it. There are similarities with the trust fund in the common law system. Trust is a concept of separation of ownership between the legal owner and beneficiary owner. Nevertheless, there are differences between *waqf* and trust. In the concept of trust, property is turned over to the trustee, because basically trust is not giving the property permanently. Second, it can be changed or cancelled. Third, the motive may not be a form of obedience to religion (Mannan 1999 & Islam, 2015).

The concept of ownership in Indonesia does not recognise trusts. The influence of the concept of trust generated two ownerships which are legal ownership (owner listed by law) and beneficial ownership (those who enjoy the economic benefits of property) (Widjaja, 2008). In the system in Indonesia, a contractual agreement is an agreement containing the accomplishments between only two parties. No third party can enter into an agreement in the Indonesian system (Widjaja, 2008).

Nazhir is a party that accepts the *waqf* to be managed and distributed to *mauquf'alaih*. In the *Waqf* Act, the ownership of *waqf* property is divided into two forms, namely the release of property and handover. The release means that the property is not part of the rights of the *waqif*, whereas hand over is granting the authority to *nazhir* to manage the *waqf*. If within the specified time the treasure of the *waqif* will be retracted, *nazhir* should return the *waqf* property intact. Figure 1 shows the relation between *nazhir*, *Waqif* and *Mauquf'alaih* in a *waqf* bank.

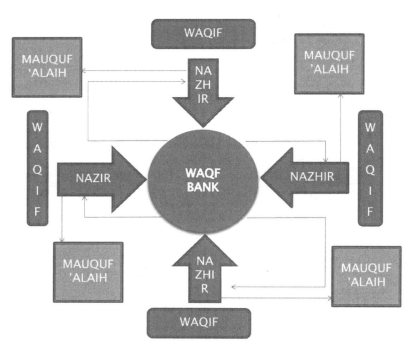

Figure 1. The relation between *nazhir*, *Waqif* and *Mauquf'alaih* in a *waqf* bank.

Waqf property ownership in the *Waqf* Act is not set. When compared with the concept of Islamic law, transfering the property of *waqf* means to restore its ownership to Allah. In the system of common law countries, the trustee has freedom to manage his property. In Indonesia, this concept is known as a trustee in financial institutions of capital markets. With this model *nazhir waqf* is a shareholder. The power of ownership of the bank is at *nazhir*.

1.2.2 Acquisition

Acquisition in company law uses the term 'takeover' while the Banking Act uses the word 'acquisition'. According to Law Number 40 of 2007 on the Limited Liability Company (Company Law) Article 1 paragraph 11, acquisition is a legal action performed by legal entities or individual persons to take over the company's shares which result in the shift of control over the company. The Islamic Banking Law Article 1 Number 31 explains that the definition of the acquisition is a legal act carried out by legal entities or individuals to take over the shares, which results in the shift of control over the bank.

This model can be used as an alternative to both the establishment of the *waqf* bank. However, the thing to note is that the acquisition of ownership may result in the transfer of control of the bank. It is stipulated in the Decree of Directors of Bank Indonesia Number 32/51/KEP/DIR dated 14 May 1999 which states that the control is only realised if at least 51% of the shares of the company, which is the target, is taken over. Therefore, if less than 51% is taken over, a *waqf* bank cannot be established (Bank Indonesia, 1999).

1.2.3 Conversion

Conversion is a change in status of a bank into another bank. In Indonesia, the change happens with conventional banks changing to Islamic banks, not vice versa. Conversion of conventional banks into Islamic banks is regulated in Bank Indonesia [BI] Regulation Number 8/3/PBI/2006 concerning Conversion of Business of Conventional Commercial Banks to Commercial Banks Conducting Business based on *Sharia* Principles by Conventional Banks as amended by BI Regulation Number 9/7/PBI/2007, which has been amended to the issuance of Regulation No. 11/15/PBI/2009 concerning Conversion of Business Activities from Commercial Bank to *Sharia* Bank (Anshori, 2010).

This alternative can be chosen in the establishment of a *waqf* bank. Banks that are considered financially potential can be converted from conventional banks into *waqf* banks or from Islamic banks into *waqf* banks. This alternative can only happen if it is strongly desired and there is a high sense of social responsibility.

1.2.4 Windows waqf in islamic banking

In the Law on Islamic Banking Article 1 Number 10, the reference to *sharia* states that the 'unit of the central office of a Conventional Commercial Bank serves as the head office of the office or unit conducting business based on *sharia*, or a unit of a branch office from a bank domiciled abroad conducting conventional business that serves as the head office of *sharia* branch offices and/or *sharia* units'.

Sharia Business Unit (UUS) established by a Conventional Commercial Bank must obtain an operating licence from Bank Indonesia. Amount of founding capital is set in the Regulation No. 11/10/PBI/2009 Article 4. Assets are established and maintained of at least Rp. 100,000,000,000.00 (one hundred billion rupiahs). As a start-up, this alternative can be used as an option to establish a *waqf* bank. It requires cooperation with Islamic banks in the form of capital relations. When the capital adequacy ratio has been met, then the unit of *waqf* can become its own *waqf* bank.

2 CONCLUSION AND RECOMMENDATIONS

The establishment of the *waqf* bank in the system of laws and regulations in Indonesia has opportunities and constraints. There are two factors inhibiting the establishment of a *waqf* bank. First, the position of banks in PP 42 of 2006 Article 23 as a custodian, which means the bank cannot be allowed as *nazhir*. Second, *waqf nazhir* as stipulated in Article 10 paragraph (3) letter c requires that the agency as *nazhir* be engaged in social, educational, social and/or religious activities.

Meanwhile, the factors that support the establishment of *waqf* banks are: first, the legalization of cash *waqf* as one of *waqf* properties that is regulated in the *Waqf* Act; second, the permissibility of legal entities as *nazhir* could be an opportunity to establish *waqf* banks; third, *Waqf* Act orders to productively manage *waqf* property.

In order to establish a *waqf* bank, there are changes required for the provisions so as not to hamper the establishment of *waqf* banks. However, to fall within the requirements of a *waqf* bank, financial institutions require new, more powerful special rules related to the establishment of this *waqf* bank. In addition, it needs strict rules on the ownership of *waqf* property and its relationship with the *nazhir*. We argue that the concept of trusts needs to be adopted into the laws and regulations in Indonesia.

REFERENCES

Al-Syarbini, M.K. (1994). *Mughni al-Muhtaj Ila Ma'rifati Ma'ani al-Fadz al-Manhaj, Juz 2*, Beirut: Dar Al-Kutub Al-Ilmiyah.

Al-Zuhaili, W. (2002). *Al-Fiqh Al-Islamy Wa Adillatuhu Juz 4*. Beirut: Dar al-Fikr. [Islamic Jurisprudence and its Argumentation].

Anshori, A.G. (2010). *Pembentukan Bank Syariah melalui Akuisisi dan Konversi: Pendekatan Hukum Positif dan Hukum Islam*. Yogyakarta: UII Press. [Development of Sharia Banks through Acquisition and Conversion: Positive Law and Islamic Law Approach].

Bank Indonesia (1999). BI: *The Decree of Directors of Bank Indonesia Number 32/51/KEP/DIR/1999 Dated 14 May 1999 regarding the Requirements and Merger, Consolidation and Acquisition Procedures of Commercial Banks*.

Hasanah, U. (2011). Cash Waqf Potential to Develop Public Housing. *Jurnal Al-Awqaf*, Special Edition, 40–50.

Indonesia (2004). *Undang-undang No. 41 Tahun 2004 tentang Wakaf* [The Law Number 41 of 2004 regarding *Waqf*].

Indonesia (2006). *Peraturan Bank Indonesia No. 8/3/2006 tentang Perubahan Kegiatan Usaha Bank Umum Konvensional Menjadi Bank Umum Yang Melaksanakan Kegiatan Usaha Berdasarkan Prinsip Syariah Dan Pembukaan Kantor Bank Yang Melaksanakan Kegiatan Usaha Berdasarkan Prinsip Syariah Oleh Bank Umum Konvensional* [Bank Indonesia Regulation Number 8/3/2006 regarding the Conversion of Business of Conventional Commercial Banks into Commercial Banks Conducting Business based on Sharia Principles and the Opening of Bank Offices Conducting Business based on *Sharia* Principles].

Indonesia (2006). *Peraturan Pemerintah No. 42 Tahun 2006 tentang Pelaksanaan Undang-undang No. 41 Tahun 2004 tentang Wakaf* [Government Regulation Number 42 Year 2006 regarding the Implementation of The Law Number 41 of 2004 regarding *Waqf*].

Indonesia (2007). *Peraturan Bank Indonesia No. 9/7/2007 tentang Perubahan Peraturan Bank Indonesia No. 8/3/2006 tentang Perubahan Kegiatan Usaha Bank Umum Konvensional Menjadi Bank Umum Yang Melaksanakan Kegiatan Usaha Berdasarkan Prinsip Syariah Dan Pembukaan Kantor Bank Yang Melaksanakan Kegiatan Usaha Berdasarkan Prinsip Syariah Oleh Bank Umum Konvensional* [Bank Indonesia Regulation Number 9/7/2007 regarding Amendment to Bank Indonesia Regulation Number 8/3/2006 regarding the Conversion of Business of Conventional Commercial Banks into Commercial Banks Conducting Business based on Sharia Principles and the Opening of Bank Offices Conducting Business based on *Sharia* Principles].

Indonesia (2007). *Undang-undang No. 40 Tahun 2007 tentang Perseroan Terbatas* [The Law Number 40 of 2007 regarding Limited Liability Companies].

Indonesia (2008). *Undang-undang No. 21 Tahun 2008 tentang Perbankan Syariah* [The Law Number 21 of 2008 regarding Sharia Banking].

Indonesia (2009). *Peraturan Badan Wakaf Indonesia No. 1 Tahun 2009 tentang Pedoman Pengelolaan dan Pengembangan Harta Benda Wakaf Bergerak Berupa Uang* [Indonesian Waqf Board Regulation Number 1 of 2009 regarding Cash Waqf Management and Development Guidance].

Indonesia (2009). *Peraturan Bank Indonesia No. 11/10/2009 tentang Unit Usaha Syariah* [Bank Indonesia Regulation Number 11/10/2009 regarding Sharia Business Unit].

Indonesia (2009). *Peraturan Bank Indonesia No. 11/15/2009 tentang Perubahan Kegiatan Usaha Bank Konvensional Menjadi Bank Syariah* [Bank Indonesia Regulation Number 11/15/2009 regarding Conversion of Business of Conventional Commercial Banks into Sharia Commercial Bank].

Indonesia (2009). *Peraturan Bank Indonesia No. 11/3/2009 tentang Bank Umum Syariah* [Bank Indonesia Regulation Number 11/3/2009 regarding Sharia Commercial Banking].

Indonesia (2013). *Undang-undang No. 17 Tahun 2013 tentang Organisasi Kemasyarakatan* [The Law Number. 17 of 2013 regarding Community Organization].

Islam, M.M. (2015). Cash-Waqf: New financial Instrument for SMEs Development in Bangladesh. *World Vision Research Journal*, 9(1): 111–120.

Kementerian Agama Republik Indonesia (2006). *Fiqih Wakaf*. Jakarta: Ditjen Bimas Islam Direktorat Pemberdayaan Wakaf Kementerian Agama Republik Indonesia. [Islamic Jurisprudence of *Waqf*].

Khairandy, R. (2009). *Perseroan Terbatas: Doktrin Peraturan Perundang-undangan, dan Yurisprudensi*, Yogyakarta: Kreasi Total Media. [Limited Liability: Doctrine of Legislation and Jurisprudence].

Mannan, M.A.A. (1999). Cash *Waqf* Certificate—An Innovation in Islamic Financial Instrument: Global Opportunities for Developing Social Capital Market in the 21st Century Voluntary Sector Banking. *Proceedings of the Third Harvard University Forum on Islamic Finance, 1 October 1999, Cambridge, USA*. Cambridge: Harvard University.

Purwosutjipto, H.M.N. (2007). *Pengertian Pokok Hukum Dagang Indonesia Volume 2: Bentuk-Bentuk Perusahaan*. Jakarta: Djambatan. [Basic Understanding of Indonesian Trade Law Second Edition: the Model of Corporation].

Ubaid, A. (2014). Kemitraan Nazhir dengan bank syariah dalam mengembangkan wakaf uang: Studi kasus di Indonesia, Bangladesh, dan Yordania. *Jurnal Al- Awqaf*, 7(1): 47–56. [Nazhir partnership with Islamic Banks in Developing Cash Waqf: A Case Study in Indonesia].

Widjaja, G. (2008). *Transplantasi Trust dalam KUH Perdata, KUHD, dan UU Pasar Modal*. Jakarta: Raja Grafindo Persada. [Transplant Trust in the Civil Code, the Trade Code, and Capital Market Act].

Presentation of suspects: The paradox of presumption of innocence

A. Pangaribuan
Faculty of Law, Universitas Indonesia, Depok, Indonesia

ABSTRACT: It is a common practice in the Indonesian justice system for the criminal suspect to be presented to the public before the news media. The suspect, usually in handcuffs, will be standing behind a table on which there are spread evidentiary objects and other items seized at the time of arrest. The event usually combines the presentation of the suspect with a press conference. Sometimes, the suspect can even be questioned by the media. It is, in fact, an event orchestrated by the authorities to demonstrate to the public the effectiveness of law enforcement. Moreover, the public is more than happy to see that justice is apparently being served. This practice raises a significant question that has increasingly gone unnoticed. The presentation of the suspect triggers a risk that the orchestrated event can be manipulated to cast an impression of guilt on the suspect before their trial has even begun. In effect, the suspect is already tried by the public once they are displayed in handcuffs or a detention uniform. This is where the paradox of the presumption of innocence begins. This paper proposes that the presumption of innocence needs to be reconsidered at a practical level, especially in the case of presentation of suspects. This paper also analyses the twofold interpretation of the presumption of innocence principle: from the suspect's point of view, where the violation of rights and freedoms can be deemed a violation of the presumption of innocence, and the presumption of innocence merely as a limited normative meaning.

1 INTRODUCTION

In the period from July to October 2016, the media have extensively covered the investigation and trial of Jessica Wongso; the 27-year-old woman accused of murdering her friend with a cyanide-laced coffee was allegedly seeking "revenge" over comments made about her former boyfriend in Australia. The case attracted massive public attention and debate and the high-profile murder trial became a public drama. The media's coverage did not stop with the reporting of the proceedings, but was supplemented by comments from observers from various disciplines. Besides this cyanide murder case, the public representation of a suspected offender was also conducted in the case of methamphetamine abuse involving the Regent of Ogan Ilir, Ahmad Wazir Noviandi. Given this practice, questions arise with regard to the public display of suspects in criminal proceedings, particularly in the presence of the mass media, and its links to the observance of the principle of presumption of innocence.

As yet there is no standard term to describe the display of a criminal suspect before the media in the Indonesian justice system. Therefore, for the sake of establishing a common understanding, I will refer to such practice as *the presentation of suspects* (the Indonesian version of which being "perp[etrator] walk"). The presentation of suspects or defendants is the act of presenting and displaying individuals who have allegedly committed a crime before the media. In other words, the suspect or defendant is placed face to face with journalists, is interviewed, and is bombarded with questions that are arguably based on a presumption of guilt. Proponents of such practice argue that this activity is aimed at demonstrating to the public that law enforcement is being implemented and justice being served. However, without discounting the public's right to be informed on matters of law enforcement, the presentation of suspects may, in fact, lead to a violation of the principle of presumption of innocence.

Observance of the presumption of the innocence principle when the mass media is involved has presented a never-ending polemic for decades, even before the birth of the free press through press law just after the reformation era, which giving a way to the freedom of press in Indonesia. As a consequence, the various existing legislations have not been able to provide clear and concrete delineation of admissible practices in relation to presumption of innocence.

This paper employs the legal doctrinal research methodology to define the practice of presenting a suspect before the public and its paradoxical nature in the potential undermining of the presumption of innocence principle. In the light of a lack of clear definitions in relation to the presumption of innocence principle in Indonesia and the practice of presenting the suspect, this paper concerns itself with the formulation of pertinent legal doctrines through the analysis of various legal rules of both national and international law. I will be applying the relevant legal rules pertaining to the presumption of the innocence principle by applying the relevant legal rules to the particular facts of presenting the suspect. In short, this paper seeks to clarify ambiguities within the presumption of the innocence principle and, arguably, its paradoxical nature when seen through the practice of presenting the suspect, and places them in a logical and coherent structure, while also describing their relationship to other rules.

1.1 *The presumption of innocence principle and its meaning*

Debate about the presumption of innocence principle in relation to the presentation of a suspect before the media, globally, is not a new issue. However, it seems that, recently, this issue has gone unnoticed in the criminal justice discourse in Indonesia. Before discussing whether or not the practice of presenting a suspect curtails the suspect's rights and freedoms, one must look into the very definition of the presumption of innocence principle.

Keijzer (1997) states that in order to have a deeper understanding of the principle of presumption of innocence, one must first understand the history of its development. Presumption of innocence was recognised as early as the 11th century in the common law system, particularly in England, as later embodied in the Bill of Rights (1689). Within the criminal justice system under a common law system (where an adversarial/contest approach is employed), the principle is a primary criterion in assessing whether a legal procedure has been conducted in an honest, fair and impartial manner (Walker, 1999).

In the Netherlands, it is acknowledged that canonic law provided for the origin of presumption of innocence, which was introduced in 1010 in the decree of Bishop Burchard von Worms, Section XVI-C6, and refers to the decree of Patris Hadriamus, stipulating that: "no person in a proceeding shall be accused as the person at fault prior to any examination proving of his guilt, by admission of such person, and witnesses testimony to prove such guilt, thus leading to a ruling that declares the guilt of the defendant" (Rukmini, 2007).

In an international setting, presumption of innocence can be found in the Universal Declaration of Human Rights (UDHR) and the International Covenant on Civil and Political Rights (ICCPR) of 1966, which has been ratified by Indonesia since 2005. Article 11 of the UDHR stipulates that every person is entitled to be presumed innocent until proven guilty in a public court of law where such person has been given the opportunity to present a defence. Additionally, Article 14 Paragraph 2 of the ICCPR explicitly declares this principle in its text: "everyone charged with criminal offence shall have the right to presumed innocence until proved guilty according to law".

When looking back at the origin of the presumption of the innocence concept, some may suggest that it adopts an individualist paradigm that protects the rights and interests of the offender (offender-based protection) and disregards the collective rights and interests of the society that has to suffer as a result of the commission of the offence (Atmasasmita, 2009).

The concept of presumption of innocence in the United Nations declaration does not provide for equal protection for the two legal entities, thus propagating a reaction on the importance of fundamental rights and obligations.

In Indonesia, the legal foundation for presumption of innocence is implicitly contained in Law No. 8 of the Criminal Procedural Code (*KUHAP*) of 1981, specifically in sub-article 3 point c, which states that "any person who is accused of a crime, arrested, detained, prosecuted, or brought before a court of law must be presumed innocent until found guilty by final and binding judgement". As such, lawmakers have established the principle of presumption of innocence as a legal foundation upon which KUHAP, and the enforcement thereof, is built (Harahap, 2013).

As a comparison, unlike Indonesia that does not specifically nor expressly provide for the fundamental rights and obligations of the accused or defendant, US laws specifically set forth various principles relating to the rights of the defendant through the country's constitution (Israel & LaFave, 1993).

Prior to being implicit in KUHAP, presumption of innocence is also stated in Article 8 of the Law on the General Authority of the Judiciary No. 14 of 1970, which also stipulates that any person who is accused of a crime, arrested, detained, prosecuted, or brought before a court of law must be presumed innocent until found guilty by a final and binding judgement. Furthermore, Law Number 48 on Judicial Powers of 2009 contains exactly the same clause in its Article 8, paragraph 1.

Judged from this perspective, the principle of presumption of innocence is often referred to as the fruit of the accusatory principle. The accusatory principle positions the defendant as a subject rather than the object of proceedings, and therefore the accused or defendant is treated with dignity as a human being. According to this principle, what is made the object of the proceedings is the offence allegedly committed by the accused or the defendant (Harahap, 2013). Practice has shifted since the colonial period. During that period, the accused was deprived of any right or opportunity to defend himself, because from the outset of the case law enforcement officers had determined the guilt of the defendant, as if he had been passed a verdict of guilty by the investigators. Additionally, treatment of the suspect or defendant took no regard of any humane considerations or human rights, such as the right to present a defence or to maintain dignity or uncover the truth. As a result, a defendant had, in practice, to resign themselves to an unfortunate fate. An example is the 1974 case of Sengkon and Karta, who had to spend years in jail while, in fact, the murder attributed to them was perpetrated by another person (Harahap, 2013).

The consequence of the implementation of the presumption of innocence is that a suspect or defendant accused of a crime cannot be treated as a guilty person, despite having been arrested/detained or subjected to any other coercive action in accordance with the prevailing laws and regulations, and thus law enforcement must take into account the human rights of the defendant or suspect (Harahap, 2013).

Presumption of innocence cannot be deemed as the opposite of the presumption of guilt. Presumption of innocence is, indeed, different from the presumption of guilt, but they are not in conflict with each other. Packer (1997) provides the following illustration: a murderer, by reason that he adequately comprehends, chooses to shoot his victim in front of many people. When the police arrive he is still holding the gun while saying that he is the killer. The incident has been witnessed by many. He is then arrested and sent to prison. This is not where presumption of innocence is applied (Packer, 1997). This case may seem an extreme example, but it demonstrates that evidence may be overwhelming and it would be out of the question that the defendant had not been involved in the murder. Thus, it may be concluded that what Packer means by presumption of innocence is not a prediction of the outcome, but the direction or principle used to guide officers on how to treat the suspect.

Some of the underlying elements of the presumption of innocence are: (a) protection from arbitrary action by public officials; (b) the court to determine the guilt or innocence of the defendant; (c) open court proceedings (i.e. must not be held in secret); (d) the defendant or suspect to be assured of the opportunity to present a defence to the fullest extent (Reksodiputro, 1994).

The right of a person to be deemed as innocent until found guilty by a court verdict (presumption of innocence) is not an absolute entitlement, from either a formal or material standpoint, as this right does not constitute one of the "non-derogable rights" defined by the ICCPR, such as the right to life or the right not to be prosecuted under retroactive criminal legislation.

1.2 *Presentation of the suspect and the paradox of presumption of innocence*

The presentation of suspects, better known in the common law system as 'the perpetrator walk', is defined by the American Heritage Dictionary of the English Language as 'the deliberate escorting of an arrested suspect by police in front of the news media, especially as a means of pressuring or humiliating the suspect'. The shortened term 'perp walk' has been used for at least five decades by New York police and media photographers (Lidge, 2006). On the basis of this definition it can be inferred that the taking of pictures or videos of a person accused of a crime is intended to pressure or shame the suspect. I suspect the act of parading the suspect may be partly inspired by ancient Roman military culture, called *'fustuarium'*, which is a punishment of being displayed in public to bring a sense of shame to the suspect.

Aside from pressuring and humiliating the suspect, another intended purpose of the presentation of the suspect is to make transparent to the public the operation of law enforcement. In addition to these purposes, there are also some who believe that the perp walk aims to show to the public that police and prosecutors are working to protect them and/or to create the public notion that the suspect has committed a crime or offence (Lidge, 2006).

In regard to the justification or argument for the display of a suspect's identity and picture in the media, according to van Veen (Adji, 1977), some cases that have caught the public's attention from the outset can no longer be deemed as private. Adji (1977) expresses the view that disclosing the identity of the potential offender is seen as giving society some satisfaction. In England, the display of a suspect's full identity is intended to avoid any misidentification of people who may share the same name. Others opine that the name and face of a suspect should be shown to the public only if the transgression allegedly committed consists of a series of offences involving multiple victims, such that by disseminating the identity of the suspect or defendant the collection of data relating to other victims may become easier.

In regard to the presentation of the suspect, one prominent story began when Dominique Strauss-Kahn, then director of the International Monetary Fund (IMF), was alleged to have committed sexual harassment in a hotel in New York in 2011. The accusation was ultimately dismissed due to lack of evidence. However, due to the perp walk conducted at the time, pictures of Strauss-Kahn being led in handcuffs by New York police officers can still be found everywhere and, as such, he has condemned the action of the New York police in forcing him to undertake the perp walk (Wringe, 2015).

In Mexico, a perp walk or presentation of a suspect, known locally as *presentacion*, is only enforced for extraordinary or serious crimes, particularly those involving drugs. During a *presentacion* session, the suspect is made to wear a prison uniform and the evidence is laid out while reporters ask questions of the suspect. The reason Mexican authorities carry out such practices is to convey to Mexican society that the government is serious in eliminating drugs. With their hands shackled, the alleged offenders are asked questions by the reporters, such as "Did you really commit the crime?" or "Why did you do it?", which are questions that imply that the suspects are already considered guilty, before a verdict has been passed by the court.

Another example relates to 2009, when two Bear Stearns executives were arrested at their homes, handcuffed, and were taken to the office of the Federal Bureau of Investigation (FBI) where they were greeted by reporters who were already waiting for them. They were then photographed by journalists with their hands in handcuffs and found their faces pasted all over the media. Although practised in the US since the 1930s, the perp walk only became prevalent for such white-collar criminal cases in the 1980s (Goldstein, 2010).

It is widely known that presentation of suspects is used by police officials and prosecutors as a means to further their careers (Lidge, 2006) or to gain appreciation from the public. Such public display of suspects that aims to pressure or shame suspects reflects a basic aspect of human nature that takes pleasure in seeing the suffering of others. In other words, a 'perp walk' can become a form of entertainment for those who watch it, and also a means for them to vent their anger and frustration on a suspect who has not even been found guilty by the courts.

The term 'perp walk' is not well known in Indonesia, but it is extensively practised in the country and the activity sometimes goes beyond simply photographing the suspect being escorted by police. For example, it is common in Indonesia, regardless the crimes, the suspect is placed before reporters with their hands cuffed or in a detention suit, and is bombarded with questions from the journalists. In looking at that scenario, it is reasonable to assume that the presumption of innocence principle is absent from the Indonesian justice system.

The principle of presumption of innocence should be embedded in the various legislations as previously mentioned, including in the regulations governing the press. In journalism, the presumption of innocence should be treated as an acknowledgement that, until a person is conclusively found to be guilty of the crime, to judge that person is to tarnish the values of democracy that uphold freedom (Huda, 2010).

Thus, aside from the criminal procedural perspective, in debating the practice of perp walk one must look into the press legislation in the country. Article 5 of the Indonesian Press Law requires members of the press to adopt the presumption of innocence when reporting events and voicing opinions. Rules on the obligation to apply this principle are contained in Article 3 of the Journalism Code of Ethics, which stipulates that members of the Indonesian press must test the information that they collect, conduct balanced reporting, refrain from mixing facts with judgemental opinions, and work under the presumption of innocence. This rule reflects the obligation under which the press must report and present information in a non-judgemental manner and not convey the presumption of guilt. Coverage by the press of suspects wearing a prison uniform, wearing handcuffs, and/or standing next to the evidence does not reflect a presumption of innocence, but rather is a statement to the public that the suspects or defendants deserve to receive such treatment.

It is interesting to further examine the practice of displaying suspects, given that a person alleged to have committed a crime should be presumed innocent until found guilty by virtue of a court verdict having permanent legal force. This raises a further question as to the lines that the press must not cross in reporting suspects in a criminal case. An example can be found in a court trial. If the hearing is conducted openly, the public have unrestricted access to all aspects of the proceedings, from the way the judge presides over the case and the attitude of the public prosecutors and lawyers, to the identity of the defendant. In this regard, the press acts as the eyes and ears of the public who cannot be present in the courtroom. As such, the press and media are free to report on the defendant, complete with full identity and photograph. However, what should not be conveyed to the public by the press is any statement that may appear to pass judgement on the defendant (Sukardi, 2010).

Although presumption of innocence is stipulated in the Journalism Code of Ethics, it is subjected to different interpretations and thus gives rise to varying approaches to reporting, particularly when it relates to crime. According to Siregar (1989), journalists can be grouped into several types in the context of presumption of innocence. The first type are those who comply with the principle in regular criminal cases. They do not reveal the full identity of the suspect, only reporting their initials instead. They also do not post photographs of the person. For cases that receive wide public attention, however, they do disclose the full identity or a picture of the suspects. The second type of journalist establishes a set criterion to determine whether a suspect's identity and picture can be disclosed to public. Some journalists do not feel the need to protect the identity of suspects who have committed a heinous crime that has seriously transgressed the boundaries of humanity. For many of them, embezzlement of public funds is particularly vicious, and thus corrupt public officials who have stolen from the people are deemed not to deserve to have their identity concealed.

The third type of journalist reveals the identity of the suspect regardless of the nature of the case, and based instead on the person allegedly involved. If the suspect is a public figure, they do not deem it necessary to identify the person by initials only, but insist he or she should be named in full on the basis of the argument that the public figure belongs to the public, and therefore it is the right of the people to be informed of all of the public figure's conduct, including when such person has allegedly committed a crime (Siregar, 1989).

On the basis of this discussion, it can be understood that there is no clear delineation set by any legislation regarding the presentation of suspects. The various interpretations of the established boundaries in the field of journalism show that the degree to which presumption of innocence is implemented is still abstract and, if allowed to continue, will adversely affect the suspect, particularly if it is subsequently found by a court of law that such person is innocent and yet the public has already passed judgement. What also needs to be taken into account is what happens when the identity and picture of the suspect displayed in the media are seen by their family, particularly the suspect's children.

2 CONCLUSION

It is clear that, in practice, the principle of presumption of innocence has become a real paradox. There is a question of whether it should be seen purely from a normative point of view, or whether it should be interpreted in a more fundamental fashion in order to observe human rights? We often see on television a suspect being displayed in handcuffs, wearing a prison uniform and even being interviewed. The author is merely trying to imagine the situation in which a suspect is eventually found to be innocent, yet the public have already labelled such person a criminal. Moreover, it creates a room for violation of human rights by inviting public opinion to decide the guilt of a person since the investigation stage. Thus it is also giving a label to the suspect who has the right to be presumed innocent until found guilty.

It would not be easy to shed such a label. The perp walk or presentation of a suspect, which intends to shame a suspect, indicates how our legal system has taken a step backwards and merely intends to seek revenge, whereas in fact the purpose of prosecution should go beyond retaliation and should also provide rehabilitation and protection.

Indonesian national law does not specifically regulate the presentation of suspects in the media. The principle of presumption of innocence still remains an abstract concept, the implementation of which relies on the authorities in charge.

In an effort to resolve this paradox of the presumption of innocence, a normative legal instrument is needed to regulate boundaries within which suspects can be displayed. In this regard, the presumption of innocence must be interpreted in a fundamental manner in order to protect the rights of the suspect and avoid false accusations, and to serve as a check and balance mechanism in law enforcement measures undertaken by the state and in the protection of the human rights of individuals.

REFERENCES

Adji, O.S. (1977). *Pers: Aspek-Aspek Hukum [Legal Aspects]*. Jakarta, Indonesia: Erlangga.

Atmasasmita, R. (2009, December 14). Logika Hukum Asas Praduga Tak Bersalah: Reaksi atas Paradigma Individualistik [Presumption of Innocence Legal Logic: Reaction towards Individualism Paradigm]. *HukumOnline*. Retrieved from http://www.hukumonline.com/berita/baca/lt4b25f96c2ed41/logika-hukum-asas-praduga-tak-bersalah-reaksi-atas-paradigma-individualistik-br-oleh-romli-atmasasmita-.

Goldstein, H.W. (2010). 'Perp Walks' undermine presumption of innocence. *New York Law Journal*, January 7, 2010. Retrieved from http://www.friedfrank.com/siteFiles/Publications/C1D29343 A45043B03B9E-284BEFA229A3.pdf.

Gordon, J.S. (2009, March 30). A sad case of trial by 'Perp Walk'. *Barron's*. Retrieved from http://www.barrons.com/articles/SB123819760412261783.

Harahap, M.Y. (2013). *Pembahasan Permasalahan dan Penerapan KUHAP: Penyidikan dan Penuntutan [Discussion on the Issues and Implementation of KUHAP]*. Jakarta, Indonesia: Sinar Grafika.

Huda, C. (2010). Makna asas praduga tidak bersalah dan pemakaiannya dalam praktek pers [The Definition of the Presumption of Innocence Principle and Its practice in Press]. *Jurnal Dewan Pers, 2*, 33–44.

Indonesia (1971). *Undang-undang No.14 Tahun 1971 tentang Ketentuan-ketentuan Pokok Kekuasaan Kehakiman [Law Number 8 of 1971 regarding the Principal Provisions of Judiciary Power]*. State Gazette 1971 No. 20, State Secretariat, Jakarta.

Indonesia (1981). *Undang-undang No. 8 Tahun 1981 tentang Hukum Acara Pidana* [*Law Number 8 of 1981 regarding the Criminal Procedural Code*]. State Gazette 1981 No. 36, State Secretariat, Jakarta.
Indonesia (2009). *Undang-undang No. 48 Tahun 2009 tentang Kekuasaan Kehakiman* [*Law Number 48 of 2009 regarding Judicial Powers*]. State Gazette 1981 No. 36, State Secretariat, Jakarta.
Israel, J.H. & LaFave, W.R. (1993). *Criminal procedure, constitutional limitations in a nutshell*. St. Paul, MN: West Publishing.
Keijzer, N. (1997). The presumption of innocence. In *Majalah Hukum Triwulan Universitas Parahyangan*. Bandung, Indonesia: Universitas Parahyangan.
Lidge, E.F., III. (2006). 'Perp Walks' and prosecutorial ethics. *Nevada Law Journal, 7*(1), 55–72.
Packer, H. (1997). Two models of the criminal process. In S. Wasserman & C. Snyder (Eds.), *A criminal anthology* (pp. 3–9). Cincinnati, OH: Anderson.
Reksodiputro, M. (1994). Hak asasi manusia dalam sistem peradilan pidana [Human Rights in Criminal Justice System]. In *Kumpulan Karangan Buku Ketiga*. Jakarta, Indonesia: Lembaga Kriminologi UI.
Rukmini, M. (2007). *Perlindungan HAM melalui Asas Praduga Tidak Bersalah dan Asas Persamaan Kedudukan Dalam Hukum Pada Sistem Peradilan Pidana [Human Rights Protection through Presumption of Innocence Principle and Equality Principle in Criminal Justice System]*. Bandung, Indonesia: Alumni Press.
Simorangkir, J. (1980). *Hukum dan Kebebasan Pers [Law and Freedom of Press]*. Jakarta, Indonesia: Badan Pembinaan Hukum Nasional.
Siregar, R.H. (1989). Beberapa catatan Kode Etik Jurnalistik PWI dan asas praduga tak bersalah [Several Journalism Ethical Notes PWI and the Presumption of Innocence Principles]. In *Seminar Diskusi Asas Praduga Tak Bersalah dan Trial by Press dalam Kode Etik Jurnalistik, 25 Maret 1989, Jakarta, Indonesia*. Jakarta, Indonesia: Dewan Pers.
Sukardi, W.A. (2010). Menghindari tuduhan pelanggaran asas praduga tidak bersalah [Dismissing the Alleged Violation of Presumption of Innocence Principle]. *Jurnal Dewan Pers, 2*, 19–31.
United Nations. (1976). International covenant on civil and political rights. In *UN Treaty Series Vol. 999 No.14668. Multilateral, International Covenant on Civil and Political Rights and Optional Protocol to the above-mentioned Covenant. Adopted by General Assembly of the UN on 19 December 1966, New York, USA* (pp. 171–346). New York, NY: United Nations.
UNOHCHR. (1996). *Universal Declaration of Human Rights*. Geneva, Switzerland: United Nations Office of High Commissioner for Human Rights.
Walker, C. (1999). Miscarriages of justice in principle and practice. In C. Walker & K. Starmer (Eds.), *Miscarriages of justice: A review of justice in error*. New York, NY: Oxford University Press.
Wringe, B. (2015). 'Perp Walks' as punishment. *Ethical Theory and Moral Practice, 18*, 615–629.

The risk of joining the Trans-Pacific partnership for Indonesia: An investment perspective

W. Setiawati
Faculty of Law, Universitas Indonesia, Depok, Indonesia

ABSTRACT: International investment and trade have a new purpose: having previously focused on profits, they are now about domination and power. This new goal places developing countries in a constant struggle to catch up with the modernisation of economic competition in the wider world. The liberalisation of trade not only comes in the form of a set of rules produced by the World Trade Organisation (WTO) but also in the form of regional or bilateral agreements. Recent debates in Indonesia have concerned the declaration by the country's president of the intention to join the Trans-Pacific Partnership (TPP). One pertinent issue in the TPP involves the investment chapter, which has started to raise concerns from countries with bad experiences of bilateral investment treaties. Joining the TPP could present a future risk for Indonesia if we consider its recent experiences of the settlement of investment disputes. This paper will explain the risks for Indonesia in joining the TPP in relation to the investment chapter and the application of the same investor–state dispute settlement, making the outcome of future dispute predictable. This research was conducted by using a qualitative method and a deductive approach to provide a comprehensive understanding of the issue presented.

1 INTRODUCTION

In the last decade, the Trans-Pacific Partnership (TPP) agreement has become the focal debate among people from many backgrounds and countries. The pros and cons of the TPP are discussed in relation to an array of issues, including labour, intellectual property, information technology and procurement. One issue that recently prompted debate of the TPP concerns the investment chapter in which an Investor–State Dispute Settlement (ISDS) mechanism is adopted by the agreement. This paper will first explain how the negotiation of TPP chapters is carried out from the beginning until the agreement is signed and how the US' policy decided not to continue with it thus TPP then considered failed. Then, the investment chapter in the TPP and the adoption of the ISDS mechanism for investment disputes will be discussed. Finally, the paper will consider the dangers of ISDS implementation for Indonesia.

1.1 *The development of the trans-pacific partnership agreement*

Since the failure of the Doha round of World Trade Organisation (WTO) negotiation in 2001, countries have shifted their economic interests to bilateral and regional free trade agreements as a quicker route to free trade between countries (Pakpahan, 2012). The Trans-Pacific Partnership is one example of free trade agreements between countries that seek economic benefits from free trade cooperation. Constructed by 12 countries—Chile, New Zealand, Singapore, Brunei Darussalam, the United States, Australia, Peru, Vietnam, Malaysia, Mexico, Canada and Japan (Hilpert, 2015; Cheong, 2013) – it is said that the TPP is a mega Regional Trade Agreement (RTA) in which the GDP of the 12 member countries represents 40 per cent of global GDP and 20 per cent of global trade. This is obviously larger than other high-achieving RTAs such as the North American Free Trade Agreement (NAFTA) and the EU Single Market (World Bank, 2016, p. 221).

The published final text of the TPP agreement, concluded on 4 February 2016, consists of 30 chapters and four annexes, as well as several additional annexes associated with specific chapters, with wide coverage of trade and investment issues (USTR, n.d.). Following the signing of TPP, the next stage is the ratification process, which is to be completed within two years of signing, with the agreement coming into force 60 days after the expiry of this period provided that there are at least six of the original 2013 signatories that together account for at least 85 per cent of the combined GDP of the original signatories (USTR, n.d.).

This requirement has led to an expectation being placed on the US and Japan to ratify it first by other members. However, given that the newly elected president of the US, Donald Trump, has already stated that he is against the TPP, which he described as the greatest danger, stating that NAFTA was the worst deal in the history of the country (Cirilli & Knowles, 2016), the future of the TPP seems bleak. As president-elect, Trump vowed to issue a note of intent to withdraw from the TPP (Woolf et al., 2016).

Japan's legislature, the National Diet, began an extraordinary session on 26 September 2016 with the TPP as one of the key items on the agenda (Nadeau, 2016). Meanwhile, Singapore has not started the ratification process, although Prime Minister Lee Hsien Loong seems to prefer the TPP to the Regional Comprehensive Economic Partnership (RCEP), especially after the ASEAN South China Sea dispute of June 2016 (Potkin, 2016). Prior to the 2016 US election, Malaysia was going to amend 18 laws as part of its effort to implement its TPP commitment before it took effect in 2018. Once these laws have been amended, Malaysia will decide on ratification (Rasid, 2016). In Vietnam, as of mid-September 2016, TPP ratification was said to be on the agenda for its next parliamentary session (Potkin, 2016), but by the end of that month Vietnam's National Assembly had decided to exclude TPP from its agenda for the upcoming session (Doan, 2016). Australia is one of the countries that is eager to see the successful implementation of TPP and the Australian parliamentary inquiry into TPP was set to report in February 2017 (Loussikian, 2016). Until then, public hearings on TPP ratification were to be convened in every state. Other member countries, such as Chile and Peru, have already sent the TPP agreement to their congresses around mid-2016 to be discussed and eventually approved (ICTSD, 2016a, 2016b).

However, with the developing situation in the US, all of these efforts on TPP ratification became muted. Japan's Prime Minister, Shinzo Abe, still intends to ratify TPP, although Japanese parliamentary opposition criticises the TPP as a "total failure" (Simoyan, 2016). Singapore, according to Prime Minister Lee, will take a stand of waiting to see on what policies Trump's administration will settle (Othman, 2016). Meanwhile, Australia and Malaysia have shifted their interests to a China-led trade deal in the form of the RCEP (Murphy, 2016; Sipalan, 2016).

1.2 *Investment coverage in the trans-pacific partnership agreement*

Although the TPP may be considered as a failing cause, it is still interesting to discuss the provisions made in the agreement, and there is always a possibility of the US and/or other countries making a very similar trade deal in the future. In the context of trade deals, the US has already had NAFTA, providing a regional free trade agreement with Mexico and Canada since 1994. When NAFTA and TPP texts on investment are compared, one of the similarities of both texts is the adoption, with a few additional clauses, of a Bilateral Investment Treaty (BIT), the most recent US model being revised in 2012. A discussion of the most important issues in the investment chapter follows.

In the definition article, investment is defined as (USTR, n.d.):

> ***Every asset*** *that an investor owns or controls, directly or indirectly, that has the characteristics of an investment, including such characteristics as the commitment of capital or other resources, the expectation of gain or profit, or the assumption of risk.*

This definition of investment uses a non-exhaustive asset-based test that, according to UNCTAD (2011, p. 21), is meant to safeguard the interests of investors, with those interests being everything of economic value with no limitation. The asset-based test is traditionally used in an international investment agreement, and in the TPP the use of an asset-based test, followed by a non-exhaustive illustrative list of forms of investment, means that it will cover future evolving forms of investment, and it will not only form a set in the list (OECD, 2008a, p. 9).

Another key issue in this agreement is in Article 9.2 concerning the Scope of the agreement; in the second paragraph it is stated:

A Party's obligations under this Chapter shall apply to measures adopted or maintained by:

a. *the central, regional or local governments or authorities of that Party; and*
b. *any person, including a state enterprise or any other body, when it exercises any governmental authority delegated to it by central, regional, or local governments or authorities of that Party.*

Within this scope, it means every action either of the central government or the regional or local government, as in the lower level of government, will fall under this agreement with the same obligation. In addition, the obligation extends to the action of any state-owned enterprise or government body or even a person with delegated authority from the government (central, regional, local).

For national treatment provisions, the clause uses both pre-entry and post-entry protection, as stated in Article 9.4:

1. *Each Party shall accord to investors of another Party treatment no less favorable than it accords, **in like circumstances**, to its own investors with respect to the establishment, acquisition, expansion, management, conduct, operation, and sale or another disposition of investments in its territory.*
2. *Each Party shall accord to covered investments treatment no less favorable than it accords, **in like circumstances**, ...*
3. *For greater certainty, the treatment to be accorded by a Party under paragraphs 1 and 2 means, with respect to a regional level of government, treatment no less favorable than the most favorable treatment accorded, in like circumstances, by that **regional level of government** to investors, and to investments of investors, of the Party of which it forms a part.*

Under the "Calvo doctrine", aliens and their property are entitled to the same treatment accorded to nationals of the host country under its national laws (UNCTAD, 1999a, p. 7), and this non-discriminatory treatment is known as national treatment. The TPP uses the pre- and post-entry model, an approach used in US treaty practice and also in the 1994 model BIT (UNCTAD, 2010), and consistently until the 2012 model BIT, and in NAFTA. This combined approach seeks to attain a compromise between the possibilities offered in pre- or post-entry investment, and also to cover *de jure* and *de facto* differential treatment that a host country might exercise (UNCTAD, 2004, p. 90). Paragraph 3 prescribes that national treatment is also an obligation for the regional level or local government, which means the conduct and measure of the regional or local government must abide by the same standards as the national-level government (Mann et al., 2006, p. 13). Another issue worthy of attention is the use of the phrase "in like circumstances" in the provision, which is not easy in practice and can give rise to very unclear outcomes when a dispute arises (UNCTAD, 1999a, p. 33).

Most-Favoured-Nation Treatment is stipulated in Article 9.5, and the clause is just the same as the national treatment provision, except that this article regulates the treatment between party and non-party, which means the host country will not discriminate one country from another, and will provide the same protection for investors from different countries. Again, because the "in like circumstances" clause is also used in this provision, it will have the same effect as its use in the national treatment provision.

Fair and equitable treatment in TPP is included in Article 9.6: Minimum Standard of Treatment, which states:

Each Party shall accord to covered investments treatment in accordance with applicable customary international law principles, including fair and equitable treatment and full protection and security.

According to UNCTAD (2012, p. 10), "fair and equitable treatment" can be viewed from two different perspectives: the plain meaning approach, and equating fair and equitable treatment with the international minimum standard. The provisions in TPP regarding fair and equitable treatment have adopted the latter approach, which includes full protection and security within its scope (UNCTAD, 2012, p. 10). Interestingly, part of the second paragraph means that the obligation not to deny justice in legal proceedings becomes the requirement for the fair and equitable standard. The requirement set for the full protection and security standard is the obligation to provide police protection required under customary international law; this is a very specific criteria which serves as a good measurement for a host country (Malik, 2011).

The Expropriation and Compensation provision in TPP prescribes that:

> [A] *party shall not expropriate or nationalize a covered investment either directly or indirectly except: for public purpose; and being done in a non-discriminatory manner; and the compensation paid is prompt, adequate and effective; and in accordance with due process of law.*

This is a clear formulation of an expropriation provision where expropriation and nationalisation are allowed as long as they are for public purpose.

The provision on Performance Requirements is the next key issue that needs attention. Performance requirements are conditions imposed by host countries on investors in order to achieve government policy goals (APEC/UNCTAD, 2012, p. 86). These performance requirements are prohibited under Trade-Related Investment Measures (TRIMs). However, in the TPP, technology transfer is included as a performance requirement, as stipulated below:

> *(f) to transfer a particular technology, a production process or other proprietary knowledge to a person in its territory.*

The open door for foreign investment is maximally utilised by Trans-National Corporations (TNCs), and becomes the major factor in the integration of the world economy (UNCTAD, 1999b, p. 9). The role of TNCs is a double-edged sword, where on the one side they are a means of improving the well-being of societies and development in the economic sense, but on the other side, they are monopolistic entities that grow through the exploitation of their competitive advantage in technology and know-how at the expense of the host country, and create a gap between themselves and the domestic capability (UNCTAD, 1999b, p. 23). The OECD Guidelines for Multinational Enterprises (OECD, 2008b, p. 23) recommend the adoption of technology transfer and rapid diffusion of technologies and know-how, with due regard to the protection of intellectual property rights. However, the stipulation of transfer technology as a performance requirement means the TPP disregards the OECD Guidelines, and this will not be a benefit for developing countries.

The last key issue in the TPP agreement is the adoption of Investor–State Dispute Settlement (ISDS), as prescribed in Article 9.17. Specifically, this provision gives investors that directly invest in the host country, or directly or indirectly own or control an enterprise which is a juridical person in the host country, to bring a claim under the ISDS clause. This stipulation does not oblige exhaustion of local remedies as a dispute settlement mechanism prior to an arbitration process, but only consultation and negotiation with a third party.

1.3 *The dark side of the investor–state dispute settlement mechanism for investment disputes*

ISDS in itself has already become a problem for many countries, and to date, has placed risk on the host country, making it a growing concern in every developing country that needs foreign investment but has only limited resources when a dispute arises that relates to ISDS.

As shown in the diagram in Figure 1, the number of cases brought to the International Centre for Settlement of Investment Disputes (ICSID) under the ISDS clause is quite modest. In terms of concluded cases, 36.7% were decided in favour of the state and 26.5% in favour of investors. This seems promising for the ISDS mechanism, but is it a fair reflection of reality? Mann (2015) has investigated the truth about these numbers, and in the 144 cases

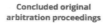
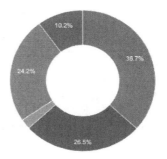

Figure 1. Number of cases brought to ICSID (Source: http://investmentpolicyhub.unctad.org/ISDS (UNCTAD, n.d.)).

decided in favour of the state (out of 255 concluded cases), it turns out that the decisions were in a jurisdiction that terminated the arbitration, thereby reducing the state's wins to 71 cases or 28%, with the significant news that investors effectively won on 72% of occasions (Mann, 2015).

There are also trends in treaty-shopping, forum-shopping and nationality-shopping when it comes to dispute settlement. Known as cherry-picking, treaty-shopping means investors can choose the most beneficial provisions among all of the investment agreements entered into by a country and apply them in a dispute (von Moltke, 2004, p. 9). Meanwhile, forum-shopping is known for the term 'fork-in-the-road', which refers to an investment treaty that stipulates the availability of two types of litigation for the same investment, that is, international arbitration and local court (Yannaca-Small, 2006, p. 205). In terms of nationality-shopping, this means investors shifting their registration to acquire nationality of convenience, without having "effective economic activities", in order to take advantage of the most investor-friendly treaty (Subedi, 2008, p. 181). These trends could be described as the drawback of international investment agreements that affect developing countries the most, because the gradual rise of the trends has left developing countries in devastating conditions from investment disputes. For example, Argentina is the country with most claims (53) in which it is a respondent in investment disputes, and Venezuela comes second with 36 claims (Miller & Hicks, 2015).

1.4 *Indonesia's past experiences in investment disputes*

There are several investment dispute cases associated with Indonesia (BKPM, 2013, p. 28):

– 1981 – *Amco Asia Corporation and others v. Republic of Indonesia* (ICSID Case No. ARB/81/1). Concluded on 5 June 1990 in favour of investor. Claimed and awarded USD 3.2 million.
– 2004 – *Cemex Asia Holdings Ltd v. Republic of Indonesia* (ICSID Case No. ARB/04/3). Concluded by agreement of the parties on 23 February 2007. Claimed USD 400 million; awarded USD 337 million.
– 2007 – *Government of the Province of East Kalimantan (C) v. PT Kaltim Prima Coal and others* (ICSID Case No. ARB/07/3). Concluded on 28 December 2009 with no award because the claimant failed to prove the authority to represent the central government.

- 2011 – *Rafat Ali Rizvi v. Republic of Indonesia* (ICSID Case No. ARB/11/13). (BIT INA-UK). Concluded on 26 June 2013 in favour of the state. Claimed USD 75 million.
- 2012 – *Churchill Mining Plc v. Republic of Indonesia* (ICSID Case No. ARB/12/14). (BIT INA-UK). Not concluded yet. Claiming USD 1.05 billion.
- 2012 – *Planet Mining Pty. Ltd v. Republic of Indonesia* (ICSID Case No. ARB/12/40). (BIT INA-AUS). The claim to be joined with the Churchill Mining Plc v. Republic of Indonesia tribunal.
- 2014 – *Nusa Tenggara Partnership BV and PT Newmont Nusa Tenggara v. Republic of Indonesia* (ICSID Case No. ARB/14/15). (BIT INA-NED). Discontinued.
- 2016 – *Oleovest Pte. Ltd v. Republic of Indonesia* (ICSID Case No. ARB/16/26). (BIT INA-SIN). Not concluded yet. No information on the claims.

Besides the aforementioned cases, there are also cases brought under the United Nations Commission on International Trade Law (UNCITRAL) (BKPM, 2013, p. 30):

- 1999 – *Himpurna California Energy Ltd v. PT Perusahaan Listrik Negara (Persero)*. Concluded on 4 May 1999 in favour of investor. Claimed USD 315 million; awarded USD 273 million.
- 2000 – *Karaha Bodas Company LLC v. Perusahaan Pertambangan Minyak dan Gas Bumi Negara and PT PLN (Persero)*. Concluded on 18 December 2000 in favour of investor. Claimed USD 512 million; awarded USD 261 million.
- 2011 – *Hesham al-Warraq v. Republic of Indonesia.* (OIC Investment Agreement). Concluded on 15 December 2014 in favour of neither party. Claimed USD 19.6 million; awarded nothing.
- 2015 – *Indian Metals & Ferro Alloys Ltd v. Republic of Indonesia.* (BIT INA-IND). Not concluded yet. Claiming USD 581 million (Amindoni, 2015).

The cases involving Indonesia in relation to investment have had a mind-altering effect on the government, especially the USD 1.05 billion claim in the Churchill Mining case, which focused the government's attention on how disruptive the use of the ISDS mechanism can be (Warner, 2016). That sum was just the claim from the claimants, and does not include the cost of the ISDS process that, according to an OECD survey (Yannaca-Small, 2006, p. 17), costs an average of about USD 8 million per case and can exceed USD 30 million per case. If we compare the amount of the claims with Indonesia's 2015 GDP of USD 861.9 billion, the amount of the claims seems minuscule, but if we take into consideration the GDP in terms of the population of 257.6 million (World Bank, n.d.), the claims are costly for Indonesia. Taking into account the costs of arbitration, the claims and costs could be doubled, and this is a big risk for Indonesia's future development.

ISDS is certainly a very expensive process, and while developing countries expect benefits from foreign investment, the result could be a backlash for many of them. While there is no suggestion that the protection for investors should be decreased, there are options to ensure a better balance of the position between investors and a host country. In the meantime, Indonesia has decided to terminate BITs with other countries, and thus far there are 19 BITs for which Indonesia has already sent a termination notification to each partner state (UNCTAD, n.d.). This is the first step in mitigating the risk from ISDS in international investment agreements, so when the Indonesian president stated that it was the country's intention to join the TPP and that 12 laws would be amended to this end (Amindoni & Hermansyah, 2016) it was surprising news.

It is still unclear how the Indonesian government will ensure the necessary precautions in the event of Indonesia joining the TPP. Besides the amendment of various laws, there are still many things which need to be repaired and developed. Thus, for investment, there are many things to consider, such as the scope of authority of regional and local government, particularly when the evidence shows that regional government decisions can be inconsistent with central government policy, and have become a basis for several cases that have placed Indonesia in risky situations. With the relevant stipulation in the TPP, local/regional government will have the same obligation as the central government, and the same obligation will also prevail for a person or a body with delegated authority from the government. This provision would cause a major change in the governmental system, and it would not only mean regulatory changes but also administrative ones in terms of how the local/regional government and the central government will create a harmonised treatment of investments and investors. Failure to do this will bring risk to Indonesia because disputes with investors will increase in number and will almost certainly be costly for the nation.

In addition, we should keep in mind that Indonesia is in the process of terminating the remainder of the BITs it has entered into, and the possibility of investors maximising the grace periods to gain benefit from the treaties is still open and, therefore, the risk to Indonesia under the existing BITs remains present. Whether or not Indonesia should add more risk of investment disputes by joining the TPP, or entering into other trade agreements with ISDS provisions, is a question that the government needs to consider most carefully.

In addition, more questions arise when the benefit offered from the TPP is only in the form of Foreign Direct Investment (FDI) because the agreement stipulates that technology transfer is considered as a performance requirement, which is prohibited. With this provision, it will not be possible for Indonesia to gain a benefit from investment in the form of technology, and the possibility is that Indonesia will simply become a market and a source for foreign investors. Furthermore, the definition of investment in the TPP covers all forms of investment and is not limited to sustainable investment; therefore, there is no requirement for investment to be performed as a contribution in a certain duration of performance of the contract related to investment as defined in the Salini test (García-Bolívar, 2012). These are all lessons to be learned from one mega regional trade agreement, the Trans-Pacific Partnership. Hence, the lessons should become considerations when the Indonesian government enters negotiations in relation to the RCEP or any other free trade agreement.

2 CONCLUSION

The TPP agreement has already become a failed case of a mega regional trade agreement, but it does not necessarily mean there will be no such similar agreements made in the future. Indonesia still has to face investment cases brought under the ISDS clause, and given the costly procedure for each case, the government needs to rethink its strategy in relation to international investment agreements. When a trade deal brings more risks than benefits to the country, the government needs to stand firm in repudiating this and other such deals.

REFERENCES

Amindoni, A. & Hermansyah, A. (2016, February 5). TPP officially signed, Indonesia to amend 12 laws. *The Jakarta Post*. Retrieved from http://www.thejakartapost.com/news/2016/02/05/tpp-officially-signed-indonesia-amend-12-laws.html.

Amindoni, A. (2015, November 18). Indian Mining Co. sues Indonesia for $581 million. *The Jakarta Post*. Retrieved from http://www.thejakartapost.com/news/2015/11/18/indian-mining-co-sues-indonesia-581-million.html.

APEC/UNCTAD. (2012). *International investment agreements negotiators handbook: APEC/UNCTAD modules*. Singapore: APEC Secretariat.

BKPM. (2013). *Review Perjanjian Investasi Bilateral dan Multilateral [Review on Bilateral and Multilateral Investment Treaties]*. Research Report. Jakarta, Indonesia: Badan Koordinasi Penanaman Modal.

Cheong, I. (2013). *Negotiations for the Trans-Pacific Partnership Agreement: Evaluation and implications for East Asian regionalism*. ADBI Working Paper 428. Tokyo, Japan: Asian Development Bank Institute. Retrieved from http://www.adb.org/sites/default/files/publication/156283/adbi-wp428.pdf.

Cirilli, K. & Knowles, D. (2016, June 29). Trump likens Trans-Pacific Partnership trade deal to rape. *Bloomberg*. Retrieved from http://www.bloomberg.com/politics/articles/2016-06-28/trump-channels-brexit-in-anti-trade-speech-at-pennsylvania-factory.

Doan, X.L. (2016, November 25). Has Trump's election affected Vietnam's foreign policy? *Asia Times*. Retrieved from http://www.atimes.com/trumps-election-affected-vietnams-foreign-policy/.

García-Bolívar, O.E. (2012). Defining an ICSID investment: Why economic development should be the core element. In *Investment Treaty News, April 13, 2012*. Geneva, Switzerland: International Institute for Sustainable Development.

Hilpert, H.G. (2015). Trans-Pacific Partnership (TPP) agreement: Increased pressure on European trade policy. In *SWP Comments 51, November 2015*. Berlin, Germany: Stiftung Wissenschaft und Politik. Retrieved from https://www.swp-berlin.org/fileadmin/contents/products/comments/2015C51_hlp.pdf.

ICTSD. (2016a, July 31). Perú remite el TPP al Congreso para avanzar en su ratificación. *International Centre for Trade and Sustainable Development*. Retrieved from http://www.ictsd.org/bridges-news/puentes/news/perú-remite-el-tpp-al-congreso-para-avanzar-en-su-ratificación.

ICTSD. (2016b, September 19). Ratificación del TPP: Batalla cuesta arriba. *International Centre for Trade and Sustainable Development*. Retrieved from http://www.ictsd.org/bridges-news/puentes/news/ratificación-del-tpp-batalla-cuesta-arriba.

Loussikian, K. (2016, October 3). BCA urges action on TPP as opposition mounts in the US. *The Australian*. Retrieved from http://www.theaustralian.com.au/business/bca-urges-action-on-tpp-as-opposition-mounts-in-the-us/news-story/b60e135c0890318ed877a2e77b4e0aba.

Malik, M. (2011). *The full protection and security standard comes of age: Yet another challenge for states in investment treaty arbitration*. Winnipeg, Canada: International Institute for Sustainable Development. Retrieved from http://www.iisd.org/pdf/2011/full_protection.pdf.

Mann, H. (2015). ISDS: Who wins more, investors or states? *Investment Treaty News, June 2015*. Winnipeg, Canada: International Institute for Sustainable Development.

Mann, H., von Moltke, K., Peterson, L.E. & Cosbey, A. (2006). *Model international agreement on investment for sustainable development, negotiator's handbook* (2nd ed.). Winnipeg, Canada: International Institute for Sustainable Development.

Miller, S. & Hicks, G.N. (2015). *Investor–state dispute settlement: A reality check*. A Report of the CSIS Scholl Chair in International Business, January 2015. Lanham, MD: Rowman & Littlefield.

Murphy, K. (2016, November 17). Australia signals support for Chinese-led trade deals to replace TPP. *The Guardian*. Retrieved from https://www.theguardian.com/world/2016/nov/17/australia-signals-support-for-chinese-led-trade-deals-to-replace-tpp.

Nadeau, P. (2016, October 1). Can the United States and Japan ratify TPP? *The Diplomat*. Retrieved from http://thediplomat.com/2016/10/can-the-united-states-and-japan-ratify-tpp/.

OECD. (2008a). *International investment law: Understanding concepts and tracking innovations*. Paris, France: OECD Publishing. Retrieved from http://www.oecd.org/investment/internationalinvestmentagreements/internationalinvestmentlawunderstandingconceptsandtrackinginnovations.htm.

OECD. (2008b). *OECD guidelines for multinational enterprises*. Paris, France: OECD Publishing. Retrieved from http://www.oecd.org/corporate/mne/1922428.pdf.

Othman, L. (2016, November 21). TPP without the US would mean a new agreement, but would be a great loss: PM Lee. *Channel News Asia*. Retrieved from http://www.channelnewsasia.com/news/singapore/tpp-without-the-us-would-mean-a-new-agreement-but-would-be-a/3305712.html.

Pakpahan, B. (2012). Deadlock in the WTO: What is next? WTO Public Forum 10 September 2012. Geneva, Switzerland: World Trade Organization. Retrieved from https://www.wto.org/english/forums_e/public_forum12_e/art_pf12_e/art19.htm.

Potkin, F. (2016, September 29). Singapore, Japan mark 50 years of diplomatic ties with dig at Beijing, TPP talk. *Forbes*. Retrieved from http://www.forbes.com/sites/fannypotkin/2016/09/29/singapore-and-japan-talk-trans-pacific-trade-deal-and-train-diplomacy/#159502d94f09.

Rasid, A.H. (2016, September 29). Malaysia amending 18 laws in preparation for TPPA ratification. *New Straits Times Online*. Retrieved from http://www.nst.com.my/news/2016/09/175727/update-malaysia-amending-18-laws-preparation-tppa-ratification.

Simoyan, M. (2016, November 28). Japan's Abe wants to talk Trump out of quitting TPP. *RT (Russia Today)*. Retrieved from https://www.rt.com/business/368413-us-abe-trump-ttp/.

Sipalan, J. (2016, November 15). As TPP falters, Malaysia looks to China trade pact to bolster trade. *Reuters*. Retrieved from http://www.reuters.com/article/us-malaysia-tpp-idUSKBN13A0H6.

Subedi, S.P. (2008). *International investment law: Reconciling policy and principle*. Oxford, UK: Hart Publishing.

UNCTAD. (1999a). *National treatment*. UNCTAD Series on Issues in International Investment Agreements. New York, NY: United Nations. Retrieved from http://unctad.org/en/Docs/psiteiitd11v4.en.pdf.

UNCTAD. (1999b). *Trends in international investment agreements: An overview*. UNCTAD Series on Issues in International Investment Agreements. New York, NY: United Nations. Retrieved from http://unctad.org/en/Docs/iteiit13_en.pdf.

UNCTAD. (2004). *International investment agreements: Key issues volume I*. Geneva, Switzerland: United Nations. Retrieved from http://unctad.org/en/Docs/iteiit200410_en.pdf.

UNCTAD. (2010). *Most-favored nation treatment*. UNCTAD Series on Issues in International Investment Agreement II. Geneva, Switzerland: United Nations. Retrieved from http://unctad.org/en/Docs/diaeia20101_en.pdf.

UNCTAD. (2011). *Scope and definition*. UNCTAD Series in Issues in International Investment Agreements II. Geneva, Switzerland: United Nations. Retrieved from http://unctad.org/en/Docs/diaeia20102_en.pdf.

UNCTAD. (2012). *Fair and equitable treatment. UNCTAD Series on Issues in International Investment Agreements II*. New York, NY: United Nations. Retrieved from http://unctad.org/en/Docs/unctaddiaeia2011d5_en.pdf.

UNCTAD. (n.d.). IIA Mapping Project. In *International investment agreements navigator*. Geneva, Switzerland: United Nations. Retrieved from http://investmentpolicyhub.unctad.org/IIA/mappedContent.

USTR. (n.d.). *TPP Final Table of Contents*. Washington, DC: Office of the United States Trade Representative. Retrieved from https://ustr.gov/trade-agreements/free-trade-agreements/trans-pacific-partnership/tpp-full-text.

von Moltke, K (2004). *A model international investment agreement for the promotion of sustainable development*. Winnipeg, Canada: International Institute for Sustainable Development.

Warner, J. (2016, April 13). Indonesia may exit case against Churchill Mining after failing to pay. *Morningstar*. Retrieved from http://www.morningstar.co.uk/uk/news/ AN_1460547406582918900/indonesia-may-exit-case-against-churchill-mining-after-failing-to-pay.aspx.

Woolf, N., McCurry, J. & Haas, B. (2016, November 22). Trump to withdraw from Trans-Pacific Partnership on first day in office. *The Guardian*. Retrieved from https://www.theguardian.com/us-news/2016/nov/21/donald-trump-100-days-plans-video-trans-pacific-partnership-withdraw.

World Bank. (2016). *Global economic prospects, January 2016: Spillovers amid weak growth*. Washington, DC: World Bank.

World Bank. (n.d.). *Indonesia*. Washington, DC: World Bank. Retrieved from http://www.worldbank.org/en/country/indonesia.

Yannaca-Small, C. (2006). Improving the system of investor–state dispute settlement: An overview. In *International Investment Perspectives Part II*: Chapter 7: 2006 Edition. Retrieved from https://www.oecd.org/investment/internationalinvestmentagreements/40079647.pdf.

The establishment of small claims court in Indonesia: Expectation and challenge in encountering the globalisation era

A. Afriana & E.L. Fakhriah
Faculty of Law University of Padjadjaran, Bandung, Indonesia

ABSTRACT: The Indonesian civil judicial system has now been improved by Supreme Court Regulation No. 2 of 2015 regarding the settlement of claims in simple disputes. The government expects to implement the principle of a fast, simple and affordable court process, to provide access to justice and ease business practice, through the establishment of a small claims court mechanism. This article elaborates the implementation in Indonesia using a juridical-normative method. It can be safely concluded that in facing the ASEAN Free Trade Area, besides an amendment, judicial reform is also needed to expedite judicial proceedings. The adoption of a small claims court into Indonesian judicial proceedings in Supreme Court Regulation No. 2 of 2015 is expected to prevent a vacuum arising in law, by providing investors with a mechanism for fast dispute settlement. However, for the best interests of the civil procedural law, it is also important to codify the unification of the national law in this area.

1 INTRODUCTION

In Indonesia, a civil claim can be settled both through litigation and non-litigation mechanisms. In practice, most parties to disputes tend to choose the non-litigation mechanisms, such as mediation, consolidation and arbitration, because they are deemed to be more effective in terms of time, cost and confidentiality (Fakhriah, 2013). However, dispute settlement through this option requires consensus and good faith from the relevant parties. Unfortunately, it is ineffective if the "defeated" party does not perform his/her duty according to the judgement/award made in the non-litigation proceeding; hence, its execution cannot be forced.

Meanwhile, the litigation mechanism needs lengthy procedural stages, starting with the filing of the claim, subpoenas, and so on, before it can reach a final judgement. Current positive laws in Indonesia, namely the Herzien Indlandsch Reglement (HIR) and the Rechts Reglementvoor Buitengewesten (RBg), do not differentiate the procedures in civil actions according to the amount of claim. This condition is relatively unfair for disputing parties that have simple claims involving a small amount. Further, disputes that can be settled through the Consumer Dispute Settlement Body [*Badan Penyelesaian Sengketa Konsumen* (BPSK)] are limited to certain cases in which the plaintiff has to be the end-consumer.

For the foregoing reasons, Indonesia is currently updating its judicial bodies through the blueprint of the Indonesian Supreme Court Judicial Reform 2010–2035, in which there are several things that represent a primary focus of renewal, including the Small Claims Court (SCC), which is intended to settle simple and/or small civil claims.

SCCs already exist in many countries, both those with common law systems and those with a civil law system. Recognised as the "people's court", the SCC focuses on informal procedures (Ferraz, 2008, 2010) and aims to give access to justice for the disputing parties within a short period of settlement as well as acting as a bridge between litigation and non-litigation proceedings. A further introduction to the SCC is described under Indonesian Supreme Court [*Mahkamah Agung Republik Indonesia*] Regulation No. 2 of 2015 regarding the Procedure for Small Claims Settlement (Republic of Indonesia, 2015).

As a country that enacts civil law, adopting an SCC according to Supreme Court Regulation No.2/2015 creates a significant difference in the code of conduct between it and common judicial proceedings. In fact, one of the main principles underlying the civil law system is that law acquires its binding power due to the existence of regulations in its legislation that are systematically arranged in a certain codification or compilation (Suherman, 2004). This basic principle is maintained, given the fact that the main objective of the law in this system is certainty.

The SCC exists in Indonesia with the purpose of reforming the judiciary system by giving access to justice for disputing parties involving small claim amounts. The SCC is meant to give a fast and economical solution for dispute settlement, particularly in cases with small to medium amounts of losses, without incurring too much cost.

The SCC also aims to acknowledge the 2015 commencement of the ASEAN Free Trade Area, in which experts expect that there will be many business conflicts and disputes of a small scale that will end up in court proceedings. Supreme Court Regulation No. 2/2015 appears to expedite the dispute settlement process in accordance with the principle of simple, fast, and affordable courts, noting that, up until now, people who seek justice have complained that it takes a long time to settle their disputes in court. As a form of globalisation, free trade in the ASEAN region leads to the removal of economic borders between countries. As one of the most active trading countries in the world, Indonesia cannot detach itself from this influence of globalisation as a social phenomenon, and it has also influenced other aspects of life, both economically and in the regime of law. Therefore, because of the urgency of the trading world, Indonesia needs faster proceedings at the SCC, where investors with small claim amounts can settle their disputes more quickly than before.

This article will elaborate the effectiveness of the SCC in Indonesia in its encounter with the era of globalisation, and the regulation model that is most likely to follow thereafter.

1.1 *Research questions*

Using the judicial-normative as a research basis, which prioritises the use of secondary data to support the primary data, this article discusses the following questions:

1. How effective is the implementation of the SCC in Indonesia in dealing with the era of globalisation?
2. What kind of model and regulation of the SCC can be made in order to reform the Civil Procedural Law in Indonesia?

2 DISCUSSION

2.1 *The effectiveness of the SCC in Indonesia in the globalisation era*

As a country actively participating in world trade activity, Indonesia cannot be free from the influence of globalisation. It penetrates almost every aspect of life, including social, economic and, of course, legal. Globalisation has been contributing towards the changes that have occurred in regional, national and international regimes. In business activity, globalisation has contributed to changes in paradigms of the law, with every change in the economy also altering the law and legal practice. Customs that previously could only be found in the common law system are now formally implemented in Indonesia as a result of the power of globalisation.

In spite of its practice in civil law countries, the SCC was actually founded and developed in common law countries. Practices in Australia, Singapore, the United States and Japan show that the SCC is used to examine and to adjudicate legal matters with small claim amounts of, for instance, not more than 100 million rupiah. Thus, in Singapore, a claim of no more than 200 million rupiah can be settled straightaway just by a written consensus between the parties to the dispute. Meanwhile, in Europe, the Small Claims Procedure (SCP) has been introduced since 2009 to the courts of all countries, except Denmark, in the European Union to settle disputes with a maximum claim amount of 2,000 euros (Mańko, 2014). The philosophy behind the establishment of these procedures involves an institution to provide access to justice for all parties, creating proportionality between the fast speed of the judgement and the small size of the claim (Durray, 2002).

According to the 2013 Annual Report of the Supreme Court of the Republic of Indonesia, the access to justice policy is embodied as an amendment to the public service policy for the poor and/or the incapable in court (Mahkamah Agung RI, 2014), providing:

1. Exemption from payable fees in Court, access to the legal aid post, and trial outside of the Court building;
2. Mediation;
3. The right to a comprehensive service of legal identity;
4. A simple mechanism of dispute settlement (i.e. the SCC).

In this globalisation era, the fields of economics and business are growing sharply, so it is the right time for Indonesia to improve the ease of doing business for its people (Sutedi, 2014). This is not only closely related to investment issues, but is also related to the manifestation of access to justice for disputing parties, as well as legal certainty in settling disputes in business practice.

The Indonesian Chief Justice, Mr Hatta Ali, emphasised that in this Free Trade Area, international society has placed a spotlight on Indonesia, thanks to the absence of a SCC within its legal system. Hence, in responding to this issue, Supreme Court Regulation No. 2/2015 was issued in relation to the Procedure for Small Claims Settlement, which is adopted conceptually from the SCC, in order to manifest a modern democratic country and enhance the public service provided to society and, in particular, those seeking justice (embaharuan Peradilan, 2014).

According to the indicator created by the World Bank, Indonesia ranked 170th for its enforcement of contracts within business practice. This indicator was created in 2015 through examination and research by the Ease of Doing Business (EODB) team. The problem with this indicator is that the standard used for the assessment is based on a common law system, but Indonesia is implementing a civil law system. Furthermore, Indonesia has changed since that assessment, due to the existence of new regulations included in the assessment indicator: for instance, Supreme Court Regulation No. 2/2015 and Supreme Court Regulation No. 1 of 2016 [Perma No. 1/ 2016] (Afriana, 2008). The main criteria for cases to be settled by the simple claim mechanism are breach of contract and/or other unlawful acts, with a maximum amount of claim of 200 million rupiah.

Civil proceedings before a court in Indonesia consist of a long series of stages, starting with the court of the first instance, then appeals, cassation and judicial review. This takes a lot of time, regardless of society's need for a simple and fast process of law. This is why an effective and efficient judicial system design is needed (Kementrian Koordinator Bidang Perekenomian Republik Indonesia, n.d.)

As a matter of fact, the main focus of a civil dispute settlement is time, both in litigation and non-litigation proceedings. However, Indonesian litigation proceedings have a lot of procedural mechanisms, which are all concluded in its formal-juridical provisions and thus cannot be superseded. Dispute settlement within a judicial body uses the final and binding national civil procedural law, so the applicable provisions and principles therein cannot be evaded. One of the most important principles is the fast, simple, and affordable principle.

Ironically, however, this most important final and binding principle is frequently unfulfilled, with the parties to the dispute incurring excessive bills over long periods (Tjukup et al., 2015). In order to enforce the law, the court places a certain price on a registrar's service, the summons of the parties, and the stamp cost on a required signature (Wijayanta, 2012). However, this should not be taken lightly, as the parties often have to pay such bills on numerous occasions thanks to the long process, accompanied by the anxiety associated with waiting for legal certainty. The court has been experiencing difficulties in its effectiveness in maintaining court administration in good order and, as a result, a backlog of cases is building up.

For the aforementioned reasons, a new effective mechanism in all instances was urgently needed. In particular, the existence of huge costs might actually hamper people with small claims from filing their cases in court. Hence, the SCC was adopted and subsequently regulated under Supreme Court Regulation No. 2/2015 and, with this new mechanism, the fast, simple and affordable principle can be achieved effectively and efficiently. It is fast, cheap and just, as well as highly suitable for small claims.

Since its enactment, Supreme Court Regulation No. 2/2015 has helped simple claims to be adjudicated effectively, both for breach of contract and unlawful act issues, and whether involving an individual or a corporation (Mahkamah Agung RI, 2016).

This regulation is also commonly used by banks to sue debtors, particularly those in connection with securities given by the debtors to the banks, along with the right to sell in cases of breach of contract. Thus has it been demonstrated that the SCC has been one of the most effective law reforms in Indonesia, because it has answered the growing need of society to settle its disputes efficiently.

2.2 *The SCC regulatory model in Indonesia in reform of the national civil law*

It can clearly be seen from Presidential Regulation No. 5 of 2010, concerning the National Medium Term Development Plan (RPJMN) 2010–2014, that the government is prioritising national development in the fields of law and the economy. These two fields have to be developed simultaneously to create harmonised legal and economic systems.

One of the most prominent points in the development of national law is the legislative process or law reform. As a civil law country, Indonesia upholds legal certainty through its laws and regulations. It is developed through systemised unification within a certain codification or compilation. The legal system in Indonesia is still oriented towards its former occupying country, the Netherlands. Most Indonesian civil law regulations were adopted from the Dutch and some have not even been amended to this day. Due to the legalism principle observed by civil law countries, Indonesia highly respects written laws and regulations as sources of authority, including its civil law code of conduct. The civil law code of conduct regulates matters in relation to the conduct of responsibilities, which are formal and binding, so that they cannot be evaded and have to be obeyed by every person. Since it has been ages since Indonesian civil law codification was first created, it now needs to be reformed to respond to the dynamic development of society.

Philosophical, sociological and juridical aspects are all important in the amendment of law. Philosophically, a sense of justice and righteousness has to be upheld, while from the sociological perspective, the concordance between the law and living cultures has to be taken into account. In terms of the juridical aspect, every new regulation must be in accordance with the prevailing laws and regulations in the hierarchy.

Currently, the government is preparing a bill on civil procedural law. However, in this bill, the SCC has not yet been stipulated. The amendment of national law is best carried out in full, not partially. This means that all rigid provisions enforced within civil procedural law have to be unified. Because the preparation of the bill on civil procedural law is still in progress, the use of the Supreme Court Regulation as the grounds to establish the SCC as one of the alternatives for rapid dispute settlement is important. Supreme Court Regulation No. 2/2015 is meant to fill the gap, and to avoid a vacuum of law, as regulated in Article 79 of Law No. 14 of 1985, which states that "the Indonesian Supreme Court may further regulate necessary provisions in order to keep circulated judicial arrangement if there is insufficiency in this Law."

As a comparison, in countries that have already fully established SCCs, whether they are civil law or common law countries, the SCC is promulgated within the national law; for instance, the "Implementation Act" in the Netherlands and the "SCT Act" in Singapore. Thus, it is expected that SCC promulgation be integrated into Indonesian national law as part of the bill on civil procedural law, in order to ensure its unification.

3 CONCLUSION

As a country adopting a system of civil law, the implementation of an SCC in Indonesia is effective in meeting the needs of society, because it offers a fast, simple and affordable procedure, which surpasses existing proceedings. As a result of the globalisation era, there is an increased possibility of disputes arising from the commercial field in, for instance, the ASEAN Free Trade Area. Hence, this will need the amendment of laws and judicial reform in order to expedite trial proceedings. The adoption of an SCC in Supreme Court Regulation No. 2/2015 is deemed to be supportive of the commercial field, and it also gives benefits to investors in disputes involving small claim amounts, enabling their disputes to be resolved in a rapid manner.

The establishment of the SCC in Supreme Court Regulation No. 2/2015 has been one of the more effective solutions in preventing a vacuum of law, as well as compensating for the absence of national civil procedural law regarding such matters. Nonetheless, without the integration and further establishment of the essence of the SCC in the form of national law, the core interest will not be met. Hence, amendment of Indonesian national civil law is needed, through the bill on civil procedural law, to achieve the necessary unification of formal law in Indonesia.

REFERENCES

Afriana, A. (2008). Penerapan Peraturan PerMA No. 1 Tahun 2008 tentang prosedur mediasi di pengadilan dalam praktik di Pengadilan Negeri Bandung. *Jurnal Al-Adalah, 7*(1), 1–9.

Durray, A. (2002). Some thoughts on current issues of natural justice and tribunals. In *5th Annual AIJA Tribunal's Conference 6–7 June 2002, Melbourne, Australia*. Melbourne, Australia: The Australian Institute of Judicial Administration.

Fakhriah, E.L. (2013). Mekanisme *Small Claims Court* dalam mewujudkan tercapainya peradilan sederhana, cepat, dan biaya ringan. *Jurnal Mimbar Hukum UGM, 25*(2), 258–270.

Ferraz, L.S. (2008). *Small claims courts and qualified access to justice: An empirical research* (Doctoral thesis, University of São Paulo, São Paulo, Brazil).

Ferraz, L.S. (2010). Access to justice and mediation in small claims courts: Empirical research and statistical analysis in nine Brazilian states. In *5th Annual Conference on Empirical Legal Studies, 5–6 November 2010, New Haven, Connecticut, USA*. New Haven, CT: Yale Law School. Retrieved from https://papers.ssrn.com/sol3/Delivery.cfm/SSRN_ID1636454_code988099.pdf?abstractid=1636454&mirid=1.

Kementrian Koordinator Bidang Perekonomian Republik Indonesia. (n.d.). EODB: Ease of Doing Business. Retrieved from http://eodb.ekon.go.id.

Mahkamah Agung [Supreme Court] Republik Indonesia. (2014). *Annual Report 2013*.

Mahkamah Agung [Supreme Court] Republik Indonesia. (2016). *Direktori Putusan: Perdata*. Retrieved from https://putusan.mahkamahagung.go.id/pengadilan/mahkamah-agung/direktori/perdata.

Mańko, R. (2014). *European Small Claims Procedure: Legal analysis of the Commission's proposal to remedy weaknesses in the current system*. Brussels, Belgium: European Parliamentary Research Service. Retrieved from http://www.europarl.europa.eu/RegData/etudes/IDAN/2014/542137/EPRS_IDA(2014)542137_REV1_EN.pdf.

Republic of Indonesia. (1985). *Undang-undang No.14 Tahun 1985 tentang Mahkamah Agung* [*Law Number 14 of 1985 regarding Supreme Court*].

Republic of Indonesia. (2010). *Peraturan Presiden Republik Indonesia No. 5 Tahun 2010 tentang Rencana Pembangunan Jangka Menengah Nasional Tahun 2010–2014* [*President Republic of Indonesia Regulation Number 5 of 2010 regarding National Medium Term Development Planning for 2010–2014*].

Republic of Indonesia. (2015). *Peraturan Mahkamah Agung Republik Indonesia No. 2 Tahun 2015 tentang Pelayanan Terpadu Sidang Keliling Pengadilan Negeri dan Pengadilan Agama* [*Supreme Court Republic of Indonesia Regulation Number 2 of 2015 regarding Integrated Service of District/Sharia Circuit Courts*].

Republic of Indonesia. (2016). *Peraturan Mahkamah Agung Republik Indonesia No. 1 Tahun 2016 tentang Prosedur Mediasi di Pengadilan* [*Supreme Court Republic of Indonesia Regulation Number 1 of 2016 regarding Mediation Procedures in Courts*].

Suherman, A.M. (2004). *Pengantar Perbandingan Sistem Hukum*. Jakarta, Indonesia: Raja Grafindo Persada.

Sutedi, A. (2014). *Hukum Perbankan: Suatu Tinjauan Pencucian Uang, Merger, Likuidasi, dan Kepailitan*. Jakarta, Indonesia: Sinar Grafika.

Tjukup, K. *et al.* (2015). Penerapan Teori Hukum Pembangunan dalam Mewujudkan Peradilan Sederhana, Cepat, dan Biaya Murah. *Law Journal in Civil Law, Conduct ADHAPER, 1*(1), 145–160.

Wijayanta, T. (2012). Bantuan Hukum Golongan Tidak Mampu dalam Perkara Perdata di Pengadilan Negeri Yogyakarta. *Mimbar Hukum Journal, 24*(1), 111–119. doi:10.22146/jmh.372.

Causation in the context of environmental pollution crime

A. Sofian & S. Fitriasih
Faculty of Law, Universitas Indonesia, Depok, Indonesia

ABSTRACT: The number of cases of environmental pollution committed by corporations is on the rise in Indonesia. A limited number of offenders were able to be prosecuted and held accountable before the courts, but the courts have found it difficult to measure the extent to which the defendants must be held accountable with regard to the environmental pollution. This research aims to determine actions that cause pollution to the environment and measure the accountability of offenders. To answer these questions, the research employs a normative approach. The study found that there are two types of criminal offenses related to environmental pollution, namely formal and material offenses. In a formal offense, the causation doctrine is not used to determine the relationship between action and consequences. In a material offense, on the other hand, causality is needed to determine the action (or actions) that gives rise to the resultant effect, which in this case is damage to the environment. In examining material offense of environmental pollution, the courts have difficulty in determining the actions that have caused damage to the environment. The causation doctrine is not employed by the courts due to the difficulty of defining the actions that result in the prohibited consequences. Given the situation, this research recommends that the causation doctrine be used as a filter to identify the action that gives rise to the effect. The courts have to keep up with developments in causation doctrines that are relevant in determining the action (actions) that leads to environmental pollution, and to ultimately measure the extent to which the offender is accountable.

1 INTRODUCTION

This paper aims to measure the attribution of corporate criminal liability for environmental damage or pollution. It is important to apply the doctrine of causation to measure the attribution of the criminal liability. More specifically, the paper aims to answer the research question whether or not judges tend to use the doctrine of causation when they handle cases causing environmental damage or pollution and when there are some defendants who are convicted of commiting a crime causing a prohibited consequence. It also aims to review causal uncertainty in the context of crime causing environmental damage or pollution. In this context, whether or not judges will measure it by applying a probabilistic approach when they find it difficult to establish an act causing the prohibited consequence. The probabilistic approach to environmental pollution cases has been deeply studied by some scientists, such as Shavell (1984) and Faure (2001). In this context, Shavell (1984) explains measurements for polluter liability. Meanwhile, Faure (2001) is concerned about efforts to attribute a consequence to an act(s). Meanwhile, Robinson (1985) adopts a probabilistic approach to causation when the causality is less than 100 (Bentata, 2013, p.12).

Judges need to understand causation in the context of an environmental crime causing a prohibited consequence such as environmental damage or pollution because the existing laws do not provide any measurement for an act which causes a consequence. Given the situation above, this research actually aims to answer the question whether or not the doctrine of causation has been used by courts in environmental crimes.

The researchers will apply a normative juridical research method to answer the question by examining a number of court decisions, journals, and literature related to environmental pollution crimes, the Law on Environmental Protection and Management as well as the Draft Criminal Code.

1.1 *Doctrine of causation*

Causation (cause and effect) is a relationship or a process between two or more events or conditions of an event in which a factor causes another factor or other factors. Let us take a simple example of a causal statement: If the switch is up, the light is on. It can be said that the causal (cause and effect) relationship between the two events is that switching up the switch causes the light on. Causation (cause and effect) has always been a central topic in the history of philosophy. This is because the concept of causation encompasses our way of thinking about ourselves, our environment, and the world in which we live and our relation to the world itself (Blackburn, 1993, p. 366). In fact, someone can say that our awareness of the world and our representation in the world depend on a causal relationship. Discovering the causal relationship gives us an insight of the structure of causation in the world and forms our basis to learn how to act smartly in the world. Exploring what really becomes a cause will enable us to build the structure of causation and this will enable us to predict rationally, make a decision, and act in the world.

Causation has a long history. It is more popular in the fields of natural sciences and philosophy (Bemmelen, 1987, p. 154). In natural sciences, the doctrine of causation is used to explain the structure of things, or the system of elements, which are interrelated, form a causal relationship, and have a causal principle. Natural sciences which are used to explain natural law use the principle of cause and effect. For example, a metal frame will expand when it is heated. The relationship between heat and expansion is a cause and effect relationship (Kelsen, 2007, p. 86).

Causation which is often called a causal relationship is the most difficult issue in the philosophical dictionary. The main difficulty lies in the formulation of the cause and effect. Cause is often formulated as an agent which causes a change or prevents a change, while effect is formulated as something resulting from the cause. In short, it can be said that determinism of this teaching perceives that any occurrence is solely an effect of a cause (Kattsoff, 2004, pp. 55–56).

Daniel Little in his book, *Varieties and Social Explanation: An Introduction to the Philosophy of Social Science*, explains that theory of causation is very suitable for a social explanation, although we need to consider a few things: First, the theory of causation cannot be generalized for all social conditions. Second, the theory of causation greatly depends on the causal mechanism which relates cause and effect, and third, the theory of causation involves a reference to beliefs, wishes, strengths, and obstacles which influence an individual in a social fact (Little, 1991, pp. 13–14). From this explanation, it is clear that causation is actually a logic thinking to explain an event in which the event occurs because there are factors influencing it. In explaining the event, there are many influencing factors, not just one factor. In causation, it is these factors which are explained. In Daniel Little's words, there is a social explanation that must be explainable.

It is important to give a meaning of "cause" to better understand social explanation. According to him, there are three main ideas to explain cause, namely: Causal Mechanism (CM), Inductive Regularity (IR), and Necessary and Sufficient Condition (NCS).

Daniel's causal mechanism formula can be explained through the following example: If C is Cause and E is Effect, CM of C and E is: $C \rightarrow C1 \rightarrow C2 \rightarrow C3 \rightarrow C4 \rightarrow E$. A more concrete example: C = A bolt is left loose on an automobile wheel, E = Accident. The causal mechanism is: A bolt is left loose on an automobile wheel (C). The vibration of the moving wheel causes the loose bolt to fall off completely (C1). The increased vibration causes the remaining bolts to loosen and detach (C2). Once the bolts were completely gone, the wheel was released (C3). Since the wheel was released, the automobile was out of control (C4). Finally, the accident occurred (E) (Little, 1991, p. 14).

Little (1991, p. 16) says that according to the theory of Necessary and Sufficient Condition (NSC), C has a causal relation with E only if C is a necessary and a sufficient condition for the occurrence of E. Therefore if C does not present, E will not occur. He takes the relation between solar radiation and fire as an example by saying that the presence of excessive solar radiation on the dark surface of an object may cause fire although it is a fact that a single condition is almost never a sufficient condition for the occurrence of another event. It means that a single factor, excessive solar radiation, may not be sufficient to cause an object be caught on fire because other conditions such as oxygen, temperature, etc., will be jointly needed to cause fire.

He further says that necessary and sufficient conditions play a vital role in measuring a causal relationship. Necessary is a condition which is absolutely present in a series of causes because without their presence, certain events will not occur. The presence of oxygen is a necessary condition for the occurrence of combustion. Therefore, if oxygen does not exist, there will not be a combustion. This shows that oxygen is a necessary for combustion. Meanwhile, sufficient is a condition which is deemed sufficient for the occurrence of an event. He says that the presence of oxygen is not a condition for the occurrence of combustion because oxygen may present without the occurrence of combustion. Little (1991, pp. 25–26) says that it is rarely true that any single condition is sufficient for the occurrence of any other, instead a group of conditions may be jointly sufficient.

Daniel Little prioritizes regularity of each event. Unlike Daniel, John Mackie explains that elements of act are equally necessary to produce an effect. However, these elements should not work regularly (Honoré, 1999, p. 95). It is possible that the occurrence of the event is a combination of various causes (plurality of causes). However, although these conditions or events are different, each of them must be a sufficient element to enable them to produce a consequence (Honoré, 1999, p. 96).

These plural conditions are jointly sufficient to produce a consequence. However, it is possible that they are not so necessary. There are a number of alternative conditions which jointly produce the same consequence. Mackie states such conditions as something which is necessary of a sufficient collection (but necessary) for bringing about a type of consequence which he calls as an INUS condition (Insufficient but Necessary part of an Unnecessary but Sufficient) (Honoré, 1999, p. 97).

He gives an example of a traffic accident which causes death. The traffic accident occurred because the driver drove the car carelessly (one condition). Another condition was the driver drove the car in a high speed on a highway. These two conditions, driving the car carelessly and driving the car in a high speed, are the INUS conditions which cause the accident. The two conditions are different, but they jointly produce a necessary condition for the traffic accident (Honoré, 1999, p. 96).

Theory of causation can only be applied to a crime which causes a prohibited result. In the context of common law, it can be only applied to offenses which fall under the category "result crimes". Meanwhile, there are some offences in civil law, such as material offence, result-qualified offence, and non pure omission. However, these three categories of offences actually are still categorized into "result crimes".

1.2 *Environmental regulations*

Environmental issues in Indonesia are regulated under the Law Number 32 Year 2009 on Environmental Protection and Management. The law contain legal provisions related to state administration, criminal law, and civil law. The law gives proper protection to the interests of the environment because those polluting or damaging environment shall be subject to not only administration sanctions but also criminal sanctions. It also explicitly states that the personnel in charge of a company causing environmental damage or pollution will be criminally responsible for environmental damage or pollution caused by the company. It also regulates corporate liability so that an entity and/or its board can be held criminally responsible for environmental crimes (Hardjasoemantri, 2002, p. 12).

In addition to the Law on Environmental Protection and Management, other laws also give protection to environment in Indonesia, such as the Law Number 7 Year 2004 on Water Resource, the Law Number 26 Year 2007 on Spatial Management, the Law Number 18 Year 2008 on Waste Management, the Law Number 18 Year 2013 on the Prevention and Eradication of Forest Destruction, etc. Of these laws, this paper will only discuss the Law Number 32 Year 2009 on Environmental Protection and Management and the Draft Criminal Law which regulate environmental crimes.

In the Law Number 32 Year 2009, criminal provisions are regulated from Article 97 to Article 120. Of a total of 23 articles, there are only three articles which are formulated materially, while the rests are formulated formally.

Since the two articles are formulated materially, the offences are perfect when a prohibited result has occurred. In Article 98, the prohibited result occurs when the act has exceeded the set of "standard quality" or "standard criteria for environmental damage". However, when the act has not yet exceeded the "standard quality" or "standard criteria for environmental damage", it is not yet categorized as an offense. The measurement of "standard quality" and "standard criteria for environmental damage", of course, is very academic and scientific because scientists need to conduct a scientific measurement. Such measurement, of course, is much influenced by various factors, such as skills of scientists who measure them, sampling techniques, laboratory tools, chemical substances used, calculation formula used, etc. to ensure the accuracy of its result(s). If, based on the result of the measurement, there is a fact that the act has exceeded the "standard quality" or "standard criteria for environmental damage, it has brought about a prohibited consequence or vice versa.

In practice, however, different scientists may have different measurement results. For example, by using the same measurement method, scientists A and B may have different results. Based on the results of measurement conducted by scientist A, the act has exceeded the standard quality or standard criteria for environmental damage, while based on the results of measurement conducted by scientist B the act has not yet exceeded the standard quality or standard criteria for environmental damage.

In this context, judges, of course, will use their belief as a basis for decision making, although it is clear that a subjective belief cannot be used in this case because instead of using their subjective belief, they have to use scientific measurement results as a strong basis for their decision. Therefore, these two different measurement results may lead the judges into a confusion: Scientist A says it has exceeded the standard quality or standar criteria for environmental damage, while scientist B says it has not yet exceeded the standard quality or standard criteria for environmental damage.

Formulation of material offense also has to consider other factors which may also contribute to the occurrence of a prohibited consequence. In the context of an act causing a prohibited consequence (exceeding the standard quality or standard criteria for environmental damage), it is also important to consider other factors that may contribute to the occurrence of the prohibited consequence. In many cases, there are other factors or actors that independently contribute to the occurrence of the consequence. For example, there are three companies that separately dump their waste into a river. Company A dumps 150 liters of waste into the river per day; company B dumps 160 liters of waste into the river per day; and company C dumps 170 liters of waste per day into the river. Hundreds of people living along the river also discharge their domestic waste into the river. There are other factors which also have deteriorated the condition of the river because a number of hospitals and hotels also dump their waste into the river. Some government's offices also dump their waste into the river. Therefore, the environment is polluted. The question is: Which act has caused the environmental pollution? After a water sample from the river is taken and tested, it is found that it has exceeded the set water standard quality.

The difference between Articles 98 and 99 lies in the level of guilt. Article 98 is categorized as intentionality, while Article 99 is categorized as omission. However, the levels of guiltare are significantly determined by an act causing a prohibited consequence. Guilt can only be measured after the consequence of the act occurs. As long as its consequence does not exceed the set standard quality or standard criteria for environmental damage, the element of guilt will be ignored or need not be proven.

Article 112 explicitly mentions "negligence" or failure to act (omission) which causes a prohibited consequence. The prohibited consequence is the occurrence of environmental pollution and/or damage which causes death. Thus, there must be consequences to enable us to categorize them as an offense, namely: (1) environmental pollution and/or damage, and (2) death. If the two prohibited consequences do not occur, the omission cannot be categorized as a prohibited offense. The next category is that the one who is negligent must be an authorized government official. It means that he or she has a legal obligation to prevent the prohibited consequence.

1.3 Court decision

Environmental crimes discussed in this paper are court judgments which punish criminals causing environmental pollution or damage. The first case is a forest fires case involving PT. MAL in Riau. The defendant committed its action from 2008 to 2009. In the case, PT. MAL cleared forest for its palm oil plantation expansion. This had resulted in forest fires. The forest fires were exacerbated by its failure to take appropriate measures to extinguish the forest fires. Consequently, the forest fires covered more than 4,745 acres. The forest fires, which occurred at the same time and covered such a wide area, had caused environmental pollution and damage, such as peat land damage and the production of gases which might endanger environment, humans and other creatures.

Prosecutors prosecuted PT. MAL with one year imprisonment and a fine of IDR 100,000,000 (one hundred million rupiah) and a subsidiary punishment of two months imprisonment in case it could not afford to pay the fine. In its *pledoi*, the lawyer said that the company had taken appropriate measures to extinguish the fires by providing tractors, firefighter units, excavators, fire watch towers, etc.

In their decision, the judges of Pekan Baru District Court stated that the defendant had unlawfully committed an action causing environmental damage as had been prosecuted by the prosecutors. Then, the prosecutors appealed the case because the judges did not impose imprisonment on the defendant.

In the appellate court decision, the judges stated that in environmental crime cases, there was an environmental effect i.e. harms, such as soil or land damage, the death of microorganisms, health problems, flight interruptions, global warming, etc. In their considerations, the judges also acknowledged that they faced difficulties to prove the case because they did not have technology to enable them to measure the harms or losses caused by the act. Based on these considerations, the judges of the appellate court punished the defendant to 6 (six) months imprisonment and 1 (one) year of probation and a fine of IDR 100,000,000 (one hundred million rupiah). Since the defendant did not appeal the decision to the Supreme Court, we do not know the views of judges in the Supreme Court.

In the case above, the prosecutors also presented an expert to the court to assess the environmental pollution and damage. Explanations from the expert had become judges' considerations to punish the defendant. Assessment conducted by the expert found that there were environmental pollution and damage. However, the assessment was only conducted based on the results of a short observation and observable effects. He did not explain the act and causal factors more comprehensively.

Neither the prosecutors nor the judges used doctrine of causation in establishing act(s) which caused the effects (environmental pollution and damage). The judges also did not carefully consider factors contributing to the occurrence of the environmental pollution and damage. Similarly, the lawyer did not consider other factors that might contribute to the occurrence of the forest fires and environmental pollution. Both the prosecutors and judges only focused on an act i.e. PT. MAL.

The prosecutors said that the only act which had caused the forest fires and environmental pollution was the act of PT. MAL (forest clearance) for the expansion of palm oil plantation. The company also did not take serious measures to extinguish the fires. However, its lawyers denied this by saying that the company had provided heavy equipment and firefighter units, although they were not enough to extinguish the forest fires. The judges insisted that the forest damage was caused by the act of PT. MAL. However, when measuring the level of guilt, they said that the company did not have any intention to cause the forest fires so that they did not impose imprisonment on the defendant but only a fine.

The judges should consider logical consequences of forest fires. Any possible consequence of the act should be considered. A series of acts committed by PT. MAL showed that there was a systematic effort to set the forest on fire and the company did not carefully consider possible risks of forest fires. The risk of forest fires was a logical consequence of the acts so that they needed to consider the number of firefighting units that they had to minimize or prevent the prohibited consequence.

According to Daniel Little's theory of causation, we have to take all of the factors into account. In this context, setting forest on fire can be categorized as C1 and failure to provide proper firefighter units can be categorized as C2. Thus, the forest fires are an accumulation of $C1 + C2 = E$. C1 is the first cause, C2 is the second cause, and E is the effect.

Meanwhile, the views of appellate court judges showed that they scientifically took into account effects caused by the act (setting forests on fire). They insisted that it was this sole act which had caused the effect. Therefore, they punished the defendant to 6 (six) months imprisonment and 1 (one) year of probation and a fine. Irrespective of the length of the imprisonment, the court could rationally take into account the effect of the defendant's act, the act had caused the prohibited effect, and the effect had caused a great loss. It is irrelevant to use the theory of fault to limit the act because they were committed by a corporation.

Decisions of the higher court and the first level court were made based on different points of view. The judges of the first level court established the attribution of the acts to the criminal liability by taking into account the defendant's fault, and they, therefore, said that the defendant was negligent. Meanwhile, the judges of the appellate court had used the theory of corporate criminal liability i.e. strict liability (there is no need to prove corporate fault). In the context of causation, it is only the criminal act that needs to be taken into account. Meanwhile, fault is a component to establish whether or not there is a criminal liability or the level of the criminal liability.

The second case was an environmental damage or pollution case handled by the Supreme Court. The case involved PT. Dongwoo Environmental Indonesia (DEI). It is a foreign company that works in hazardous and toxic waste management in Bekasi. In the case, the company's staff disposed waste, and this caused dizziness, queasy, and faintness among local communities as they were exposed to toxic gas. They put the processed waste, which took a form of black mud, into some plastic bags and containers and disposed them in an empty area in Sempu Village, North Cikarang, Bekasi. The black mud and liquid caused odor. The manager of the company knew that they disposed the waste in the area, and these staff got extra payment for that. This was done from 2005 to 2006. They disposed about 410.2 tons of waste in some areas in Bekasi.

Therefore, the public prosecutors prosecuted PT. DEI with (1) paragraph 1 of Article 41 of the Law Number 23 Year 1997 on Environmental Protection and Management jo paragraph 1 of Article 64 of the Criminal Code, (2) paragraph 1 of Article 42 jo Article 47 of the Law Number 23 Year 1997 jo paragraph 1 of Article 64 and (3) paragraph 1 of Article 44, Jo Article 45, Jo Article 47 of the Law Number 23 Year 1997 and Jo paragraph 1 of Article 64 of the Criminal Code.

Then, the public prosecutors prosecuted PT. DEI, which was represented by Kim Young Woo as the President Director of PT. DEI, with: (1) a fine of IDR 325,000,000 (three hundred and twenty-five million rupiah) subsidiary 6 (six) months of imprisonment, (2) deprivation of profit from the disposal of 410.2 tons of waste, and (3) the closing of PT. DEI.

The first level court, Bekasi District Court, fully accepted the prosecutions. Then, West Java High Court strengthened the Bekasi District Court decision.

In his plea, the lawyer of PT. DEI claimed that the judex fact did not consider other waste similar to the one disposed by PT. DEI in Sempu Village, North Cikarang, Bekasi. There were also a number of other companies which disposed their waste in the empty area so that the judex fact should consider the actions of other companies. Therefore, the judge had to be able to prove that it was waste disposed by PT. DEI which caused the environmental damage or pollution and caused the dizziness, queasy, and faintness among local people in Sempu Village. The lawyer also claimed that the public prosecutors mixed up their prosecution with two types of crimes, namely: formal and material crimes so that it was unclear.

The Supreme Court imposed the following punishments on PT. DEI: (1) a fine of IDR 650,000,000 and subsidiary of six months of imprisonment, (2) deprivation of profits from the disposal of 410.2 tons of waste, and (3) the closing of PT. DEI.

It is interesting to analyze the above case because the public prosecutors used articles for material crimes or result crimes in which the action had caused environmental pollution and damage and diseases among local communities living around the hazardous and toxic waste disposal. It means that they had applied the doctrine of causation in this case. In the case, the prohibited consequence is environmental pollution and damage caused by the action of disposing the hazardous and toxic waste in an empty area near the residential area. Then, the hazardous and toxic waste caused the environmental pollution and damage and even local communities' exposure to the hazardous and toxic waste. Causation logic given by the public prosecutors was able to convince the judges. In this context, waste disposal (C1) was the mere action. Hence, there was no *novus actus interveniens* which could break this action. Therefore, the District Court, the Appellate Court, and the Supreme Court believed in this repeated mere action.

In his appellate memoir, the defendant's lawyer claimed that there was another factor that could break PT. DEI's action i.e. other companies' actions of disposing their waste in the empty area. The actions of these other actors had also contributed to the environmental pollution and damage. Therefore, they also should be held responsible for the environmental pollution and damage. However, he could neither prove this nor give authentic evidence. Thus, these intervening factors could not directly omit the attribution of PT. DEI's action to the criminal responsibility.

2 CONCLUSION

Cases discussed in the paper show that the theory of causation had been applied to establish an act which has caused the prohibited effect. Prosecutors and judges had different views of which theories of causation that they would apply to establish an act causing the prohibited effect. The prosecutors did not carefully take into account the theories when they were handling cases of environmental crimes causing environmental pollution and damage. If judges could attribute the act and the prohibited result based on evidence brought to the court, they usually would not further take into account and apply the causation doctrines. Evidence from environmental experts and other evidence were enough to attribute the defendant's acts and the environmental pollution and damage.

A similar situation could also be found in cases handled by the courts. Judges would not take into account and apply the doctrines to cases in which acts and results had a direct relation. However, if there was a series of acts which caused the effect, judges would tend to use their logic to establish an act causing the prohibited effect and attribute the act to the prohibited effect. This logic was built based on facts presented in the court room. A choice of certain causation doctrines was not a priority for the judges. The doctrines were in an abstract room as they were in the mind of the judges.

REFERENCES

Bentata, P. (2013) Environmental regulation and civil liability under causal uncertainty: An empirical study of the French legal system. *Review of Law & Economics*, 9(2), 239–263.

Blackburn, S. (1993) Realism: Quasi or Queasy? In: Haldane, J. & Wright, C. (eds.) *Reality, Representation, and Projection*. Oxford, Oxford University Press.

Faure, M. (2001) *Economic Analysis of Environmental Law: An Introduction*. Marseille, IDEP [Institut d'économie publique]. [Online] Available from: http://economiepublique.revues.org/1592.

Hardjasoemantri, K. (2005) *Hukum Tata Lingkungan* [*Law on Environmental Management*]. Yogyakarta, Gadjah Mada University Press.
Honoré, T. (1999) *Responsibility and Fault*. Oxford, Hart Publishing.
Indonesia (2004). *Undang-Undang Nomor 7 Tahun 2004 tentang Sumber Daya Air*. [*The Law Number 7 of 2004 regarding Water Resources*].
Indonesia (2007). *Undang-Undang Nomor 26 Tahun 2007 tentang Penataan Ruang* [*The Law Number 26 of 2007 regarding Spatial Management*].
Indonesia (2008). *Undang-Undang Nomor 18 Tahun 2008 tentang Pengelolaan Sampah* [*The Law Number 18 of 2008 regarding Waste Management*].
Indonesia (2009). *Undang-Undang Nomor 32 Tahun 2009 tentang Perlindungan dan Pengelolaan Lingkungan Hidup* [*The Law Number 32 of 2009 regarding Environmental Protection and Management*].
Indonesia (2013). *Undang-Undang Nomor 18 Tahun 2013 tentang Pencegahan dan Pemberantasan Perusakan Hutan* [*The Law Number 18 of 2013 regarding The Prevention and Eradication of Forest Destruction*].
Kattsoff, L.O. (2004) *Pengantar Filsafat* [*Elements of Philosophy*]. Translated by Soemargono, S. Yogyakarta, Tiara Wacana.
Kelsen, H. (2007) *Teori Hukum Murni: Dasar-dasar Ilmu Hukum Normatif* [*Pure Theory of Law*]. Translated by Muttaqien, R. Berkeley, University of California Press.
Little, D. (1991) *Varieties of Social Explanation: An Introduction to the Philosophy of Social Science*. San Francisco, Westview Press.
Robinson, G.O. (1985) Probabilistic causation and compensation for tortious risk. *Journal of Legal Studies*, 14, 779–798.
Shavel, S. (1984) A model of the optimal use of liability and safety regulation. *Rand Journal of Economics*, 15, 271–280.
van Bemmelen, J.M. (1987) *Hukum Pidana 1: Hukum Pidana Material Bagian Umum* [*Criminal Law 1: Criminal Law Material General Part*]. Translated by Hasnan. Bandung, Bina Cipta.

Public official recruitment system: Towards creation of a rule of law based on *Pancasila*

S. Anam & J. Asshiddiqie
Faculty of Law, Universitas Indonesia, Depok, Indonesia

ABSTRACT: *Pancasila* is the basis of all sources of law in Indonesia. *Pancasila*'s philosophy should be reflected in the legislation-making process in Indonesia, which includes the system of appointing public officials. The system and mechanisms of staffing and appointment of public officials still have weaknesses and issues, in terms of both the concept of institutional structures, and the culture and character of individual public officials in Indonesia. The problems of decadency in structural, personal and cultural standards occur at almost all levels of public officials in Indonesia; therefore, the implementation of the values and principles of *Pancasila* are drifting further away from the ideals and expectations of the nation. The deterioration of morals among public officials and the organisational culture have an adverse effect on the function of good governance, and this is due to the non-implementation of *Pancasila* values in the recruitment system of public officials in Indonesia. This study uses normative legal research and sociological methods in order to devise a better recruitment system of public officials, one in accordance with the values and principles of a state of law of *Pancasila*.

1 INTRODUCTION

The system and mechanisms of selection of Indonesian public officials still have some weaknesses and problematic issues, in terms of both the concept of the institutional structures themselves (Asshiddiqie, 2010), and the personnel acting as public officials in Indonesian culture. This has been influenced by the absence of technical rules concerning the standards that should provide a detailed and comprehensive system and mechanisms for selection, one that is primarily correlated to the concept of institutions, requirements and the choice of public officials (Black, 1999). In addition, there are multiple interpretations of the basis on which an institution is authorised to conduct the selection of public officials in Indonesia (Asshiddiqie, 2005).

The problems of decadency (Matanggui, 2009) in structural, personal, and cultural standards have occurred at every level among public officials in Indonesia. The declines in the personal morality of public officials and in the organisational culture have been impeding the optimal application of good governance. The decline in the quality of personnel and a poor organisational culture have led to political interests being placed above all else. This condition increases the rates of cases of corruption, collusion and nepotism, unjust practices and authority abuse, as well as encouraging a lack of integrity and transparency in public officials in their duties and representations.

There are problems in the concept of institutions (structures); for instance, there are serious issues in determining who can be categorised as a public official, and this is caused by an unclear division of institutions whose functions can be categorised as public in nature (Article 1, Law No. 18 of 1998). This affects the systems and mechanisms of both the selection and discharge processes for public officials, and whether the same standards can be applied to every public official. It becomes very important because the system of recruitment of public officials has yet to establish a proper standard of competence in recruiting the necessary professionals to occupy certain positions of public office.

1.1 The concept of the state of law of Pancasila

In Indonesia, one of the national ideals handed down by 'the founding leaders' of Indonesia is the application of just laws based on *Pancasila* (Kusuma, 2004), the five principles held to be the basis for Indonesian identity and its national ideology. In the explanation, it is stated that Indonesia must conform to the idea of *rechtsstaat* (a law-based state), rather than *machtsstaat* (a power-based state). In the RIS Constitution of 1949, the idea of a state of law was explicitly included. Similarly, in the Provisional Constitution of 1950, the reformulation of law firmly stated that Indonesia is a law-based nation. In 2001, the Third Amendment of the Constitution of the Republic of Indonesia of 1945 also expressly restated this provision in its Article 1(3), which stated: "Indonesia is a Nation that is based on Law".

The characteristics of *rechtsstaat* are also embedded in Indonesia as a nation that is based on law. The description of Indonesia as a nation of law cannot be separated from the preamble of the 1945 Constitution of Indonesia as an ideal for a nation that is based on law (Asshiddiqie, 2010), and it is then stated in the contents and explanations of Indonesia's 1945 Constitution (before amendment). Thus, the first paragraph of the Constitution's preamble contains the word 'justice'; the second paragraph contains the word 'fairness'; the third paragraph contains the word 'Indonesia'; and the fourth paragraph contains the terms 'social justice' and 'just and civilised humanity'. All of these terms refer to the notion of a state of law, because one of the purposes of a state of law is 'to achieve justice' (Thaib, 1996).

According to the doctoral research conducted by Azhary (1995), the term *rechtsstaat* is a conceptual type (*genus begrip*) (Darmodiharjo & Shidarta, 2006), and its relationship with the 1945 Constitution of Indonesia involves a specific meaning of the term. Studies of *rechtsstaat* have frequently been conducted by Indonesian legal experts; however, their studies have not been fully able to establish whether Indonesia is a state of law in the sense of *rechtstaat* or the rule of law (Hartono, 1982). Azhary (1995) identified the principles of a state of law in Indonesian terms as follows:

1. The law of the nation is based on *Pancasila*.
2. Sovereignty is in the hands of the people.
3. It is a constitutional system.
4. There is equality before the law and government.
5. There is independent power of the judiciary.
6. The President and the House of People's Representatives are those who arrange the law.
7. The application of the system is through the People's Consultative Assembly.

According to Azhary (1995), the principle of a state of law is very interesting because it places *Pancasila* as a principle of the law of Indonesia in the highest position. The state of law of *Pancasila* does not only have a certain distinctiveness, but it is the special hallmark of any western concept of laws (*rechtsstaat* and the rule of law) and the so-called socialist legality.

According to Azhary (2010), the use of the term *rechtsstaat* in the explanation of the 1945 Constitution of Indonesia is a concept (*genus begrip*) that is still interpreted in a general sense, while the term 'the state of law of *Pancasila*' has a special meaning. In addition, the term '*rechtsstaat*' in the explanation of the 1945 Constitution of Indonesia is totally different and non-identical to the European continent's concept of law-based states, as well as from the Anglo-Saxon version of the concept of socialist legality. However, Azhary (2010) argues that the concept of the state of law of *Pancasila* is similar to the concept of Islamic nomocracy, especially as a principle of the nation of Indonesia.

In addition, Adji (1980) stated that Indonesia as a state of law does have unique characteristics because *Pancasila* should be assigned as the basic principles and the source of law, and Indonesia as a state of law could be considered as a state of law of *Pancasila*. One of the fundamental characteristics of a state of law of *Pancasila* is the guarantee of freedom of religion (Adji, 1980). According to Adji (1980), another characteristic of Indonesia is that there is no rigid and absolute separation between religion and state, because religion and state are in a harmonious relationship.

Meanwhile, according to Wahjono (1982), a state of law of *Pancasila* is grounded in the principle of family values that are contained in the 1945 Constitution of Indonesia. In the principle of family values, the main priority is "the common people, but the value and dignity of humans remain respected" (Wahjono, 1982). Wahjono (1982) understood that the law is a tool or a vehicle to organise the life of a nation, or public order and social welfare. In terms of the legal function, he confirms that there are three functions of law based on the principle of family values, namely:

1. Upholding the principles of democratisation that are in accordance with the formulation of seven principles of governance, which are stated in the explanation of the 1945 Constitution of Indonesia;
2. Actualising the social justice that is in accordance with Article 33 of the 1945 Constitution of Indonesia; and
3. Upholding the values of humanity based on the statement of 'Belief in the one and only God', which is actualised according to justice.

In the concept of a state of law of Indonesia, it is idealised that law should be the primary principle of governance for the nation, rather than politics or economics. Thus, the phrase that is commonly used in English to refer to this principle is *'the rule of law, not of man'* (Asshiddiqie, 2010). Therefore, it is necessary to distinguish between the law (*rule*), position (*ambt*), and position holders (*ambtsdrager*) (Utrecht & Djindang, 1985).

In addition, a new perspective was introduced by Asshiddiqie (2014), who stated that as well as an understanding of the rule of law, there must be a new definition in relation to the rule of ethics. In the concept of the rule of law, there is a notion concerning a code or book of laws and law courts, whereas the concept of the rule of ethics must contain a definition of the code of ethics or conduct, as well as the notion of a court of ethics. It also affects the expansion of the understanding of the meaning of constitutional law as the basic law, and the constitution should be interpreted as a source of constitutional ethics (Asshiddiqie, 2014). Thus, *Pancasila* is not only a source of law for Indonesia, but also serves as a source of ethics (Asshiddiqie, 2014). In the effort to build a healthy democracy, there must be simultaneous support from and for the rule of law and the rule of ethics (Asshiddiqie, 2015).

2 METHOD

2.1 *Actualisation of recruitment of public officials according to the principle of a state of law of Pancasila*

2.1.1 *Recruitment procedure*

The selection process starts with receipt of applications and ends with a decision on these applications. Performance is the total product of the work and effort made by an employee (Rivai, 1999). Meanwhile, according to Vroom (1964), it is the product of motivation and ability, formulated as follows:

$$\text{PERFORMANCE} = (\text{MOTIVATION} \times \text{ABILITY})$$

Based on this formula, Vroom shows that both elements have a mutual effect on each other, meaning that even the highest skill level of an employee will not produce optimum performance when it is applied with low motivation. Conversely, even the highest levels of motivation of an employee in performing their duties will not be effective without good ability. Performance is determined by individual or personal factors, such as ability and effort, and it is also determined by factors beyond our control, such as the decisions taken by others, the available resources, the system in which we work, and so on (Dharma, 2005). Personal performance is basically influenced by: (1) expectations of rewards; (2) encouragement; (3) ability, needs and character; (4) the perception of the task; (5) internal and external rewards; (6) the perception of the level of remuneration; (7) job satisfaction (Rivai, 1999).

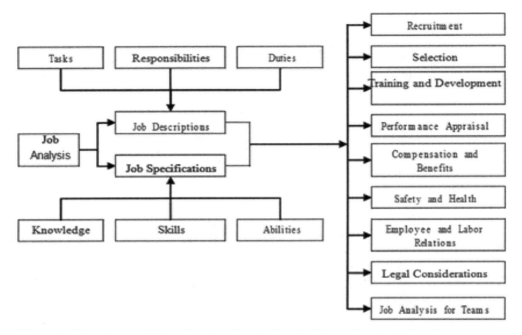

Figure 1. Flowchart explaining the role of analysis of a job as the foundation for recruitment management.

The flowchart in Figure 1 displays the thinking of Mondy and Noe (1996) in their explanation of the role of analysis of a job as the foundation for recruitment management. According to the flowchart, the thinking is actually aligned with the values of each principle of *Pancasila*. Thus, when applied to the recruitment of public officials, the first principle of *Pancasila* ("Belief in the one and only God") is expected to form a guideline for the person or institution conducting recruitment to consider obedience to, and faith in, God Almighty as the basis for the assessment of criteria for candidates, to respect the freedom of religion according to the teachings of each religion, and not to impose one's religion or beliefs on others. The second principle ("Just and civilised humanity") aims to oblige leaders to maintain the notion of proper tolerance, respecting the principles of humanity, defending righteousness and justice, and implementing the good principles of humanity (social activities, community work, and helping each other).

The third principle of *Pancasila* ("The unity of Indonesia") aims to be a benchmark that can be used as a guideline in the process of recruitment, which includes patriotism, wilful sacrifice to the nation, and the value of the integrity of the nation as part of the Unitary State of the Republic of Indonesia. The fourth principle of *Pancasila* ("Democracy guided by the inner wisdom in the unanimity arising out of deliberations among representatives") can be regarded as a guideline for the recruitment process of public officials who must be able to put the interests of the nation/society above any other. Indonesian people should never force their will onto others and should prioritise the culture of discussion to achieve common goals. Furthermore, the fifth principle ("Social justice for all of the people of Indonesia") is aimed at selecting or recruiting leaders that are capable of upholding the values of helping others, and appreciating others' ideas and rights, as well as being willing to conduct helpful actions in the interest of the people and the country, actualising justice for all.

2.1.2 *Institutional structure*

In addition to the issues outlined above, there is also another more complicated issue, and it is associated with the status of public officials in the system of Indonesian state administration both as state officials and not as state officials.

Legally, the standard criteria for state officials is not set out in existing legislation, but it is explicitly stated in some laws and regulations, which include:

a. Article 1, Point 1 of Law No. 28 of 1999 concerning Clean Government Executives Free from Corruption, Collusion, and Nepotism;
b. Article 2 of Law No. 28 of 1999 concerning Clean Government Executives Free from Corruption, Collusion, and Nepotism;
c. Article 1, Point 4 of Law No. 43 of 1999 concerning changes in Act No. 8 of 1974 concerning Basic Law on Civil Servants;
d. Article 11, Paragraph 1 of Law No. 43 of 1999 in the Explanation;
e. Article 122 of Law No. 5 of 2014 concerning State Civil Apparatus.

On the basis of the provisions above, we will not find any standard criteria for any state institution to be treated as employing either state/public officials or non-public officials. Some legislation only states that specific institutions could be managed by public officials or by non-public officials. Some legislation also abstractly provides authority to legislators to determine the institutions that could be managed either by public/state officials or by non-public/state officials. This issue is also abstractly regulated by the Law on the Ordinance of Civil Service and the Law on State Civil Apparatus.

The ideas of Hans Kelsen may provide us with insight into the deeper meaning of a state institution, and we can learn from the concept of the state organ explained in his book of *General Theory of Law and State* (Kelsen, 1961), in which Kelsen explains that "Whoever fulfils a function determined by the legal order is an organ". Then, literally, the term "public office" defines the position of a public leader who is able to manage and direct a public institution and people as a whole (a position affecting the people as a whole). However, as a legal term, the phrase "public office" contains both specific and broader meanings. In a more specific definition, this term is usually associated with the notion of the "state official", which is limitedly and administratively defined as based on having administrative rights in the form of financial support and rights of precedence. In a broader meaning, "public office" is usually interpreted as "a position or occupation established by law or by the act of a government body, for the purpose of exercising the authority of the government in the service of the public" (Bell, 2010).

However, in its practice in the United States, as reflected in various court judgements, the definition of "public office" can also be understood with a more limited scope of meaning. First, a "public officer" is different from a "public employee". Second, the posts of public officers are determined according to their authority to make decisions on behalf of the state or public interests. If the office is intended merely to be an 'advisory' body, which provides non-binding judgement or a non-mandatory recommendation in the process of decision-making, then it will not be regarded as a "public office".

Therefore, the definition of public officials could be divided into three parts: first, state or public officials; second, administrative officials; third, professional officials (Asshiddiqie, 2006). State or public officials are "political appointees", who are appointed or selected for political consideration, while administrative officials or "administrative appointees" are appointed purely for administrative reasons. By contrast, professional officials are appointed for the inevitable market needs, because they also deal with public administration despite the fact that their budget and facilities do not come from the government.

3 CONCLUSION

The implementation of a public official recruitment system that is in accordance with the principle of a state of law of *Pancasila* could be actualised by at least two things. First, the recruitment procedure must be based on the values contained in every principle of *Pancasila*. The assessment criteria for potential officials or candidates must be based on the principles of belief in one and only God, just and civilised humanity, the unity of Indonesia, a democracy guided by the inner wisdom of the unanimity arising out of deliberations among representatives, and social justice for all of the people of Indonesia. Second, institutional structures have to be divided according to clear and exact criteria. For example, public or state officials should be categorised as 'political appointees' who are appointed on the basis of political considerations, while administrative officials are those appointed purely for administrative reasons, and professional officials are those appointed to meet inevitable market demands.

REFERENCES

2001, the Third Amendment of the Constitution of the Republic of Indonesia of 1945.
Adji, O.S. (1980). *Free Trial State Law*. Jakarta, Indonesia: Erlangga.
Asshiddiqie, J. (2005). *Dispute over the Constitutional Authority of the State Institution*. Jakarta, Indonesia: Konpress.
Asshiddiqie, J. (2006). *Development and Consolidation of State Institutions Post Reforms* (2nd ed.). Jakarta, Indonesia: Setjen Kepaniteraan Mahkamah Konstitusi Republik Indonesia.
Asshiddiqie, J. (2010). *Indonesian Law Country*. Public Lecture, 23 January 2010. Jakarta, Indonesia: Dewan Pimpinan Pusat Ikatan Alumni Universitas Jayabaya.
Asshiddiqie, J. (2014). *Constitutional Ethics and Ethics Trial: A New Perspective on the Rule of Law and the Rule of Ethics and the Constitutional Law and Constitutional Ethich* (Rev. ed.). Jakarta, Indonesia: Sinar Grafika.
Asshiddiqie, J. (2015). *Strengthening Governance and Justice System* (1st ed.). Jakarta, Indonesia: Sinar Grafika.
Azhary, M.T. (2010). *State of Law: A Study of Its Principles Seen from the aspect of Islamic Law, its Implementation of the Medina Period and the Current Period*. Jakarta, Indonesia: Prenada Media Group.
Azhary. (1995). *Indonesian Law Country: Normative Juridical Analysis of its Elements*. Jakarta, Indonesia: University of Indonesia Press.
Bell, F. (2010, February 17). *What's a public office?* [Web log message]. Retrieved from https://canons.sog.unc.edu/whats-a-public-office/.
Black, H.C. (1999). *Black's law dictionary* (7th ed.). St. Paul, MN: West Publishing.
Budiardjo, M. (1998). *Fundamentals of Political Science*. Jakarta, Indonesia: Gramedia Pustaka Utama.
Darmodiharjo, D. & Shidarta. (2006). *Principles of Philosophy of Law: What and How of Indonesian Law Philosophy*. Jakarta, Indonesia: Gramedia Pustaka Utama.
Dharma, S. (2005). *Performance Management: Theory Philosophy and Its Implementation*. Yogyakarta, Indonesia: Pustaka Pelajar.
Hartono, S. (1982). *What is The Rule of Law*. Bandung, Indonesia: Alumni Press.
Kelsen, H. (1961). *General theory of law and state* (A. Wedberg, Trans.). New York, NY: Russell & Russell.
Kusuma, A.B. (2004). *The birth of the 1945 Constitution*. Jakarta, Indonesia: Pusat Studi Hukum Tata Negara Fakultas Hukum Universitas Indonesia.
Law No. 18 of 1998.
Law No. 28 of 1999 concerning the Clean Government Executives Free from Corruption, Collusion, and Nepotis.
Law No. 43 of 1999 concerning changes in the Act No. 8 of 1974 concerning Basic Law on Civil Servants.
Law No. 5 of 2014 concerning State Civil Apparatus.
Matanggui, J.H. (2009). *Synonym Dictionary*. Jakarta, Indonesia: Grasindo.
Mondy, R.W. & Noe, R.M. (1996). *Human resource management*. Upper Saddle River, NJ: Prentice Hall.
Provisional Constitution of 1950.
RIS Constitution of 1949.
Rivai, V. (1999). *Human Resource Management for Companies and Theories and Practices*. Yogyakarta, Indonesia: Raja Grafindo.
Thaib, D. (1996). In Manan, B. (Ed.), *People's Sovereignty State of Law and Human Rights: Collection of Posts within the framework of 70 years of Sri Soemantri Martosoewignjo*. Jakarta, Indonesia: Gaya Media Pratama.
Utrecht, E. & Djindang, M.S. (1985). *Introduction to Indonesian State Administration Law* [*An Introduction to Indonesian State Administrative Law*]. Jakarta, Indonesia: Ichtiar Baru.
Vroom, V.H. (1964). *Work and motivation*. New York, NY: John Wiley & Sons.
Wahjono, P. (1982). *Juridical Concept of State of the Republic of Indonesia*. Jakarta, Indonesia: Rajawali.

A *waqf* and *musyarakah* implementation model in *takaful ijtima'i* as an alternative *sharia* insurance system: An experiment to maximise the realisation of the social justice principle in *sharia* insurance in Indonesia

Z. Abdullah, U. Hasanah & Gemala Dewi
Faculty of Law, Universitas Indonesia, Depok, Indonesia

ABSTRACT: *At-takaful* institutions embody the spirit of helping each other, and that spirit has become the basic principle for a *sharia* insurance institution (*asuransi syariah* or *at-takaful al-ijtima'i*). In general, the current *sharia* insurance that has been practised still uses a hybrid contract (*multiakad*) model, involving *akad tabarru'* and *akad mudharabah*. However, such a hybrid contract model still does not reflect fairness. This paper proposes the use of a *waqf* and *musyarakah* model as an alternative solution to overcome the challenge of this issue of social justice. First, a *waqf* model will be used because it has a similarity with the *sharia* insurance system, with both sharing similar goals of overcoming poverty and creating prosperity. This similarity could become an entry point for *waqf* to create a synthesis with already-established insurance institutions. Second, *akad syirkah* (*musyarakah*) is implemented in a civil partnership entity that runs insurance institutions that endorse a mutual principle. According to the mutual principle, profit sharing must have a positive impact on people's prosperity (social justice). This article focuses on *takaful ijtima'i* with a *waqf* and *musyarakah* model, and positions it as an alternative *sharia* insurance system for the future, especially in Indonesia's context. A doctrinal approach is used in discussing the topic.

1 INTRODUCTION

Social justice in Islam is known as '*takaful ijtima'i*'. The meaning of *takaful ijtima'i* is to provide a decent living standard for every citizen. Social justice is not just about food, but also about housing, education, health and clothing (Rahman, 2003). According to Harry Calvert, a British social security law scholar, social justice is the main legitimate mechanism associated with the provision of security of sufficient individual income if its implementation is conducted by using other social services, to ensure someone meets the minimum standard of living that is culturally acceptable (Rahman, 2003).

In terms of law classification, social justice is included in *fardhu kifayah* (Al-Haritsi, 2006), which means that every Muslim has an obligation to help others as far as their ability permits it, while the state plays the role of ensuring its implementation. These roles reflect the state's capacity as an authority to oblige every individual within their power to implement Islamic law. Therefore, social justice, relieving the burden of fellow human beings, is not merely the state's duty but also the duty of every Muslim.

In the *fiqh* of Umar, Al-Haritsi (2006) wrote that the parties dealing with social justice affairs shall consist of individuals, society, and the state. The state's main duty is to manage this social justice system, but individuals and society can take parts in the implementation. In relation to this matter, the Hadith of the Prophet explains, "Muslim is like a building that establishes each other" (Al-Haritsi, 2006; Rahman, 2003).

According to Ramesh Mishra, the state has an obligation to fulfil the basic rights of its citizens in the economic and social fields. Meanwhile, the public and business sectors are not an obligation but the responsibility of wider humanity. As an institution that has public legitimacy and is appointed by the people, the state has an obligation to fulfil, to protect, and to respect people's basic economic rights. Therefore, the function of arrangement should be owned by the state (Mishra, 1999).

From the above discussion, it can be concluded that *takaful ijtima'i* or social justice can be performed by parties other than the state. Insurance as part of *takaful ijtima'i* can be provided by private parties with the purpose of helping the state in manifesting social justice for the people. According to Syekh Abu Zahrah, *at-takaful al-ijtima'i* means every society member is guaranteed under their society itself (Zahrah, n.d.). As a result, it can be understood that *at-takaful* refers to the characteristic of protecting each other in the society. This characteristic has been used as the basic principle in the institution of *sharia* insurance or *at-takaful (al-ijtima'i)*.

The *takaful ijtima'i* concept is very dynamic so, in relation to current practices, they could be modified and mixed with other instruments of economics and law. This includes the modification and mixing of the current *sharia* insurance system to provide the best alternatives in social justice programmes for the people.

2 *TAKAFUL IJTIMA'I*: CONCEPT AND HISTORICAL BACKGROUND

One of the missions of the messengers in Islam is to maintain social justice in the world, and it is reflected in the Qur'an, Surah Al-Hadid [57:25]. Based on this mission, therefore, the Prophet introduced the social security system in Islam, which was then known by His successors (Khalifah) as *takaful ijtima'i*.

From a historical perspective, a pattern of helping similar to *takaful ijtima'i* had already existed in Arabic culture before the emergence of Islam. This concept was then also implemented by Khulafa ar-Rasyidin until Khalifah Umar bin Abdul Aziz. However, they only started to establish a *takaful ijtima'i* institution in the era of Khalifah Umar ibn Al-Khattab.

The *takaful* concept was first used by the Prophet in dealing with a murder case of two persons of the Huzail tribe and a law was imposed, *al-aqilah*. *Al-aqilah* as a law required the killer to pay some money, known as 'blood money' (*diyat*), as a compensation to the victim's family (Haekal, 1992; Aly, 2008 At that time, the Prophet considered *al-aqilah*, which was applied in the *jahiliyyah* era, to be positive and to be an example to everyone. In addition, the Prophet even adopted *al-aqilah* in the Medina Charter (Maslehuddin, 1999).

The social justice practised in the era of the Messenger can be seen in the economic policies that were established in *Baitul Mal*, such as the *zakat* and *ushr* distributions to those who were entitled to receive them in accordance with the Qur'an: the payment for the Muslims who were slaves, the payment for the people who were accidently killed by Muslim troops, and the allowance for the poor.

After the death of the Messenger, Caliph Abu Bakar Ash-Shiddiq, Umar ibn Al-Khatab, Ustman Ibn Affan and Ali Ibn Abi Thalib (Khulafa ar-Rasyidin) were still practising the social security system as the Prophet had done. However, all of them had different styles of implementing it (Al-Qardhawi, 1972; Ja'Farian, 2004; Amalia, 2010).

In terms of the history of *takaful ijtima'i*, insurance in the form of an institution started to be known around the 19th century, when western colonisation encroached on the Islamic world. However, this does not mean that the subject is invalid and is not allowed in Islamic law because, in a *muamalah* matter, there is a principle to give dispensation in determining edicts on matters that bring advantages to human life.

3 *SHARIA* INSURANCE PRINCIPLES

The purpose of insurance is to protect the participants from any risk that might happen. *Sharia* insurance, which is an adaptation of conventional insurance agencies, has its own ultimate principles, namely *ta'awanu 'ala al bir, wa al-takwa* (mutual cooperation in virtue and righteousness), and *al-ta'min* (provision of security) (Djazuli, 2007). These principles effectively make the participants part of a big family, who are all committed to taking risk for one another's sake. With this goal, *sharia* insurance employs the principle of *takafuli* (taking each other's risks) in its *akad* (transaction).

Besides the *ta'awun* (helping) principle, there are some other relevant principles, namely, corporations, mutual responsibility, mutual protection, fairness, and also *gharar-*, *maysir-* and *riba*-free elements.

Gharar means uncertainty, risk or danger. According to Ali Adnan Ibrahim (2008), *gharar* is:

> *Excessive uncertainty or ambiguity [and] generally includes lack of complete information, deceit, risk, and inherent uncertainty as to the subject matter of the contract. Uncertainty about future events and quality of goods may also be included, whether any particular incompleteness is unilateral, bilateral, multilateral, intentional, or otherwise.*

Islam forbids any transaction that contains imbalance between the risk and the profits. In other words, Islam prohibits every transaction intended to gain benefits without the willingness of bearing loss (Al-Kamil, n.d.).

In insurance, *gharar* can be seen in the form of a contract (*akad*) that provides the basis of a closure policy. In conventional life insurance, a *tabadduli* (exchange) contract is used. In *sharia* insurance, the exchange contract must clearly state the amount which has to be paid and the amount that will be received. This condition made it confusing (*gharar*).

Maysir or a gambling element must be avoided too. *Maysir* is any *muamalah* (transaction) that involves people in the possibility of making losses or gaining profits (Hosen, 1987). Allah forbids every gambling action, as mentioned in Surah Al-Baqarah [2:219] and Al-Maidah [5:90], where Allah called *maysir* the work of Satan.

Maysir emerges as an effect of uncertainty. In insurance, there are three areas of possibility where *maysir* may occur (Antonio, 2004):

1. If a policyholder experiences misfortune after having only paid a limited amount of money, the insurance company has to bear the difference between the amount that has been paid and the amount insured. In this matter, the client gains benefit.
2. If, by the end of the agreement, nothing has happened for which the client can receive a pay out, the insurance company gains benefit.
3. If the client resigns from the agreement before a given time limit (the reservation period), he may only receive a refund of very small amounts, or even none at all with some insurance companies.

Riba is also prohibited in Islam. According to Barbara L. Seniawski, *riba* means "excess, increase, augmentation, expansion or growth" (Seniawski, 2001). In financial terminology, *riba* is defined as profits without any contract of value required by one of the parties who holds the contract in exchanging two types of goods with the same values (Schacht, 2010).

Riba could emerge from the investments made by insurance companies. Basically, insurance companies are similar to banks, in that both collect people's funds. These funds will be invested in order to make profits. However, the problem with the investment instruments of conventional insurance is that the company may not pay any attention to the halal or haram aspects associated with such investments, giving rise to a concern that they might involve an investment that is based on interest (*riba*).

4 HYBRID CONTRACTS IN *SHARIA* INSURANCE

According to the fatawa of Dewan Syariah Nasional Majelis Ulama Indonesia (DSN-MUI) on *sharia* insurance, namely Fatwa No. 21/DSN-MUI/X/2001 on *Sharia* Insurance Basic Guidance, Fatwa No. 51/DSN-MUI/III/2006 on *Akad Mudharabah Musytarakah*, Fatwa No. 52/DSN-MUI/III/2006 on *Akad Wakalah bil Ujrah*, and Fatwa No. 53/DSN-MUI/III/2006 on *Akad Tabarru'*, there are four known insurance contracts that produce four types of insurance. Those contracts are:

1. *Tabarru'*;
2. *Mudharabah*;
3. *Mudharabah musytarakah*;
4. *Wakalah bil ujrah*.

The four *sharia* insurance contracts apply a hybrid or double (*al-uqud al-murakkabah*) form, namely the merger of two types of *akad* that are different due to the law; in this case, the combination between *akad tijari* (*tijari* contract) and *akad tabarru'* (*tabarru'* contract).

Tijari contract is represented by the *akad mudharabah musytarakah, mudharabah*, or *akad wakalah bil ujrah*, while *akad tabarru'* is represented by *akad hibah*. *Akad tijari* (*mudharabah musytarakah, mudharabah or wakalah bil ujrah*) and *akad tabarru'* (*hibah*) have different characteristics and purposes.

Akad tabarru' produces a *hibah* (grants) type, and has the characteristics and purpose of helping fellow insurance members; therefore, it must always be present in every transaction of *sharia* insurance. *Akad tabarru'* has a characteristic of not seeking any profit, so the *akad tabarru'* can always be combined with an *akad tijari*, namely *wakalah bil ujrah* or *mudharabah musytarakah*. This *akad tijari* has the characteristics and the purpose of seeking profits (profit-oriented).

The different characteristics and purposes of the two *akad* have implications for the financial management that has to be done by the insurance company, namely, fund separations.

Tabarru' means charity or doing good deeds without demanding any requirements (Sumanto et al., 2009). *Akad tabarru'* is the main *akad*, and has to be inherent in every *sharia* insurance, as stipulated in Fatwa DSN-MUI No. 53/DSN-MUI/III/2006, Part Two.

Akad tabarru' is a contract made between the participants in insurance that stipulates that each participant must agree to help the other in taking the risk if any unfortunate event befalls one of the participants. The means for each participant to make contributions is called *tabarru'* funds. The concept of handling risk like this is called risk sharing. This type of *akad* can eliminate *gharar* and *maysir* factors in *sharia* insurance.

Application of *akad tabarru'* can produce a *hibah* (grants) model, but the use of this *hibah* concept in *sharia* insurance is still debatable. The debate revolves around the issue of whether it is allowable for funds that have been granted to be readmitted, in the form of benefits or funds. This is applied in the *sharia* insurance, where the insurance participants can still get benefits from the grants in the form of restitution if they suffer any unfortunate event or in the form of repayment of some of the grants if no unfortunate event happens.

The provision regarding the grants that is contained in the 1997 Compilation of Islamic Law (*Kompilasi Hukum Islam* (KHI)), Article 171 point g, states: "*Hibah* is a provision of an object voluntarily and without a reward of a person to others who are still alive to possess". Article 212 of KHI also regulates that, as also stated in the Hadith: "*Hibah* cannot retract, except *hibah* from parents to children".

This concept is different from the *hibah* concept that is practised in *sharia* insurance in Indonesia. There are several types of *hibah*. One is *hibah* with returns, rather than *hibah* that only expects pleasure (reward) from Allah without expecting anything. This type of *hibah* is allowed in Islam but is of a different "degree" from the *hibah* that is merely for Allah. Nevertheless, it is still in the line of Islamic law and consists of *rukhsoh* (waivers) in the implementation of Islamic law. This type of *hibah* is the result of at least two factors: first, the competition with conventional insurance; and second, because people are yet to develop a good understanding of *sharia* economy, the main consideration is always the economic benefit factor.

The *akad tijari* that has been practised is still using *akad mudharabah* and *akad wakalah* (*bil ujrah*). These two *akad* have advantages and disadvantages in their implementation. The advantage is that these two contracts (*akad*) are most suitable for use in limited companies, which act as the *sharia* insurance operator, because the operator does not have any risk of loss in relation to the funds they manage. However, this is also the disadvantage of *sharia* insurance in Indonesia, which is associated with profit-oriented business entities. Such profit-oriented business entities only want to share any profits with the insurance participants, but do not want to handle any losses.

5 PROPOSING A *WAQF* AND *MUSYARAKAH* MODEL FOR *TAKAFUL IJTIMA'I*

From the foregoing descriptions, it is understandable that existing *sharia* insurance institutions in Indonesia are yet to achieve cooperative purpose, or the purpose of helping each other, based on the principle of fairness. To maximise the achievement of the helping principle in *sharia* insurance, there are several things that can be done: among others, making use of the two *akad* that embody a spirit of togetherness and mutual helping, namely, *akad waqf* and *akad musyarakah*.

Akad waqf is chosen because the *waqf* has a similar purpose to insurance, namely, to help each other and to attain common prosperity. This similarity could act as an open door to a *waqf* system in the insurance instrument.

Waqf is also included in *akad tabarru'*, and this *akad* is not for making profit, but is to help people. *Hibah* and *waqf* are included in *akad tabarru'*, and both of them have some similarities and differences.

Majelis Tarjih and Tajdid Muhammadiyah outline several similarities and differences between *hibah* and *waqf* (Fatwa Tarjih, 2009), including:

1. In both *waqf* and *hibah* there are people who give some of their wealth (*wakif* and *wahib*), goods are given, and people receive them.
2. *Waqf* goods are every goods, that have durability that is not only disposable, or can be used in many times, and valued according to Islamic value, while *hibah* can be anything, disposable or durable. It is not permitted for *waqf* or *hibah* to be in the form of goods that it is forbidden to sell, such as liability goods or haram objects.
3. *Waqf* goods can only be given to a group of people for the benefit of the people, while *hibah* goods can be given either to an individual or a group, for the benefit of the people or only for the individual concerned.

In *sharia* insurance, the object of goods given for *waqf* for *tabarru'* is money, as mentioned in *akad hibah*, and not goods. The possibility for *waqf* objects to take the form of money is regulated in Article 16 clause 3 of Law No. 41 of 2004 on *Waqf*. In addition, DSN-MUI also issued a fatwa about cash *waqf* in 2002 that stated (among other things): the law for cash *waqf* is *jawaz* (allowed); cash *waqf* can only be distributed and used for matters that are allowed by *syari'i*; the basic value of the cash *waqf* has to be sustainable and should not be saleable nor become the object of *hibah* and/or heritable.

Sharia insurance institutions can function as the *nazhir* of the cash *waqf*. The basic concept of *waqf* is that the *waqf* property should not decrease nor perish, but should be productive and able to generate benefits; therefore, the main obligation of a *sharia* insurance institution is similar to that of a *nazhir*, which is to manage and develop *waqf* property (Sula, 2010).

Thus, the *waqf* funds should not be used for any other purpose, but first have to be invested and managed, and the profit can then be used thereafter. The use of investment funds is for claim funds and operational funds in the *nisbah* (proportion) that has been agreed upon. Furthermore, the profits generated from the investment should not be given to the participants, but have to be distributed and used for those who have the rights (*mauquf alaih*) according to the participants' will.

The position of customers while they pay the premiums is that of someone that gives a *waqf* (*muwakif*). *Waqf* depends on a purpose and, in this insurance, *waqf* is for the welfare of *ummah* (the community), or *waqf* for *ta'awun*. Therefore, the role of "direction" from the intention of the *muwakif* is very important, and the most "appropriate" is *waqf* for ummah welfare (*al-waqf limaslahatil ummah*) or *waqf* for *ta'awun* (*al-waqf lit ta'awun*). Thus, the *waqf* funds can be located for investment in asset provisions to accomplish *ummah*'s welfare, e.g. for claiming payment, operational cost, building hospital, school, or even rented building.

The concept can be used for *sharia* insurance with social-based or non-profit-oriented insurance, or even for collective benefits. However, there is a disadvantage in this *waqf* institution, especially in price competition. With the use of two accounts, a *tabarru'* account and a savings account, the participant will automatically pay a greater amount than *sharia* insurance that uses *akad hibah*. Nevertheless, this concept still can be implemented and is very likely to be applied.

The advantages of a *sharia* insurance concept based on *akad waqf* are as follows:

1. Assets will not decrease, and even tend to increase in parallel with the increase of customers and the passing of time.
2. During the time they pay the premiums, the customer will receive benefits both now and for the afterlife, because when they pay the premiums they also do *waqf* for the welfare of *ummah*.
3. The product of the *waqf* fund investment will add to the reserve of *tabarru'* funds, and some parts can also be used to contribute to the *sharia* insurance company's operational cost (*nazhir*).
4. The *waqf* funds collected can be used for *waqf* assets, such as a *waqf* building that can be used for the *sharia* insurance office, or can even be invested into rented properties, such as office buildings, provided that the investment results are for the customer interest.

To apply this concept in Indonesia involves several prerequisites; not only normative preparation, but also the readiness of the people to accept the consequences. One of the changes is that the operator should not take the form of a limited company, but should be a mutual business entity. A mutual business entity constitutes a form resembling a partnership or what is called *syirkah*. *Syirkah* or *musyakarah* means to mix one to another, making them difficult to distinguish (Al-Dīn Muḥammad s.a.).

In this matter, *syirkah* is a form of collaboration between the owners of capital in order to build a bigger business together, or cooperation between capital owners who do not have the skills to run a business and entrepreneurs who do not have the necessary capital or who need additional capital.

In such a *syirkah* form of business, the *akad* used is also *syirkah* or *akad musyarakah*. This *akad syirkah* is part of *akad tijari*. This *akad* also becomes the distinguishing factor between a limited company (*perseroan terbatas*) and a mutual business partnership (*persekutuan perdata*). This *akad musyarakah* arrangement is regulated in Fatwa DSN-MUI No.8/DSN-MUI/IV/2000 on *Musyarakah* Financing.

Musyarakah or *syirkah* can be divided into several types and, among these, *syirkah inan* is the one most similar to the mutual insurance principle. *Syirkah al-inan* is a form of cooperation in which two or more people participate in a joint business but they do not have the same level of capital and benefit portions (Hassan, 1986; Djamil, 2012).

Antonio (2004) explained that in *syirkah al-inan* every party submits their property as capital, and each member is able to manage and develop the capital. Profits and shared and risks are borne together. As well as being the form most similar to mutual *sharia* insurance, the *syirkah al-inan* model is also in line with the business entity that is regulated in Article 1618 of the Indonesian Civil Law (KUHPer).

In its Fatwa No. 08/DSN-MUI/IV/2000 on *Musyarakah* Financing, DSN-MUI states in its preamble that, "… *musyarakah* financing that has the strength in togetherness and fairness, both in profit sharing and in loss risks, now is carried out by the *Sharia* Finance Institution (Lembaga Keuangan Syariah)". Although the fatwa is not aimed specifically at *sharia* insurance matters, it can be used as general guidance in running a mutual *sharia* insurance business. The element of fairness, as mentioned in the considerations, is proved by the existence of an equal position for each party, with equality in their obligations and benefits as implied in the articles within Fatwa No. 08/DSN-MUI/IV/2000.

6 CONCLUSION

The institution of *sharia* insurance in Indonesia has not yet accomplished any cooperative goal or a condition in which people help each other on the basis of the fairness principle. To maximise the achievement of helping and the fairness principle in *sharia* insurance, there are several things that can be done, including, among others, the use of *akad waqf* and *akad musyarakah*. Some of the advantages of *sharia* insurance that is based on *akad waqf* include: first, the customer will obtain worldly and afterlife benefits for the time they pay the premiums, because in paying the premiums they are also doing *waqf* for the welfare of ummah; second, *waqf* funds that are collected can be used for *waqf* assets, such as investment in leased properties, and the profits from the investment are in the customers' interests.

Obviously, to apply this concept in Indonesia requires not only normative preparation, but also the readiness of the people to accept the consequences. One of the changes is that the operator should not take the form of a limited company (*perseroan terbatas*), but rather should be a mutual business entity called *syirkah*, particularly *syirkah al-inan*. *Syirkah al-inan* is most similar to mutual insurance, being a form of cooperation in which two or more people participate in a joint business, but do not have the same levels of capital and benefit portions. In *syirkah al-inan*, every party submits their property as capital and is allowed to manage the capital. Profits are shared and risks are borne together.

In the *syirkah* form of business, the *akad* that is used is also *syirkah* or *akad musyarakah*. The *syirkah* model as regulated in Fatwa DSN-MUI No. 08/DSN-MUI/IV/2000 on *Musyarakah* Financing has adopted and fulfilled the fairness criteria, which is reflected in several aspects. First, the manager is also a mutual insurance participant; therefore, the manager can receive interest gained from the capital and from the work (business) revenue. Second, in the *syirkah* concept, both profits and losses are shared, and not just profits, so the manager is likely to be careful in managing the insurance funds.

REFERENCES

Al-Haritsi, J.A. (2006). *Fiqih Ekonomi Umar bin Al-Khatab* [*Umar bin Al-Khatab's Economic Fiqh*]. Jakarta, Indonesia: Khalifa.

Al-Kamil, U.A. (n.d.). *Al-Qowa'idul Fiqhiyyah al-Kubro wa Atsaruha fil Mu'amalat al-Maaliyah* (Doctoral thesis, Jami'ah al-Azhar as-Syarif, Egypt).

Aly Adnan Ibrahim, "Financial Innovations in the Muslim World," *American University International Law Review* (2008).

Al-Qardhawi, Y. (1972). *Hukum Zakat* [*Zakah Law*]. Jakarta, Indonesia: Lintera Antar Nusa.

Amalia, E. (2010). *Sejarah Pemikiran Ekonomi Islam: Dari Masa Klasik hingga Kontemporer* [*The History of Islamic Economics Thought: From Classic to Contemporary Era*]. Depok, Indonesia: Gramata Publishing.

Antonio, M.S. (2004). Spekulasi dalam Asuransi Syariah [Speculation inside Sharia Insurance]. *Republika, October 7, 2002.*

Departemen Agama [Department of Religious Affairs] Republik Indonesia. (1997). *Kompilasi Hukum Islam* [*The Compilation of Islamic Laws in Indonesia*]. Jakarta, Indonesia: Departemen Agama Republik Indonesia.

Dewan Syariah Nasional Majelis Ulama Indonesia. (2000). *Fatwa No. 8/DSN-MUI/IV/2000 tentang Pembiayaan Musyarakah* [*Musyarakah financing*].

Dewan Syariah Nasional Majelis Ulama Indonesia. (2001). *Fatwa No. 21/DSN-MUI/X/2001 tentang Pedoman Umum Asuransi Syariah* [*Sharia insurance basic guidance*].

Dewan Syariah Nasional Majelis Ulama Indonesia. (2006a). *Fatwa No. 51/DSN-MUI/III/2006 tentang Akad Mudharabah pada Asuransi Syariah* [*Mudharabah contract in sharia insurance*].

Dewan Syariah Nasional Majelis Ulama Indonesia. (2006b). *Fatwa No. 52/DSN-MUI/III/2006 tentang Akad Wakalah bil Ujrah pada Asuransi Syariah dan Reasuransi Syariah* [*Wakalah bil Ujrah contract in sharia insurance and reinsurance*].

Dewan Syariah Nasional Majelis Ulama Indonesia. (2006c). *Fatwa No. 53/DSN-MUI/III/2006 tentang Akad Tabarru' pada Asuransi Syariah* [*Tabarru' contract in sharia insurance*].

Djamil, F. (2012). *Hukum Perjanjian dalam Transaksi di Lembaga Keuangan Syariah* [*Agreement Law inside Sharia Finance Institution's Transaction*]. Jakarta, Indonesia: Sinar Grafika.

Djazuli, A. (2007). *Fiqh Siyasah: Implementasi Kemaslahatan Umat dalam Rambu-Rambu Syari'ah* [*Fiqh Siyasah: Implementation of Social Welfare in Sharia Values*]. Jakarta, Indonesia: Kencana.

Fatwa Tarjih. (2009, November 6). Wakaf dan Hibah [Waqf and grants]. *Hasil Sidang Tim Fatwa Majelis Tarjih and Tajdid Pimpinan Pusat Muhammadiyah*. Retrieved from http://www.fatwatarjih.com/2012/11/wakaf-dan-hibah.html.

Haekal, M.H. (1992). *Sejarah Hidup Muhammad* [*The life of Muhammad*] (A. Audah, Trans.). Jakarta, Indonesia: Intermasa.

Hassan, A.A.H. (1986). *Sales and contracts in early Islamic commercial law*. Islamabad, Pakistan: Islamic Research Institute.

Hosen, I. (1987). *Apakah Judi Itu* [*What is the Gambling*]. Jakarta, Indonesia: Lembaga Kajian Ilmiah Institut Ilmu Al-Qur'an.

Ibn Manzur, M.M. (2000). *Lisan al-Arab I* (3rd ed.). Beirut, Lebanon: Dar Sader.

Ibrahim, A.A. (2008) Financial innovations in the Muslim world. *American University International Law Review, 23*(4), 661–732.

Ja'Farian, R. (2004). *Sejarah Islam: Sejak Wafat Nabi SAW Hingga Runtuhnya Dinasti Bani Umayah (11–132 AH)* [*History of the caliphs: From the death of the messenger(s) to the decline of the Umayyad dynasty (11–132 AH)*]. Jakarta, Indonesia: Lentera.

Khorshid, A. (2004). *Islamic insurance: A modern approach to Islamic banking*. New York, NY: Routledge Curzon.

Maslehuddin, M. (1999). *Menggugat Asuransi Modern* [*Questioning Modern Insurance*]. Jakarta, Indonesia: Lentera.

Mishra, R. (1999). *Globalization and the welfare state*. Cheltenham, UK: Edward Elgar Publishing.

Rahman, A. (2003). *Doktrin Ekonomi Islam* [*Doctrine of Islamic Economic*]. Yogyakarta, Indonesia: Dana Bhakti Wakaf.

Republic of Indonesia. (2004). *Undang-undang No. 41 Tahun 2004 tentang Wakaf* [*Law No. 41 of 2004 regarding Waqf*].

Schacht, J. (2010). *Pengantar Hukum Islam* [*An Introduction to Islamic Law*] (J. Supomo, Trans.). Bandung, Indonesia: Nuansa.

Seniawski, B.L. (2001). Riba today: Social equity, the economy, and doing business under Islamic law. *Columbia Journal of Transnational Law, 39*, 701–728.

Sula, M.S. (2010, December 30). Sinergi Wakaf dengan Instrumen Asuransi Syariah. *Badan Wakaf Indonesia* [*Synergizing Waqf with the Sharia Insurance Instruments*]. Retrieved from http://bwi.or.id/index.php/in/publikasi/artikel/715-sinergi-wakaf-dengan-asuransi-syariah.html.

Sumanto, A.E., Priarto, E., Zamachsyari, M., Trihadi, P., Asmuri, R. & Maulan, R. (2009). *Solusi Berasuransi: Lebih Indah dengan Syariah* [*Solution in Insurance: It would be Better with Sharia*]. Bandung, Indonesia: Salamadani.

Zahrah, M.A. (n.d.). *Fi al-Mujtama al-Islamiy*. Cairo, Egypt: Dar al-Fikr al Arabiy.

The implementation of Indonesia's penal sanctions to the increased illegal trade of endangered species

A.G. Wibisana & W.P. Nuning
Faculty of Law, Universitas Indonesia, Depok, Indonesia

ABSTRACT: Indonesia is a mega diversity country. It has around 10 to 20% of plant and animal species that exist in the world. However, the existence of natural resources and ecosystems in Indonesia are being threatened due to several factors, one of which is the illegal trade in endangered species. Indonesia has ratified CITES since 1978. This article attempts to answer the questions of how Indonesia has translated the Convention's provisions on the use of criminal sanctions into its national laws and how these laws have been put into practice. To answer these questions, the article will explore various laws and regulations addressing the classification of endangered species in Indonesia. This article will also discuss the use of penal sanctions for illegal trading of endangered species, and analyze several factors likely to undermine the effectiveness of law enforcement. In analyzing the effectiveness of law enforcement for illegal trading of endangered species in Indonesia, this paper will focus on the trading of endangered animals. This article observes not only that Indonesia's law on conservation of biological resources has provided limited sanctions, but also that law enforcers do not fully understand the seriousness of trafficking in endangered species in Indonesia. The article provides some recommendations for the improvement of law enforcement against illegal trade in endangered species.

1 INTRODUCTION

Indonesia is one of the world's richest nations in terms of wildlife inhabitation and biodiversity. Indonesia is composed of more than 17,000 islands, and has approximately 88.17 million hectares (ha) of forest, or about 46.33% of Indonesia's land area (Simargo et al., 2011). Indonesia has 10–20% of the world's plant and animal species. It has about 10% of flower plants in the world (25,000 types), 12% of mammals (515 species, 36% are endemic), 16% of reptiles, 17% of bird species of the world (1,531 species, 20% are endemic) and about 20% of fish species in the world (Ministry of National Development Planning Republic of Indonesia, 1993).

Unfortunately, the existence and conditions of natural resources and ecosystems in Indonesia are currently under serious threats. The deteriorating conditions of wildlife might be caused by several factors, including human activities that induce the destruction of the ecosystem in the pursuit of economic gains. The situation might also be worsened by the increased demands in wildlife as pets and other products (for example for food, medicines, and accessories). Exploitation of wildlife in Indonesia will likely lead to serious conservation problems, such as the extinction of species (Soehartono & Mardiastuti, 2003, pp. 3–4).

This paper attempts to answer the questions of how Indonesia has translated CITES provisions on criminal sanction into its national laws and how these laws have been put into practice. To answer these questions, the paper will first explore various laws and regulations addressing the classification of endangered species in Indonesia. In this regard, it attempts to describe whether theoretically Indonesia has a good set of implementing regulations appropriate to effectively comply with CITES mandates. Furthermore, the paper will also discuss the use of penal sanctions for illegal trade in endangered species, and analyze several factors likely to undermine the effectiveness of law enforcement. In analyzing the effectiveness of law enforcement for illegal trade in endangered species in Indonesia, this paper will focus on the trade in endangered animals.

After this introduction, the paper will provide a general overview of wildlife trade in Indonesia (Section II). In this section, information from various studies will be used to form a picture on how serious wildlife trade in Indonesia is. Afterwards, Section III will provide discussion on sanctions under CITES. These provisions will then be compared with penal sanctions in Indonesia (Section IV) and how these sanctions are put into practice by both public prosecutors and the courts (Section V). Some concluding remarks will be provided in Section VI.

2 GENERAL REVIEW OF WILDLIFE TRADE IN INDONESIA

Indonesia might be the largest crime scene of illegal wildlife trade in Southeast Asia, for both domestic and export markets. Protected animals in Indonesia, such as orangutans, tigers, rhinoceroses, and elephants, are facing a serious threat from poaching and trading. Wildlife Conservation Society (WCS) as referred to by Amarullah (2008) assesses that wildlife trade has in recent years become more intense because of the weaknesses of existing laws in Indonesia. Due to rampant poaching, smuggling and trade in wildlife the Sumatran rhinos [*Dicerorhinus sumatrensis*] have reached the status of almost extinct (around 100–120 population), the Sumatran tigers [*Panthera tigris sumatrae*] have reached the status of almost extinct (650 population), the Asian elephants have reached the status of endangered, and the Sunda pangolin [*Manis javanica*] have also reached the status of endangered. The value of wildlife trade in Indonesia could approximately reach the value of US$ 1 billion per year (Amarullah, 2008). The most common countries of destination for smuggling from Indonesia are China, Taiwan, Hong Kong, and the Netherlands. Surprisingly, most of the animals are smuggled based on the request of people who have superstitions about certain animals or certain parts of it (Budiari, 2015).

Aside from conventional illegal wildlife trade, nowadays illegal wildlife trade can be found online. Protection of Forest & Fauna [ProFauna] notes that from January up to mid-December 2015 there were at least 5,000 cases of wildlife trade on the web (including on social media like Facebook). The number increases considerably compared to the data compiled in 2014, when there were approximately 3,640 online ads offering a wide range of wild animal species. The rate and volume of wildlife trade in social media have increased since it is very easy for users to upload their items, develop an unlimited network, and the security is far better than the conventional trade. The trend of illegal wildlife trade has been dominated by the online method for the past five years. In recent years, there have been a few massive cases where a large number of animals and huge amount of money are involved. These cases include the trade in 96 alive pangolins, 5,000 kg frozen pangolin meat, and 77 kg pangolin scales foiled in Medan, North Sumatra in April 2015; a smuggling attempt of 10 kg Manta Ray gills, 4 sacks of mixed shark and ray bones, 2 sacks of shark bones, and 4 shark fins in East Flores in July 2015; and a smuggling attempt of a 40 feet-container of Horned Helmet [*Cassis cornuta*] shells worth IDR 20.422 billion in August 2015, which was about to be shipped to China (Protection of Flora and Fauna, 2015).

High profit and the weak law enforcement are believed to contribute to the increased illegal trade in wildlife in Indonesia. Although a large number of perpetrators have been convicted and punished, deterrence is still questionable since the current Law, namely the Law No. 5 of 1990 regarding Conservation of Biological Resources and Ecosystems (hereinafter referred to as the Law No. 5/1990), offers low penal sanctions namely an imprisonment of maximum five years and a fine of maximum IDR 100 million.

3 SANCTIONS UNDER CONVENTION ON INTERNATIONAL TRADE IN ENDANGERED SPECIES (CITES)

The Convention on International Trade in Endangered Species of Wild Fauna and Flora (CITES) is an international agreement aimed to ensure that international trade in specimens of wild animals and plants does not threaten their survival. CITES was drafted as a result of a resolution adopted in 1963 at a meeting of members of the World Conservation Union (IUCN).

The text of Convention was finally agreed upon at a meeting of representatives of 80 countries in Washington, D.C., the United States of America, on 3 March 1973. Since 1975, when the CITES entered into force, the Parties have possessed a powerful tool to control the trade.

As the consequence of joining the Convention, each Party is required to perform all their obligations which are stipulated in CITES (the convention itself, resolution, and its decision). The formal legal basis for CITES sanctions is in Article XIV.1(a) of the Convention, which expressly reserves the right of states to take stricter domestic measures regarding the conditions for trade, taking, possession, or transport of specimens of species included in Appendices I, II and III, or the complete prohibition thereof (CITES, Article XIV.1(a)). Hence, the article also implicitly authorizes the use of unilateral economic sanctions by way of trade restrictions or trade bans (embargoes) against other states, provided that these sanctions are compatible with the applicable general rules of international law (Sand, 2013, p. 253).

More importantly, the Convention also mandates the states to implement criminal sanctions against illegal trading in endangered species. The Convention asks states to take appropriate measures to prohibit trade in specimens, including through the penalization of trade in or possession of endangered species (CITES, Article VIII par. 1(a)).

If a noncompliance dispute cannot be resolved by direct communication between the party concerned and the CITES Secretariat (pursuant to Article XIII) or the Standing Committee (pursuant to its functions as delegated by the Conference of the Parties (CoP), the CoP or the Standing Committee may, as a last resort (where a party's compliance matter is unresolved and persistent and the party is showing no intention to achieve compliance), recommend an embargo in terms of the suspension of commercial or all trade of specimens of one or more CITES-listed species (Sand, id).

Recommendations to suspend trade in specimens of CITES-listed species are made by the CoP and the Standing Committee. A recommendation to suspend trade provides a period of time during which the relevant country can move from non-compliance to compliance by inter alia making progress in the enactment of adequate legislation, combating and reducing illegal trade, submitting missing annual reports or responding to specific recommendations of the Standing Committee concerning the implementation of Article IV of the Convention in the context of the Review of Significant Trade. Recommendations to suspend trade are withdrawn immediately upon a country's return to compliance.

4 PENAL SANCTIONS FOR ILLEGAL WILDLIFE TRADE IN INDONESIA

Indonesia has ratified CITES through the Presidential Decree [Keputusan Presiden [Kepres]] No. 43 of 1978. One of the CITES obligations that must be carried out by Indonesia is to draw up a national legislation for the implementation of CITES. This requirement is met by the enactment of the Law No. 5 of 1990 regarding Conservation of Biological Resources and Ecosystems. The Law No. 5 of 1990 constitutes the main legal basis for CITES implementation in Indonesia. In this regard, the Law No. 5 of 1990 prescribes several criminal sanctions on illegal activities related to trade in protected animals.

Art. 40 par. 2 of the law provides sanctions of a maximum imprisonment of 5 years and a maximum fine of IDR 100 million for those who intentionally violate the provisions of Art. 21 paras. 1 and 2, and Art. 33 par. 3 of the Law. In this regard, it should be noted that Art. 21 par. 1 of the law prohibits the taking, cutting, owning, destroying, eliminating, keeping, transporting, trading, and transferring from one place to another of the living or dead protected plants or parts thereof. However, Art. 21 par. 2 of the law prohibits a) catching, injuring, killing, keeping, possessing, transporting, and trading in the living protected animals; b) keeping, possessing, transporting, and trading of dead protected animals; c) transfer of a protected animal from one place to another, within or outside Indonesia; d) trading, keeping, or possessing the pelts, bodies or other parts of a protected animal, or goods made of parts of the animal; and e) taking, destroying, exterminating, trading, keeping, or possessing an egg and/or nest of a protected animal. Art. 33 par. 3 prohibits carrying out of an activity that is not in line with the function of utilization and other zones of the National Park, Grand

Forest Park, and Natural Recreation Park. In contrast with Art. 40 par. 2, Art. 40 par. 4 of the law punishes negligent violation of similar provisions with an imprisonment of maximum 1 year and a fine of maximum IDR 50 million.

Penal sanctions laid down in Article 40 paragraph (2) and (4) of the Law No. 5 of 1990 are only for protected wildlife. Hence, it could be concluded that the Law No. 5 of 1990 does not prescribe any penal sanction for illegal activities related to unprotected wildlife. For illegal activities involving unprotected species, one might refer to the Law No. 41 of 1999 regarding Forestry, Art. 78 par. 12, which states that those who willfully violate the provision referred to in Art. 50 par. 3(m), namely removing, carrying, and transporting plants and wildlife species which are not protected under any law, from forest area without any approval from the respective authorities, are subject to an imprisonment for a maximum of 1 year and a fine for a maximum of IDR 50 million.

Furthermore, penal sanctions for unprotected wildlife trading can refer to the Government Regulation [Peraturan Pemerintah [PP]] No. 8 of 1999 regarding Wild Flora and Fauna Exploitation. In this case, illegal trade in unprotected wildlife can be subject to imprisonment for smuggling, attempts to damage the environment, or even theft. In this case, Art. 57 of the PP No. 8 of 1999 stipulates that those who are not entitled to trade in wild flora and fauna are subject to criminal sanction of smuggling. The regulation only authorizes certain business entities and local community to conduct the trade. In addition, Art. 60 par. 1 of the PP No. 8 of 1999 stipulates that the appearance of wildlife without permission shall be punishable as an attempt to damage the environment. Furthermore, Art. 63 par. 1 of the PP No. 8 of 1999 states that the parties who send or transport wildlife without proper documents shall be sentenced for participation in smuggling and/or theft and/or the attempt to damage the environment.

If one refers to provisions on theft under the Indonesian Criminal Code, an activity resulting in environmental degradation under the Law No. 32 of 2009 regarding Environmental Protection and Management, and smuggling according to the Law No. 17 of 2006 on Customs, one might find quite high sanctions. For the examples, for import smuggling (Article 102 of the Law No. 17 of 2006) and export (Article 102 paragraph (A) of the Law No. 17 of 2006), the minimum sanctions are the imprisonment of 1 year and a fine of IDR 50 million, while the maximum sanctions are the imprisonment of 10 years and a fine of IDR 5 billion. Clearly, those maximum sanctions are higher than the maximum sanctions provided for in the Law No. 5 of 1990. The result is quite surprising: penal sanctions for illegal trade in unprotected species could be higher than those for the protected ones!

5 LAW ENFORCEMENT OF PENAL SANCTION ON ILLEGAL WILDLIFE TRADE IN INDONESIA

In order to know how law enforcement for illegal trade in endangered species works, one could look at various provisions on the investigation and enforcement procedures and on the court decisions. With respect to the procedures, there are several institutions involved in law enforcement for illegal trade in wildlife in Indonesia. These institutions are the Forest Ranger, Forestry Civil Investigator, Forest Security Unit, and other Forest Security personnel, which are part of the Ministry of Forestry. In addition, there are also other institutions outside the Ministry, which are relevant to the law enforcement of the illegal wildlife trade, such as the police and military, customs, the Ministry of Transportation, the Ministry of Agriculture, navy, immigration, and of course public prosecutors. It is very likely that these various institutions have overlapping authorities in enforcing a particular case. This might lead to a situation where every institution considers the case to fall within its authority, or otherwise no institution is willing to enforce the case out the fear of trespassing the authority of other institutions. One obviously needs a strong coordinating institution, which unfortunately is often not in place.

With respect to the court ruling, one could first refer to Becker (1968) who argues that the deterrent effects of punishment might result from the probability of a criminal conduct being punished and the severity of sanctions. Intuitively, the effect is the result of law enforcement (that is the ability of the law enforcers to enforce the law) and the degree of punishment (that is how the judges give sanctions to the crime in question). Hence, bringing an illegal trading of wildlife into the court is one thing, but having rulings that give the appropriate sanctions for the crime is quite another.

Table 1 shows several court decisions regarding illegal activities related to wildlife animals. Most of them are related to the possession, killing, and trading of the animals or part of their organs. The table indicates very low sentences of the illegal activity, which mostly were imprisonment of 1 year or less. No ruling has given the maximum imprisonment of 5 years, and only one case was given quite severe imprisonment (namely an imprisonment of 4 years).

The court decision above more or less show the striking quality of the law enforcement in relation to the illegal wildlife trade, at least as reflected in several court decisions above. In

Table 1. Selected court rulings on illegal activities related to wildlife animals.

No	Decision number	Wildlife or its parts	Indictment	Punishment
1	District Court of Lubuk Sikaping, No. 24/Pid.B/2014/PN Lbs, RI v. HAMDANI Pgl ANDAN (2014)	1 piece porcupine organ, 9 pieces of bear bile, 4 adult Sumatera tiger fangs, 4 young Sumatera tiger fangs, 16 adult Sumatera tiger nails, and 16 young Sumatera tiger nails.	Imprisonment of 7 months Fine of IDR 5 million	Imprisonment of 5 months Fine of IDR 5 million
2	District Court of Lubuk Sikaping, No. 25/Pid.B/2014/PN Lbs, RI v. SUKATMAN Pgl MAN (2014)	1 piece porcupine organ, 9 pieces of bear bile, 4 adult Sumatera tiger fangs, 4 young Sumatera tiger fangs, 16 adult Sumatera tiger nails, and 16 young Sumatera tiger nails.	Imprisonment of 9 months Fine of IDR 5 million	Imprisonment of 6 months Fine of IDR 5 million
3	District Court of Sarolangun, No. 36/Pid.Sus/2015/PN.Srl, RI v. MUSA bin KASIM (2015)	1 plastic bag of tiger bones 1 plastic bag of tiger pelts	Imprisonment of 2 years Fine of IDR 50 million	Imprisonment of 1 year and 5 months Fine of IDR 50 million
4	District Court of Sarolangun, No. 37/Pid.Sus/2015/PN Srl, RI v. HERMAN TONI bin SABLI (2015)	1 plastic bag of tiger bones 1 plastic bag of tiger pelts	Imprisonment of 2 years Fine of IDR 50 million	Imprisonment of 1 year and 5 months Fine of IDR 50 million
5	District Court of Arga Makmur, No. 45/Pid.B/2016/PN.Agm, RI v. Anzuar Anas Als Aan Bin Madren (2016)	Pelt of 1 tiger Bones of 1 tiger	Imprisonment of 4 years Fine of IDR 60 million	Imprisonment of 4 years Fine of IDR 60 million
6	District Court of Arga Makmur No. 62/Pid.B/2014/PN.AM., RI v. NASARUDIN Als NASAR Bin GROT (2014)	Pelt of 1 tiger 2 heads of deer	imprisonment of 7 months Fine of IDR 1 million	Imprisonment of 5 months Fine of IDR 1 million

(*Continued*)

Table 1. (Continued)

No	Decision number	Wildlife or its parts	Indictment	Punishment
7	District Court of Jambi No. 508/Pid.Sus/2015/PN Jmb., RI v. HERY SUPENO Bin TRISILO (2015)	1 plastic bag of tiger pelts 1 plastic bag of tiger bones	imprisonment of 7 months fine of IDR 1 million	imprisonment of 7 months fine of IDR 1 million
8	District Court of Jambi No. 644/Pid.Sus/2014/PN.Jmb., RI v. MAMAN FIRMANSYAH ALIAS MAMAN ALIAS FIRMAN (2014)	1 piece of tiger pelt 1 piece of tiger skull 2 pieces of tiger jaws	imprisonment of 10 months fine of IDR 5 million	imprisonment of 7 months fine of IDR 3 million
9	PN Jambi, No. 702/PID.Sus/2015/PN.Jmb., RI v. JEFENDI Bin M.LIS dan PAHMI Bin ABDULLAH (2015)	1 piece of preserved tiger	imprisonment of 2 years, for Jefendi, and of 1 year and 6 months for Pahmi Fine of IDR 2 million for each defendant	Imprisonment of 1 year and 6 months for each defendant Fine of IDR 1 million for each defendant
10	Supreme Court, No. 2064 K/Pid. Sus/2012, RI v. AFANDI Pgl. FANDI	1 piece of tiger pelt	Imprisonment of 3 years Fine of IDR 3 million	Imprisonment of 1 year and 6 months (lower than the verdict of the district court, namely imprisonment of 2 years and 4 months) Fine of IDR 3 million (upholding the verdict of the district court)
11	District Court of Lumajang, No. 206/Pid.Sus/2015/PN.Lmj., RI v. SIYONO dan YULI ANDRIANTO	2 pieces of dead monkeys	For each defendant: Imprisonment of 1 year and 6 months Fine of IDR 2 million	For each defendant: Imprisonment of 1 year Fine of IDR 1 million

all court decisions above, the judges gave sentences that were not much different from those asked by the prosecutors. Unfortunately, because the prosecutors never sought for the maximum punishment, the judges also never gave the maximum one.

Surprisingly, such low punishments occurred, even though the courts were convinced that the defendants were guilty of conducting the accused crimes, and the fact that most of them were involved in the killing of endangered animals, such as Sumatran Tigers. Strangely, such lenient sanctions for crimes against tigers were even lower than the punishment for killing two monkeys, which are not even endangered species.

The only decision with quite severe penal sanction is Anzuar Anas Bin Madren Als Aan (District Court Arga Makmur, 2016), in which the prosecutor employed cumulative charges, and the judges declared that the defendant was guilty of the accused crimes, namely: a) capturing and killing protected animals alive; and b) trading pelts and other parts of protected animals. Although both of the prosecutor and judges did not explain that the defendant committed concurring crimes (concursus realist), if one compares to other decisions that have low penalties (under one year), one might well get an impression that the defendants were prosecuted and sentenced with penal sanctions because he did more than one crime. If that was the case, then the reason for giving high sanction was then not entirely accurate. This is because the maximum sanction of Article 40 paragraph (2) or (4) of the Law No. 5 of 1990 has to be imposed without having to wait for evidence that the defendant committed several criminal acts at once. Moreover, if the defendant has conducted concurring criminal acts, then according to the theory of aggravated absorption (Utrecht, 2000, pp. 184–186), the penal sanctions could be as high as 5 years plus one third of 5 years.

From the court decisions, it can be seen that the issue of wildlife trade is not a priority for the judges. Therefore, the Law No. 5 of 1990 needs to be revised. Penal sanctions should be made with the minimum and maximum mechanisms in order to give stronger deterrent effects for offenders. However, it should be noted that low penalties above will not automatically disappear if the Law No. 5 of 1990 recognizes the minimum penalty aside from the maximum one. If the disparity in the minimum penalty and a maximum penalty is too far, then the tendency of low penalties remains large because it is feared that sanctions that will be prosecuted and sentenced are too far from the minimum sanction. Therefore, if we want to suggest the minimum penalty, then the minimum penalty has to be made not too low from its maximum penalty.

Theoretically, as Messer (2000) argues, countries with a high level of law enforcement have the best chance to reduce or even stop poaching (Messer, 2000, p. 55). Moreover, if one considers the perpetrators of illegal activities against wildlife as rational criminals, then one could think that negative incentives for criminal activities resulting from a situation where the costs of doing a crime, namely the expected sanction (or the probability of criminals being punished times the severity of sanctions), are higher than the benefits of doing the crime (Cooter & Ulen, 2012, pp. 463–465). It follows, thus, that the low probability of detection of crimes, such as the case of crimes related to wildlife, should be compensated by severe punishment. Unfortunately, this is not the case in Indonesia.

More importantly, the Law No. 5 of 1990 and the court rulings have failed to see crimes on wildlife as part of organized and transnational crimes. It seems that all the cases stop with the factual perpetrators. Not only did the rulings provide insufficient punishment to deter factual perpetrators, but also the law failed to give appropriate disincentives for intellectual perpetrators.

6 CONCLUDING REMARKS

This paper shows some of the weaknesses of the Law No. 5 In 1990, including the lack of classification of species, the lack of clarity of sanctions for illegal trade in unprotected species, the low penal sanctions, the failure to recognize corporate responsibility, and the failures to see illegal trade as a part of transnational organized crime. The paper has also observed a disturbing picture related to the application of penal sanctions in the court decisions for illegal wildlife trade in Indonesia. It turns out that the sanctions demanded by the public prosecutor are relatively low. Unfortunately, judges also imposed sanctions that were relatively similar to those demanded by the prosecutors.

The paper also shows the urgency for the revision of the Law No. 5 of 1990. One frequently heard on the revision of the Law is to increase maximum penal sanctions, as well as the introduction of minimum sanctions. One important thing to note in this regard is the minimum penalty cannot be far too low and much different from the maximum sanction.

Meanwhile, it also necessary for the revising Law to address the less endangered species. It should specify which species belong to this category, the mechanism for species utilization, and the sanctions for violations of law. In addition, the revision should also incorporate corporate responsibility and see the illegal trade in the perspective of organized crime, both national and transnational in scopes.

REFERENCES

Amarullah, A. (2008) Satwa Liar di Indonesia Terancam: Thailand, Malaysia dan Singapura Sebagai Negara Tujuan Penjualan Satwa Tersebut. *Viva News* December 23, 2008. [Online] Available from: http://nasional.news.viva.co.id/news/read/17565-satwa-liar-di-indonesia-terancam. [Accessed 1st September 2016].

Becker, G.S. (1968) Crime and punishment: An economic approach. *Journal of Political Economy* 76 (2), 169–217.

Budiari, I. (2015) Jakarta hotspot of illegal wildlife trade: BKSDA official. *Jakarta Post* December 28, 2015. [Online] Available at: http://www.thejakartapost.com/news/2015/12/28/jakarta-hotspot-illegal-wildlife-trade-bksda-official.html. [Accessed 2nd September 2016].

CITES (1973) *Convention on International Trade in Endangered Species of Wild Fauna and Flora.* [Online] Available from: https://www.cites.org/eng/disc/text.php#texttop. [Accessed 5th September 2016].

Cooter, R & Ulen, T. (2012) *Law and Economics.* 6th edition. Boston, Addison-Wesley.

District Court of Arga Makmur (2014) Decision Number 62/Pid.B/2014/PN.Am. Arga Makmur.

District Court of Arga Makmur (2016) Decision Number 45/Pid.B/2016/PN.Agm. Arga Makmur.

District Court of Jambi (2014) Decision Number 644/Pid.Sus/2014/PN.Jmb. Jambi.

District Court of Jambi (2015) Decision Number 508/Pid.Sus/2015/PN Jmb. Jambi.

District Court of Jambi (2015) Decision Number 702/PID.Sus/2015/PN.Jmb. Jambi

District Court of Lubuk Sikaping (2014) Decision Number 24/Pid.B/2014/PN Lbs. Lubuk Sikaping.

District Court of Lubuk Sikaping (2014) Decision Number 25/Pid.B/2014/PN Lbs. Lubuk Sikaping.

District Court of Lumajang (2015) Decision Number 206/Pid.Sus/2015/PN.Lmj. Lumajang.

District Court of Sarolangun (2015) Decision Number 36/Pid.Sus/2015/PN.Srl. Sarolangun.

District Court of Sarolangun (2015) Decision Number 37/Pid.Sus/2015/PN Srl. Sarolangun.

Indonesia (1990) *Undang-undang No. 5 Tahun 1990 tentang Konservasi Sumber Daya Alam Hayati dan Ekosistemnya* [*The Law Number 5 of 1990 regarding Conservation of Biological Resources and Ecosystems*].

Indonesia (1999) *Peraturan Pemerintah No. 8 Tahun 1999 tentang Pemanfaatan Jenis Tumbuhan dan Satwa Liar* [*The Law Number 8 of 1999 regarding Wild Flora and Fauna Exploitation*].

Indonesia (1999) *Undang-undang No. 41 Tahun 1999 tentang Kehutanan* [*The Law Number 41 of 1999 regarding Forestry*].

Indonesia (2006) *Undang-undang No. 17 Tahun 2006 tentang Perubahan atas Undang-undang No. 10 Tahun 1995 tentang Kepabeanan.* [*The Law Number 17 of 2006 regarding Amandement of Law Number 10 Year 1995 on Customs*].

Indonesia (2006) *Undang-undang No. 32 Tahun 2006 tentang Perlindungan dan Pengelolaan Lingkungan Hidup* [*The Law Number 32 of 2009 regarding Protection and Management of Environment*].

Mahkamah Agung [Supreme Court] Republik Indonesia (2012) Decision Number 2064 K/Pid. Sus/2012.

Messer, K. (2000) The Poacher's Dilemma: The economics of poaching and enforcement. *Endangered Species Update* 17 (3), 50–56.

Ministry of National Development Planning Republik Indonesia (1993) *Biodiversity Action Plan for Indonesia.* Jakarta, National Development Planning Agency. [Online] Available from: https://www.cbd.int/doc/world/id/id-nbsap-01-p01-en.pdf.

Presiden Republik Indonesia (1978) *Keputusan Presiden No. 43 Tahun 1978 tentang Mengesahkan Convention On International Trade In Endangered Species Of Wild Fauna And Flora* [*The Presidential Decree Number 43 of 1978 regarding Convention On International Trade In Endangered Species Of Wild Fauna And Flora*].

Protection of Forest and Fauna (2015) *ProFauna: Wildlife Crime in Indonesia 2015: 5,000 Trade and 370 Poaching Cases.* In: *Wildlife and Forest Crime December 30, 2015.* [Online] Available from: http://www.profauna.net/en/content/wildlife-crime-indonesia-2015-5000-trade-and-370-poaching-cases#.V9wGPdN97 sE. [Accessed 3rd September 2016].

Sand, P.H. (2013) Enforcing CITES: The rise and fall of trade sanctions. RECIEL: *Review of European, Comparative & International Environmental Law* 22 (3), pp. 251–263.

Simargo, W., Nanggara, S.G., Nainggolan, F.A. & Apriani, I. (2011) *Potret Keadaan Hutan Indonesia Periode Tahun 2000–2009.* Jakarta, Forest Watch Indonesia. [Online] Available from: http://fwi.or.id/wp-content/uploads/2013/02/PHKI_2000–2009_FWI_low-res.pdf.

Soehartono, T. & Mardiastuti, A. (2003) *Pelaksanaan Konvensi CITES di Indonesia.* Jakarta, Japan International Cooperation Agency.

Utrecht, E. (2000) *Hukum Pidana 2.* Surabaya, Pustaka Tinta Mas.

Analysis of the water resource regulation under the ordinance of the Greater Jakarta area

R.I. Dewi
Faculty of Law, Universitas Indonesia, Depok, Indonesia

ABSTRACT: The existence of local ordinances on water resources in Indonesia is crucial given that issues related to water are contextual in nature. Meanwhile, the conditions of the various areas in Indonesia are geographically and demographically diverse. The regulation of water resources in Indonesia, with its extensive land area and diverse population, requires legislation issued not only by the national government but also through delegation by the local governments. The *Greater Jakarta* area (*Jabodetabek*) still faces problems in meeting their water needs; in light of this, a study was carried out on water resource ordinances that are in force in this area. The result of the study showed that water resource regulation is still plagued with problems related to availability, content, and implementation. The approach employed in this study was a qualitative and it is expected that the results can serve as valuable input for the relevant stakeholders involved with water resource management. As an end result, it is hoped that through the use of water resource ordinances, this vital resource can be managed in an equitable manner for the prosperity of the people.

1 INTRODUCTION

The regulation of water resources in the *Jabodetabek* area (*Greater Jakarta* area) encompassing the adjacent administrative areas of *Bogor*, *Depok*, *Tangerang*, and *Bekasi* or *Jabodetabek* is still mired with a multitude of problems, mainly relating to the utilisation and distribution of water. This condition is yet to be adequately overcome, and thus the people's basic need for water has not been completely fulfilled which has, in turn, affected their prosperity. Meanwhile, water is one of the most vital natural resources, and its role is irreplaceable in sustaining human life (Mitchell et al., 1945, p. 628).

Availability of a water supply is an issue in all areas where human settlement exists. In fact, water is not only a cause of dispute in some communities, but can also trigger wars between nations (Gleick, 2014, p. 331). In the context of Indonesian society, the gap between water need and supply has caused problems with water resources. In this regard, a study on water resources and its regulating instruments was identified as needed, due to at least four driving factors: 1) water constitutes a basic need of human beings, and its supply is limited; 2) many problems arise within society that relate to water resources (Kreamer, 2012, pp. 88–90); 3) many issues relating to water resources require regulation of the utilisation and distribution of water, as demonstrated by European and Chinese water regulation studies (van Rijswick & Wouters, 2014, p. 85); and 4) the regulation of water resources in the context of the Republic of Indonesia, given the vast size of its territory and highly heterogeneous people, cannot be affected by the national government alone, but must also be delegated to the regional governments through local ordinances (Persoon & Weerd, 2006, p. 96).

In the running of local governance, the local administrations are given the authority to establish ordinances (local regulations), which include the municipalities of *Jakarta*, *Bogor*, *Depok*, *Tangerang*, and *Bekasi*. *Jakarta* is a province, designated as a special territory because it is the country's capital. The *Jakarta* province is surrounded by a number of local governments: *Bogor* municipality, consisting of the district of *Bogor* and *Bogor* city; *Depok* municipality; the local government of

Tangerang, consisting of the district of *Tangerang*, *Tangerang* city, and *South Tangerang* city; and the *Bekasi* area, consisting of the district of *Bekasi* and *Bekasi* city. The provincial head of government is the governor, whereas a district is led by a Regent/*Bupati*, and a city by a Mayor.

This study on local ordinances that regulate water resources within the *Greater Jakarta* area is intended to provide an analysis of regulatory issues relating to water resources in the area. The output of the study is expected to address the following researchable questions: 1) what constitutes the legal basis for the regulation of water through local ordinances? This involves the legal foundation upon which local ordinances are established as legally binding instruments; 2) what are the issues involved in the regulation of water resources by the local governments of the *Greater Jakarta* area? By answering these questions, we will be able to understand the water resource regulatory framework and issues that exist in the *Jabodetabek* area and identify measures to address them.

Discussion on water governance through local legislation in the *Greater Jakarta* area will be described through the following phases. First is a discussion of concepts relating to water resources and local regulations. The concepts related to water resources are generally known, and include the definition, nature of and the rights of access. Moreover, recent legislation, which is a legal product made by the government as one of the aspects of regional autonomy, will be discussed.

Second is the discussion about the legal basis for water regulation through legislation. The setting of water resources in Indonesia will be discussed with a description of Article 33 of the 1945 Indonesian Constitution (UUD, 1945) as the cornerstone of the constitutional arrangements for natural resources including water. Arrangements through the Water Resources Act are through Law No. 7 2004, which at this point has been cancelled by the Constitutional Court and has returned to Law No. 11 1974, along with a new law which has yet to be established and local regulations (*perda*). Discussion of the water resources setting refers to the statutory provisions as stipulated in Law No. 12 2011.

Third is the setting of regulation by the local government in *Greater Jakarta*. The discussion includes: 1) the Availability Regulation of water resources; 2) problems of regulation; and 3) efforts to overcome the problems of water resource management through a regulation.

2 THE CONCEPT OF WATER RESOURCES AND LOCAL ORDINANCES

An evaluation of the regulation of water resources through local ordinances must take into account major variables. In this study, a review is conducted on the right to water supply arising from the vital role of water for human life. The discussion on the relevant local ordinances will examine how water supply, and the right to water, are regulated by the ordinances as local regulations that are established under the regional autonomy status and its various implications.

2.1 *Water resources*

Water resources are an irreplaceable natural resource that is critical to the sustainability of human life. Without water a person simply cannot live as all aspects of human life are related to water. Vandana Shiva, an environmental activist from India, states that water is life itself (Shiva, 2002, pp. 39–40). As such, the use of water or a person's right to water must consider the essence of water, which is vital to his/her survival and the life of other people.

> The fact that the right over water has existed in all ancient laws, including our own *dharmasastras* and the Islamic laws, and also the fact that they still continue to exist as customary laws in the modern period clearly eliminates the water right as being purely a legal right, that is, the right granted by the state or law (Shiva, 2002, pp. 19–21).

One water right that is quite renowned in the course of history is the 'riparian right'. According to Shiva (2002), this right is based on common ownership and reasonable use of water. At the beginning, the riparian principles were based on the allotment of water and the preservation of collective ownership of water resources. However, due to the advent of globalisation and colonisation in its various forms, principles governing rights over the use of water have experienced a shift. Previously regarded as public goods or collective property, water now has become privatised; a liberalisation of the utilisation of water, is causing problems in meeting the water needs of the general public.

> With globalisation and privatisation of water resources, new efforts to completely erode people's rights and replace collective ownership with corporate control are under way. That communities of real people with real needs exist beyond the state and the market is often forgotten in the rush for privatisation (Shiva, 2012, pp. 19–23).

In Indonesia, under Law No. 7 2004 on Water Resources, there are two types of rights recognised by the state, namely the 'right of water utilisation' and the 'right of water for enterprise use'. This right is not much different from water right arrangements in Switzerland that is 'Based on the classic continental dichotomy property rights: water is divided into public and private ownership' (Rüegger, 2014, p. 19). Water rights according to Law No. 7 2004 shows the effect of the aforementioned globalisation, albeit not in entirely the same sense. However, Law Number 7 2004 was subsequently repealed by the Indonesia Constitutional Court, and regulation of water resources were reverted to Law No. 11 1974 on Irrigation, which emphasises the social function of water rights.

2.2 Local ordinances (sub-national regulations)

A local ordinance is a legal product issued by the local government as an aspect of regional autonomy, as is the implementation of Article 18 of the 1945 Indonesian Constitution. Indonesia is a unitary state divided into smaller sub-national administrations. This division of region is adopted due to the fact that Indonesia possesses a vast territory made up of numerous islands separated by bodies of water, but all operate as the Unitary State of the Republic of Indonesia (NKRI) (Soeprapto, 2007, pp. 95–96).

The existence of this ordinance is recognised in the Indonesian legislative hierarchy under Article 7 of Law No. 12 2011 on the 'Establishment of Legislations'. The classification and hierarchy of the Indonesian legislation are as follows: 1) Indonesia's Constitution in 1945; 2) the Decree of the People's Consultative Assembly (MPR); 3) the law/government regulation in lieu of law; 4) Government regulation; 5) presidential regulation; 6) provincial ordinance; and 7) district/city ordinance. The degree of legal power of each of the above legislations correspond to their position in the hierarchy. Local ordinances, in addition to being provided under Law No. 12 2011, are also referenced in Law No.23 2014 on local governance (specifically in its articles 236–271).

The water ordinance in question regulates two aspects: namely public service (Dewa, 2011, pp. 97–100) and matters regarding the use of water by the economic sector through the trading of water resources (Savas, 1987, p. 3). The present review of the Water Resource ordinance was carried out to establish whether the ordinance in question has accommodated public participation, thereby ensuring that its provisions address public needs. In the context of regional autonomy, this constitutes a manifestation of the representativeness of a policy (Achmadi and Soemantri, 2003, p. 30). Additionally, the discussion on the water resource ordinance will examine the position of the ordinance within the legislation hierarchy, namely that it should not be in conflict with any higher-ranking regulation, and must take a reference to the highest ranking legal provision, namely Article 33 of the 1945 Indonesian Constitution.

3 THE REGULATION OF WATER RESOURCES WITHIN THE JABODETABEK AREA THROUGH ORDINANCES

3.1 Legal basis of the water resource regulation

Article 33 of the 1945 Indonesian Constitution represents the constitutional foundation for a water resource regulation. The article stipulates that the authority is to control water resources that lie within the state. The extent of the state's control is only as the manager, rather than as the owner. Given that the highest power within the state lies with the people, water within the territory of Indonesia must be managed, as much as possible, to enhance the prosperity of the people.

Since 1974, water resources have been regulated by Law No. 11 1974 on Irrigation. Thirty years later the law was superseded by Law No. 7 2004 on Water Resources, in order to anticipate an expected water crisis. However, in reality, from its conception to its ratification, the provisions of Law No. 7 has been deemed to be in conflict with Article 33 of the 1945 Indonesian Constitution, and thus it has met with resistance from the people.

This has prompted the instigation of several petitions for judicial review on such law against the 1945 Indonesian Constitution. The first petitions, case number 058–059–060–063/PUU-II/2004 and case number 008/PUU-III/2005 were denied by the Constitutional Court. In 2013 another petition was filed, and pursuant to Constitutional Court Ruling Number 85/PUU-XI/2013, Law Number 7 2007 on Water Resource was repealed, as it was deemed to be in conflict with the 1945 Indonesian Constitution and lacked any binding power. The court also ruled that Law No. 11 1974 on Irrigation was to be reinstated. The existence of Act No. 11 1974 on Water Resources is intended to prevent a legal vacuum until the law governing water resource is formed.

Water is an element that controls the lives of many people. Therefore, it is necessary to regulate its distribution, utilisation, and management (Derrington, 2011, p. 63). The Water regulation in Indonesia must be in accordance with the provisions of Article 33 paragraph (2) and (3) of the 1945 Indonesian Constitution which is controlled by the state for the welfare of the people. The new law must consider the decision of the Constitutional Court on judicial review of the current laws. This is because the Constitutional Court's decision is binding, final and has the value of legal certainty and equity (Faqih, 2010, p. 114).

Based on the decision of the Constitutional Court case number 85/PUU-XI/2013, there should be strict limits to preserving and maintaining the availability of water. There are six restriction items, namely: 1) the utilisation of water should not interfere, override, let alone negate the people's right to water; 2) the state must meet the public's right to access water, as access to water is a human right itself; 3) environmental sustainability is a human right, as the provisions of Article 28 H paragraph (1) of the 1945 Indonesian Constitution states; 4) the state's monitoring and control over water is absolute; 5) top priority is given to water concessions by state or local government; and 6) if all of these restrictions are met and reviewed while the availability of water is still missing, the government is likely to allow the private sector to make use of water with certain conditions.

3.2 *Legal basis for the formulation of ordinances*

Managing water resources over a large area with varying local characteristics requires customised ordinances that are tailored to suit the condition of the respective areas, and as such, they are highly contextual. The legal basis for the establishment of ordinances is provided under Article 18 paragraph (6) of the 1945 Indonesian Constitution: 'The regional government is entitled to establish ordinances and other regulations to exercise autonomy and carry out supporting functions'.

The regulatory provisions that serve as a reference in the establishment of ordinances are, among others, Law No. 23 2014 on Local Governance; Law No. 12 2011 on the Establishment of Legislations; Government Regulations No. 16 2010 on Guidelines for the Establishment of a Regulation of the Regional House of Representatives on Rules of Conduct of the Regional House of Representatives; and the Regulation of the Minister of Interiors No. 53 2011 on Establishment of Sub-National Legal Instruments.

As ordinances constitute part of the legislative framework, the process by which they are enacted must follow the applicable principles governing the formulation of legislations (Article 5 of Law No. 12 2011). These principles lay down conditions that must be met in the establishment of an ordinance, namely that an ordinance: a) must have a clear purpose; b) is drafted by an institution/official authorised to draft legislations; c) contains materials that are needed by and beneficial for the people; and d) is clearly worded and focuses on the principle of openness and transparency. These requirements are to ensure that the ordinance being drafted would not be misconstrued, and it accommodates the aspiration or feedback from various members of society.

3.3 *Water resource ordinances in the Jabodetabek area*

3.3.1 *Issues with water resource ordinances*

Issues with water resource ordinances that apply within the *Greater Jakarta* area relates to: 1) availability; 2) harmony of the substance and formulation process of the ordinance; and 3) implementation of the ordinance in the community. Firstly, on the issue of availability, findings from the present research show that water resource ordinances that apply to the *Jakarta* municipality do not regulate water resource management, but merely provide for the taxation of water and the regulation of water utility companies and economic issues related to water resources. The current regulation on the use of water has only recently been introduced in the form of a 'Governor Regulation'. Moreover, in contrast to other provinces, namely *West Java* and *Banten*, *Jakarta* does not have any municipal regulation that governs the management of groundwater or other water resources. This is a cause for a concern, as *Jakarta* is the capital city that faces major water issues, such as those relating to the control of groundwater usage, floods, drought and safe water supply services.

Some of the causes of these problems are the limited number of specialised legal human resources, the multitude number of other pertinent issues, and the lack of coordination between the local government and the local people's representative body (DPRD). The *Jakarta* municipal government (*Pemda DKI*) relies on an ordinance bill to be proposed by the DPRD. On the other hand, the DPRD also waits for a draft to be submitted by the municipal government. As a result, the necessary process by which to establish ordinance to adequately govern water resources has yet to be initiated.

Similar situations were found in other areas targeted for the present research, such as *Bekasi* city, *Tangerang* city, and *South Tangerang* city, which do not have any ordinances that regulate the management of water resources. As is the case in *Jakarta*, the three administrative areas only have local regulations related to taxes and drinking water companies.

Where a government lacks a local regulation that regulates its water resources, the matter will be governed by a higher-ranking legislation. For *Jakarta*, given its status as a province, water related issues are referred to the law on water resources, that is, national legislation. For district/municipal governments that do not have a water ordinance, they must refer to the next higher class of regulations, namely provincial regulations.

A problem in this field is that a regulatory instrument governing water resources, in order to be applicable, needs to be highly contextualised. This entails that provisions regarding water sources at the level of law are still very broad in nature, and as such, if *Jakarta* relies directly on such law, it will face problems in terms of implementation. Likewise, the governments at the district/city level would face the same hurdle if they only refer to the provincial regulation to regulate matters relating to water sources in their areas. In this regard, the *West Java* province has laid down provisions in the West Java Provincial Regulation No. 3 2004 on the Management of Water Quality and Water Pollution Control.

The lack of regulations at the local level that regulate water resources has caused challenges in their use and the fulfilment of the people's need for water. Given that water constitutes a fundamental need for sustenance, failure to fulfil such need would also affect people's prosperity.

A second problem involves harmony of the water resource ordinances. The problem with harmony involves the substantive content of the ordinance as well as its process of enactment. In regard to the harmony of the subject matter of the legislation, there appears to be a lack of current ordinances. Each administrative area establishes their own water resource ordinances which only take into account their own local conditions. In other words, they do not consider the possible links such regulations have with similar regulations of their neighbouring regions. Additionally, the local governments who are the subject of this research still do not offer adequate transparency of information relating to their current water source legislations. For example, most of these governments' websites do not provide clear information on the water resource ordinances. This would certainly impede the harmonious usage and management of water resources.

The problem of lack of cohesiveness in the process of establishing Local Regulations on water resources occurs internally within the local government as well as externally with other regional governments. Within the local governments themselves, often the executive and legislative bodies fail to work together in the enactment of ordinances, particularly with respect to initiatives to draft ordinance bills. On the other hand, cost is also a problem as it can be felt as being quite substantial over the course of the enactment process (Dewi & Nurdin, 2014, p.111–112).

Similarly, the cooperation between local governments (the external aspect) in the establishment of ordinances is not conducted optimally. Each local government only relies on provisions governing the establishment of ordinances as stipulated in Law No. 23 2014 on the Regional Governance, Law No. 12 2011 on the Establishment of Legislations, and the Minister of Interiors Regulation No. 53 2011 on the Establishment of Regional Legislations.

As water can present issues that transcend administrative borders, particularly those of adjacent regions, establishment of ordinances regulating such resources should involve a degree of coordination between the local governments, either at the provincial or at the district/city level. Provisions on such cooperation are provided under Law No. 23 2014 on Regional Governance, the Government Regulation 38 2007 on the Distribution of Authority of Governance, Provincial Governance, and District/City Governance, the Government Regulation No. 50 2007 on the Procedure to Implement Inter-Region Cooperation, the Minister of Interiors Regulation No. 6 2006 on the Organisation and Work Procedure of the *Jabodetabekjur* (*Jakarta, Bogor, Depok, Tangerang, Bekasi,* and *Cianjur*) Development Cooperation Agency Secretariat, and the Minister of Interiors Regulations No. 22 and No. 23 2009 on the Process Technical Guidelines.

The third problem is the implementation of the water source ordinances in society. The result of this research on water resource ordinances shows that existing ordinances have not been fully implemented down to the lowest government subdivisions such as the *kelurahan* (a sub-district area/urban community). Of all the work programmes related to the use and management of water resources prepared by the sub-district areas that were researched in the *Jabodetabek* area (*Jakarta* province, *Bogor* District, *Bogor* city, *Depok* city, *Tangerang* district, *Tangerang* city, *South Tangerang* city, *Bekasi* District, and *Bekasi* city), only 20per cent of them provide for activities that involve water management such as river clean-up activities. These kinds of activities, carried out by communities through the citizens association (*rukun warga*) and neighborhood association (*rukun tetangga*), are the result of the people's own initiatives such as cleaning up the neighbourhood including its drainage system. When members of the community run into problems with their water supply, such as faulty pumps or dirty water coming out from the pumps, they will try to resolve the problem themselves or with the assistance of their neighbours. It is apparent here that communication between the local governments and the communities is still inadequate thus, the dissemination of information on ordinances cannot be carried out effectively.

3.3.2 *Efforts to resolve issues related to water resource ordinances*
Referring to the problems plaguing the regulation of water resources in the *Jabodetabek* through local ordinances, firstly, the lack of water resource ordinances, the sub-national government and the DPRD need to treat the issue as a priority for resolution, by introducing a water resource ordinance programme. Through this measure, water resource related ordinances can serve as the guiding instrument for the management of water in the *Jabodetabek* area. Effective political communication between the local government and the DPRD needs to be developed to make sure that the process by which ordinances are enacted can be undertaken smoothly, and mutual support can be given by the two institutions.

Another measure that can be undertaken is to upgrade the status of the relevant Governor regulation or Regent/Mayor regulation to become an ordinance. This can be an alternative if there occurs an urgent need to regulate water resources in a given region that has no local ordinance. Under this option, an existing Governor regulation or a Regent/Mayor regulation is discussed between the local government and the DPRD and developed into an ordinance.

One feature of such ordinance would be that it carries a sanction that will be imposed if any of its provisions are breached.

In order to resolve the problem of the ordinances not being aligned with each other, coordination and cooperation can be improved internally within the respective regions (between the local government and the DPRD), or externally between local governments. The existing inter-region cooperation agency needs to be utilised more intensively. It not only serves as a forum where the sub-national governments can meet but can also serve as a strategic forum to communicate information on the establishment of an ordinance.

To resolve the third problem, namely the implementation of water resource ordinances, the ordinances need to be introduced to the people and its implementation needs to be closely monitored. A work programme must be carried out, whereby the local government together with the public, make efforts to resolve water related problems. Dissemination of water resource ordination to the public can be undertaken in a variety of ways, and among others through the media, public education, trainings, and other open events. Thereby, problems related to water, which is a shared problem, can also be resolved together.

4 CONCLUSION

The constitutional basis for the regulation of water resources through the use of ordinances can be found in Article 33 of the 1945 Indonesian Constitution. According to this Article, the right to control water resources lies with the state. However, the highest governing power in the Republic of Indonesia is also with the people; therefore all natural resources, including water, must be used fully to advance the prosperity of the people. At the sub-national level, the state's responsibility to manage the resources is delegated to the local government as mandated by Law No. 23 2014 on Regional Governance. Decentralisation and regional governance are mandated by the 1945 Indonesian Constitution, which is intended to accelerate the achievement of prosperity for the people and includes the utilisation of water resources within local communities. Utilisation of water resources in these communities require ordinances to regulate it.

To tackle issues surrounding water resource ordinances arising in the *Jabodetabek* region, several measures need to be undertaken. Firstly, to establish adequate water resource ordinances, which can be pursued by making water resource issues as priority under the regional legislation programme. The ordinance enactment process requires solid cooperation between the local government and the DPRD. Additionally, under certain circumstances, the regulations of the governor/regent/mayor can have their status upgraded to become local ordinances through a joint discussion with the DPRD. Secondly, the process by which ordinances are passed in regions that are adjacent to each other, in this case in the *Jabodetabek* area which is also linked to ordinances in *West Java* and *Banten*, needs to be undertaken in an integrated and coordinated fashion. Thirdly, the dissemination of information and monitoring in the implementation of the ordinances needs to be done consistently, with the involvement of all stakeholders. Thereby, water resource ordinances can be an effective means to improve the prosperity of the people in regard to their need for water.

REFERENCES

Achmadi, A., Soemantri, Hardja (2003). *Penyusunan peraturan daerah yang partisipatif* [Compilation of participatory local goverment ordinances]. Jakarta, Indonesia: Masyarakat Transparansi Indonesia, DPRD Kota Ambon and LRP-Ausaid.

Derrington, E. (2011). Drinking water in the United States: Are we planning for a sustainable future? *Consilience: The Journal of Sustainable Development*, 6(1), 63–90.

Dewa, M.J. (2011). *Hukum administrasi negara dalam perspektif pelayanan publik* [The law of state administration in the perspective of public service]. Kendari, Indonsida: Unhalu Press.

Dewi, I. & Nurdin, A.R. (2014). *Pengaruh perda sumber daya air pada kesejahteraan masyarakat di jabodetabek* [The influence of water resources law on the welfare of the people in Jabodetabek]. Depok, Indonesia: Direktorat Riset dan Pengabdian Masyarakat Universitas Indonesia.

Faqih, M. (2010). Nilai-nilai filosofi putusan Mahkamah Konstitusi yang final dan mengikat [The philosophical values of the Constitutional Court decision are final and binding], *Jurnal Konstitusi*, 7(3), 97–118.

Gleick, P.H. (2014). Water, drought, climate change, and conflict in Syria. *American Meteorological Society*, 6, 331–340.

Indonesia. (1974). *Undang-undang no. 11 Tahun 1974 tentang Pengairan* [Law No. 11 of 1974 Irrigation].

Indonesia. (2004). *Undang-undang no. 7 Tahun 2004 tentang Sumber Daya Air* [Law No. 7 2004 Water Resources].

Indonesia. (2006) *Peraturan Menteri Dalam Negeri No. 6 Tahun 2006 tentang Organisasi dan Tata Kerja Sekretariat Badan Kerjasama Pembangunan Jakarta, Bogor, Depok, Tangerang, Bekasi dan Cianjur* [Minister of Home Affairs Regulation No. 6 2006 concerning the Organization and Work Procedure of the Jabodetabekjur (Jakarta, Bogor, Depok, Tangerang, Bekasi, and Cianjur) Development Cooperation Agency Secretariat].

Indonesia. (2007) *Peraturan Pemerintah No. 38 Tahun 2007 tentang Pembagian Urusan Pemerintahan antara Pemerintah, Pemerintah Daerah Provinsi dan Pemerintah Daerah Kabupaten/Kota* [The Government Regulation No. 38 2007 concerning the Distribution of Authority of Governance, Provincial Governance, and District/City Governance].

Indonesia. (2007) *Peraturan Pemerintah No. 50 Tahun 2007 tentang Tata Cara Pelaksanaan Kerja Sama Daerah* [The Government Regulation No. 50 2007 concerning Procedure to Implement Inter-Region Cooperation].

Indonesia. (2009) *Peraturan Menteri Dalam Negeri No. 22 Tahun 2009 tentang Petunjuk Teknis Tata Cara Kerja Sama Daerah* [Minister of Home Affairs Regulation No. 22 2009 concerning Technical Guidelines for Inter-Regional Cooperation].

Indonesia. (2009) *Peraturan Menteri Dalam Negeri No. 23 Tahun 2009 tentang Tata Cara Pembinaan dan Pengawasan Kerja Sama Antar Daerah* [Minister of Home Affairs Regulation No. 23 2009 concerning Procedure to Foster and Monitor Inter-Regional Cooperation].

Indonesia. (2010) *Peraturan Pemerintah No. 16 Tahun 2010 tentang Pedoman Penyusunan Peraturan Dewan Perwakilan Rakyat Daerah dengan Tata Tertib Dewan Perwakilan Rakyat Daerah* [The Government Regulation No. 16 2010 concerning Guidelines for the Establishment of a Regulation of the Regional House of Representatives on Rules of Conduct of the Regional House of Representatives].

Indonesia. (2011) *Peraturan Menteri Dalam Negeri No. 53 Tahun 2011 tentang Pembentukan Produk Hukum Daerah* [Minister of Home Affairs Regulation No. 53 2011 concerning Establishment of Sub-National Legal Instruments].

Indonesia. (2011) *Undang-undang No. 12 Tahun 2011 tentang Pembentukan Peraturan Perundang-undangan* [Law No. 12 2011 concerning Formation of Legislation].

Indonesia. (2014) *Undang-undang No. 23 Tahun 2014 tentang Pemerintahan Daerah.* [Law No. 23 2014 regarding Local Governments].

Kreamer, D.K. (2012). The past, present, and future of water conflict and international security. *Journal of Contemporary Water Research & Education*, 149, 88–96.

Mitchell, H.H., Hamilton, T.S., Steggerda, F.R. & Bean, H.W. (1945). The chemical composition of the adult human body and its bearing on the biochemistry of growth. *Journal of Biological Chemistry*, 158, 625–637.

Persoon, G.A. & Weerd, M. (2006). Biodiversity and natural resource management in insular southeast Asia. *Island Studies Journal*, 1(1), 81–108.

Rüegger, V. (2014). Water distribution in the public interest and the human right to water: Swiss, South African, and International Law Compared. *Law, Environment and Development Journal*, 10(1), 18–34. Retrieved from http://www.lead-journal.org/content/14016.pdf.

Savas, E.S. (1987). *Privation: The key to better government.* Chatham, England: Chatham House Publishers.

Shiva, V. (2002). *Water wars: Privatization, pollution, and profit.* Cambridge England: South End Press.

Soeprapto, M.F.I. (2007). *Ilmu perundang-undangan: Dasar-dasar pembentukannya* [The science of legislations]. Yogyakarta, Indonesia: Kanisius.

UUD. (1945). *Undang-undang Dasar Republic Indonesia 1945* [The 1945 Constitution of the Republic of Indonesia].

van Rijswick, M. & Wouters, P. (2014). Achieving sustainable and adaptive fresh water management: Selective studies of international, European, Dutch, and Chinese water law—Introduction. *Journal of Water Law*, 24(3/4), 85–91.

Review of the implementation of *murabahah* financing in *sharia* banks in the *Greater Jakarta* area

A.R. Nurdin
Faculty of Law, Universitas Indonesia, Depok, Indonesia

ABSTRACT: Banks are financial institutions that play a vital role in the economic system. In Indonesia, 'bank' may refer to a conventional bank or a *sharia* bank. There are a number of legal issues associated with the implementation of *murabahah* financing, which in certain cases are deemed to be no longer in accordance with the *sharia* principles. This situation has prompted this research, with the following topics as its main focus: 1) what is the *murabahah* concept and principles of *sharia* in *murabahah* financing? 2) what are the *murabahah* financing arrangements in Indonesia? 3) how does *murabahah* financing apply consistently in accordance with *sharia* banking principles in the *Greater Jakarta* area? The research method employed in this paper was the legal normative review. Implementation of *murabahah* financing still shows non-conformities with the *sharia* principles. These measures include the following: 1) the State through the Bank Indonesia-Central Bank/Otoritas Jasa Keuangan-Financial Service Authority) has made improvements to various regulations and enhanced external supervision of *sharia* banks; 2) *sharia* banks should self-evaluate to improve themselves and enhance the Dewan Pengawas Shariah (Sharia Supervisory Board); 3) the public as customers have to take the public as customers have to take on the responsibilities of their payment obligations, and become a part of social control on *murabahah* financing.

1 INTRODUCTION

Banks are financial institutions that play a vital role in a country's system of economy. Generally, banks are institutions that carry out the function of collecting and disbursing the public's money in order to assist society in keeping its economy running.

Looking into its history, the institution of banks has already existed since the seventh century B.C. in Babylon and/or during the Hammurabie reign. In Indonesia, banking was introduced during the Dutch Indies colonial period in the eighteenth century. During such period, in order to facilitate the Verenigde Oostindische Compagnie's (commonly known as the Dutch East Indies Company) commercial activities, the bank De Bank van Leening was established in 1746. Its name was subsequently changed into De Bank Courant en Bank van Leening in 1752. This bank was a conventional banking institution and its presence has continued to this day.

In addition to conventional banks, there are also other banking institutions that are more specific in nature, including the *sharia* banks. Looking at its history, *sharia* banks have operated since the founding of the Mit-Ghamr Bank in Egypt in 1963. Since that time, *sharia* banking has been practised in various countries (Ahmad, 2000, pp. 59–60).

The first *sharia* bank in Indonesia was established in 1992. Its growth however, has not been as rapid as conventional banks. The Deputy Commissioner for Bank Oversight of the Indonesia Financial Services Authority reported that the market share of *sharia* banks in the country with respect to other banks fell short of 5per cent (an indicator of industry development by regulators), experienced a drop in 2014, and again in 2015 (OJK RI, 2016). This condition raises a question given that the majority of Indonesia's population are Muslims. As such it was deemed necessary to conduct a legal analysis of the implementation of *sharia* banking.

A legal research on the implementation of *sharia* banking is important because of at least three reasons: 1) the role of banks is critical in Indonesia's economic system, including *sharia* banks. Sarwer et al. (2013) showed that Islamic banks have a significant effect on economic growth for a country such as in Pakistan and the presence of *sharia* banks is very important and needed in Indonesia because the vast majority of Indonesian people are of the Islamic faith (BPS-Biro Pusat Statistik (Statistic Center Bureau), 2010); 2) the growth of *sharia* banks is not as rapid as that of conventional banks due to a number of issues (Ismal, 2011, pp. 7–9) and the implementation of *sharia* banking in Indonesia, which was introduced in 1992, still faces problems relating to its supporting regulations; and 3) issues relating to 'transition' that must be undergone by the Indonesian people when they intend to move from conventional banking to *sharia* banking (Muhammad, 2011, pp. 139–140).

For members of the public who choose to adopt *sharia* banking, the currently most preferred *sharia* banking product is *murabahah*. This concerns the sale of goods for the cost of acquiring such goods, plus a margin agreed upon by the parties; however the seller has to first inform the cost of acquisition to the buyer. The high interest of the public in using the *murabahah* transaction (57per cent of total *sharia* bank financing) can be seen in the *sharia* banking statistics issued by OJK for May 2016. This is in line with research conducted by Alsoud and Abdalah (2013) within the Islamic Bank of Kuwait, which indicated that the product users of *murabahah* was the highest compared to other financing products.

In fact, during the *sharia* banks' early days of operation, their activities were more focused on primary products, namely *mudharabah (trust financing)* and/or *musyarakah (joint financing)* financing (Usmani, 2002, pp. 104–105). However, subsequently it was felt that their growth has been extremely languid, and thus there was an idea to develop a secondary product constituting a financing scheme outside the *mudharabah* and/or *musyarakah*, such as the *murabahah* with all of its variants which is deemed to be a replication of conventional banking practices.

The secondary product, *murabahah*, turned out to experience a more dominant growth and *sharia* banks' operations deviated from the form that was originally intended at the time of their inception. Currently it is possible for *sharia* banks to be managed and owned by non-Muslims, but there is concern that their operation does not strictly follow the *sharia* principle.

On the other hand, the length of *sharia* banks' historical presence in Indonesia is vastly different from that of conventional banks. *Sharia* banks began to be established in 1992, whereas conventional banks have been operating since the colonial era. This fact, among others, has caused *sharia* banking to encounter a number of impediments. From a cultural aspect, it is not easy to alter the habit or perspective of the public as customers of general/conventional banks and get them to embrace *sharia* banking.

The various problems described above, among others, have made *sharia* banks' lag behind in performance when compared to conventional banks. For that reason, *sharia* banks have taken various measures to overcome the problem, including modifying a number of products that they offer; one of which is the *murabahah* financing scheme.

Table 1. (Otoritas Jasa Keuangan Republik Indonesia, 2016).

Akad	1.1 Billions of Indonesian Rupiah (IDR)
Mudharabah	*14,856*
Musyarakah	*64,516*
Murabahah	*124,339*
Salam	*0*
Istishna	*780*
Ijarah	*9,616*
Qardh	*3,752*
Others	0
Total	217,859

In some cases, modification to the implementation of *sharia* banking was done to make the banking public interested in becoming customers. The problem however, is whether *sharia* banking as is currently being implemented, still follows the *sharia* principles as laid down in the *Al-Qur'an*, *hadiths*, and *ijma fuqaha*?

In practical terms, it is still questionable as to whether Islamic banking is consistently applying the principles of *sharia* or whether there have been irregularities (Noor, 2006, pp. 311–312). Therefore, it was identified as necessary to research the consistent implementation of corresponding *murabahah* financing principles of *sharia* banks based on the *Qur'an*, *hadith*, and *fiqh muamalah*. This research uses a qualitative method with the normative juridical approach, and also uses empirical or quantitative data from *murabahah* implementation practices as supporting information.

The purpose of this research was to analyse the implementation of *murabahah* financing in accordance with the principles of *sharia*. Samples were taken from Islamic banks in *Jakarta*, *Bogor*, *Depok*, *Tangerang* and *Bekasi* (in *Greater Jakarta* or also known as *Jabodetabek*). The reasons for selection of this region were based on the level of *murabahah* financing in the *Greater Jakarta* area which covers around 58per cent of the total national *murabahah* financing (OJK, 2014). Therefore, the research questions were: 1) what are the *murabahah* concept and principles of *sharia* in *murabahah* financing? 2) are there *murabahah* financing arrangements in Indonesia? and 3) how does *murabahah* financing apply consistently in accordance with *sharia* banking principles in the *Greater Jakarta* area?

In order to answer the above problems, this article is divided into several sections: 1) the *murabahah* concept and principles of *sharia* in *murabahah* financing; 2) arrangements of *murabahah* financing in Indonesia; and 3) *murabahah* financing in Islamic banking in the *Greater Jakarta* area. The first section will discuss the *murabahah* concept and principles of *sharia* in the implementation of *murabahah* in accordance with *fiqh muamalah*; these principles refer to legal sources namely the *Qur'an*, *hadith*, and *ijtihad*. Secondly, there will be description of the *murabahah* financing arrangement according to the *fatwa* of the National Sharia Board (*Dewan Syariah Nasional* or DSN) and regulations applicable in Indonesia. The description of the *fatwa* of DSN covers its legality and related *murabahah fatwa* products. Other descriptions of legislation include *murabahah* financing arrangements in accordance with Law No. 21 2008 concerning Islamic banking and supporting implementing regulations. Thirdly, there will be discussion on the implementation of Islamic banking in the *Greater Jakarta* area which includes: a) the background of the customers who choose *murabahah* products; b) the *sharia* implementation process and irregularities in *murabahah* financing; and c) the efforts of *sharia* application in the implementation of a consistent and beneficial *murabahah* financing system.

Based on preliminary data of this research, the hypothesis formulation is: 'If the stakeholders in the implementation of *murabahah* financing does not carry out their role in accordance with the *sharia* principles, hence there are irregularities in the implementation of the of *murabahah* financing'. Therefore, the thesis has to be tested through this research.

2 *MURABAHAH* CONCEPT AND *SHARIA* PRINCIPLES IN *MURABAHAH* FINANCING

Murabahah is one of the products of Islamic banking which is a special form of sale and purchase. The legal basis of *murabahah* financing is the *Qur'an* and the *hadith* which regulate the sale and purchase of the *murabahah*.

Etymologically, the definition of *murabahah* or origin of the word comes from the word '*ribh*' meaning 'profit' of 'capital added value'. The word *murabahah* means 'mutual'. Therefore, *murabahah* can be interpreted as 'mutual benefit'. In the study of *fiqh*, *murabahah* means 'to sell by the original capital along with the clear additional benefit'.

The concept of *murabahah* according to *fiqh muamalah* can be seen from legal sources namely the *Qur'an*. The consequence of putting the *Qur'an* as the main source and the first in determining the law, is that if there is an issue, the first thing to do is to find the answer from the *Qur'an*. If during the legal settlement of a problem an answer is found in the *Qur'an*, then it should not refer to sources outside of the *Qur'an*. The arguments in the *Qur'an Sura* (QS) which regulate the *murabahah* include: Al-Baqarah verse 275, verse 280 and An-Nisa verse 29. These verses generally indicate the permissibility of sale and purchase, and the *murabahah* is a sale and purchase agreement by both sides to a mutual benefit.

Sharia principles are the Islamic law principles in the banking business based on the *fatwa* issued by an institution having the authority in stipulating a *fatwa* in *sharia* matters (Law No. 21 2008 regarding Islamic banking). The *sharia* principles cover business activities which do not contain interest elements of, *maysir*, *gharar*, *haram*, and *zalim*. Businesses based on *sharia* principles include, but are not limited to, businesses which do not contain the elements of:

1. *usury*, a illegally obtained additional income (*batil*), among others;
2. *maysir*, a transaction dependent on an unstable condition and on luck;
3. *gharar*, a transaction the object of which is not clear, not owned, its whereabouts is not known or cannot be delivered at the time of transaction (unless otherwise regulated in *sharia*);
4. *haram*, a transaction the object of which is prohibited in *sharia*; or
5. *zalim*, a transaction that is unfair for the other party.

The implementation of *murabahah* financing means that sale and purchase should fulfil the pillars and terms according to *sharia*. The pillars of *murabahah* financing in Islamic economic perspective consist of: 1) the *akad* of parties, consisting of a) a seller and b) a buyer; 2) the object of the *akad*, is namely a) tradable objects and b) price; 3) the *akad/sighot*, consists of: a) delivery (*ijab*) and b) acceptance (*qabul*) (Antonio, 2004, p. 102). These *murabahah* pillars must be met. If there is one that is not met, then it can be said that the practical implementation of the *murabahah* is invalid.

Besides, the implementation of the pillars of *murabahah*, according to Antonio (2004), the terms must be met. Moreover, the terms of *murabahah* implementation are: 1) the *akad* of the parties is namely a) competence (by law) and b) voluntary (*ridha*) and is not under force/coercion/pressure; 2) the object to be traded a) does not include prohibited/banned goods, b) is beneficial, c) the delivery from seller to buyer can be achieved d) has full ownership of the *akad* parties, and e) is according to the specifications of the delivery by seller and receipt by buyer; and 3) *akad/sighot* should a) be clear and specifically address the parties in *akad*, b) be during the *ijab qabul* (the handover) and the specification of the products and the agreed price must be matching, c) not contain clauses that state that the validity of the transaction depends on upcoming events, and d) not limit the time, for example, 'I will sell it to you for a period of 12 months after that it should be returned to me'.

Moreover, in *murabahah* financing, one of the forms of agreement or treaty, *akad,* is part of an important process. Terminology wise, *akad* is the bond of what one person wants from another person in a way that elicits certain legal commitments (Ash-Shawi and Al-Muslih, 2004, pp. 25–29). *Akad* or agreement in a transaction, contains provisions agreed upon and are binding to the two parties. However, it is not allowed in an agreement to foretell things which are forbidden or unlawful in the religion, or which violates the provisions of the *Qur'an*. For example, an agreement should not be made to deceive others, or to perform transactions of *haram* items (Hasan, 2002, p. 101).

3 REGULATION OF *MURABAHAH* FINANCING IN INDONESIA

The regulating of *murabahah* financing rests on two legal foundations. The first is the provisions in *fiqh muamalah* originating from the *Al Qur'an*, *hadiths*, and *ijtihad*. These principles are manifested in *fatwas* of the DSN. The *fatwas* are substantiated by Law No. 21 2008 regarding Islamic banking, and thus having a binding legal force. The second is Law No. 21 2008 regarding *sharia (Islamic)* banking and its implementing regulations.

Another guiding principle of *murabahah* financing is found in the decrees issued by the DSN. This is an organisation that was established and operates under Indonesia's Council of

Ulema (MUI); it was created in response to the public's need for the development of *sharia* economics and finance. It was also the consequence of the enactment of Law No. 7 1992 regarding banking and its subsequent amendment by Law No. 10 1998, where the latter contains provisions on *sharia*-based banking.

The following are various legislations that provide the legality of the DSN decree: 1) Law No. 21 2008 regarding Islamic banking, specifically Article 1 sub-article 12 and Article 26; 2) the decree of the Board of Bank Indonesia No. 32/34/1999; 3) the Decision of DSN-MUI No. 01 2000 regarding Basic Guiding Principles of the National Sharia Board of the Indonesia's Council of Ulema (PD DSN-MUl); and 4) the Bank Indonesia Regulation No. 11/3/PBI/2009.

Decrees issued by DSN that regulate *murabahah* financing and other matters relating thereto include decrees on: 1) *murabahah*; 2) *wakalah*; 3) down payments in *murabahah*; 4) discounts in *murabahah*; 5) penalties on customers deferring on payments despite having the necessary resources; 6) Al-Qardh; 7) discounts of repayment in *murabahah*; 8) loan assignment; 9) compensation (*ta'widh*); 10) a discount to *murabahah* charges (*khashm fi al-murabahah*); 11) repayment of *murabahah* loans for customers unable to pay; 12) rescheduling of *murabahah* financing; 13) conversion of *murabahah* contracts; and 14) assignment of *murabahah* financing between *sharia* financial institutions (*Lembaga Keuangan Syariah* (LKS)).

In *murabahah* financing under DSN Fatwa No: 04/DSN-MUI/IV/2000, the general guiding principles of *murabahah* in *sharia* banking are: 1) the bank and customer must undertake *murabahah* that is free from *riba* (profit); 2) the goods being traded are not considered *haram* (forbidden) under *sharia* (Islamic) law; 3) the bank finances a portion or the entire cost of goods purchased under agreed upon terms; 4) the bank purchases the goods required by the customer under the bank's own name, and such purchase must be legal and free from *riba*; 5) the bank must disclose all matters pertaining to the purchase, such as if the purchase was made under a loan; 6) the bank shall then sell such goods to the customer (order placer) at a price that is the same value of the purchase price plus profit. In this respect the bank must inform truthfully the cost of the goods to the customer along with any expenditure necessary to be made; 7) the customer pays the agreed upon price for the goods over a mutually agreed period; 8) to prevent abuse or breach of the transaction, the bank can enter into a special agreement with the customer; and 9) if the bank wishes to represent the customer in purchasing the goods from a third party, the *murabahah* sale agreement must be entered into after the goods, in principle, becomes the property of the bank (Himpunan Fatwa Keuangan Syariah DSN-MUI 2014, p. 64).

Provisions of the law underlying the *murabahah* financing by Islamic banks are in Law No. 21 2008 regarding Islamic banking. The financing is the provision of funds or bill equivalent to the form of sale and purchase transactions in the form of *murabahah* receivables. *Murabahah* financing is based on agreements between the Islamic bank and/or Sharia Business Unit (UUS) and another party requiring the parties to be financed and/or given the facility of funds to repay the funds after a certain period of time in exchange of *ujrah*/fee (General Provisions Article 25 of the Law No. 21 2008).

In addition to containing an explanation of the concept of *murabahah*, the Islamic banking law also regulates among others:

1. An obligation for *sharia* banks and *sharia* divisions to perform the function of collecting and distributing public funds (Article 4 paragraph (1) of Law No. 21 2008).
2. The business activities of Islamic banks including:
   – to channel financing based on *murabahah akad*, *salam akad*, *ishtishna akad*, or other *akad* which do not conflict with *sharia* principles;
   – to buy, sell, or guarantee on its own risk securities issued by a third party on the basis of real transactions based on *sharia* principles, among others, *akad ijarah*, *musharaka*, *mudaraba*, *murabahah*, *kafalah*, or *hawalah* (Article 19 of Law No. 21 2008);

3. UUS business activities including:
   – to channel financing based on *murabahah akad*, *salam akad*, *ishtishna akad*, or any other agreement that does not conflict with *sharia* principles;

- to buy and sell securities issued by a third party on the basis of real transactions based on *sharia* principles, among others, *ijarah akad, musharaka, mudaraba, murabahah, kafalah,* or *hawalah* (Article 19 of Law No. 21 2008);

4. The business activities of Sharia Rural Bank, one of which is channelling funds to the public in the form of financing based on the principles of *murabahah, salam,* or *ishtishna* (Article 21 letter b number 2 of Law No. 21 2008).

In addition to Law No. 21 2008 concerning Islamic banking, regulations regarding *murabahah* financing as referenced in the *sharia* banking law consist of: laws related to *murabahah* financing, Bank Indonesia regulations, Financial Services Authority (OJK), and other technical implementing regulations, such as Bank Indonesia circulars, OJK circulars, and the *murabahah* financing standards published by OJK. These regulations are formulated by the authorised institutions/agencies to serve as the legal foundation for undertaking *murabahah* financing between *sharia* banks and the customer according to the *sharia* principles and to meet the needs prevailing in the society.

4 *MURABAHAH* FINANCING IN ISLAMIC BANKING IN GREATER JAKARTA AREA

4.1 *Implementation of murabahah financing*

In the implementation of *murabahah* financing in practice in *Greater Jakarta*, only 32per cent of customers understand about the meaning of *murabahah* financing. *Murabahah* financing (sale and purchase based on *sharia*) is the most preferred by customers (81.8per cent), while other forms of financing like *salam* only gets 15.2per cent, and *ishtishna* 3per cent. The reason customers choose *murabahah* according to the majority (87.1per cent) of customers is because *murabahah* financing is more advantageous compared to other financing methods.

Regarding whether the implementation of *murabahah* financing is in accordance with Islamic principles, most customers (41.9per cent) stated that it is in accordance with *sharia*. On the other hand, the rest of the customers (38.7per cent) said the *murabahah* financing is not in accordance with *sharia*. More customers do not understand whether or not *murabahah* financing is in accordance with *sharia*.

Problems of incompatibility with the *sharia* in the implementation of *murabahah* financing, lie firstly on the parties to a transaction, namely: a) ignorance of the customer (buyer) on *murabahah* transactions carried out due to lack of information from the bank (seller); and b) fewer banks conduct thorough research about the conditions of the customer (buyer) and so customers are not properly selected.

Based on the principles of *sharia*, the ignorance of the customer (buyer) of a *murabahah* transaction shows that it does not fulfil a requirement of *murabahah;* that is the sincerity of the parties, as well as the principle of exchange, the freedom of trade and cooperation, and the principle of justice and expediency. The lack of information from the bank as the seller, may cause *gharar*; the customers then accept it, which may lead the customers into *maysir* (gambling), in other words, engaging in conjecture/acting without calculation. Furthermore, the bank may provide *murabahah* financing without checking the suitability of the target of such financing. Although it has been assessing customers well through the analysis of the Five C's (Character, Capacity, Capital, Condition of Economic and Collateral), it can be said that *gharar* transactions (that are not clear/not in accordance with its purpose), have occurred.

Secondly, is the problem of the object of merchantability, namely: a) the designation of financing (goods) does not comply with the agreement; b) the goods were not present when the financing was approved; and c) the form of goods is not in accordance with the agreement, shows the non-fulfilment of the requirements in harmony, and exchange regarding buying and selling. This also leads to *gharar* (the transaction that are not clear) of goods bought and sold. It also leads to failure in recognition of the principle of ownership of property and determination of a fair price, which will lead to unfairness and lack of benefit from the activities carried out because one of the parties does not obtain the rights that should be received.

The third issue is that in the implementation stages of the agreement, the problems are: a) pre-contract stage—customers do not know about the transactions made; b) the approval stage of the contract—customers do not understand the contract, the goods have not been received by the customers, the contract cancelled, the customers do not get a chance to study the form of the contract and the large cost of contract; c) post-contract stage—the obligation of payment is delayed, the payment restructuring mechanism is not informed, and the bank is auctioning items that have been purchased (i.e. house) by customers due to payment difficulties without previously informing the mechanism of solution to this problem. These are not complying with the—principles and requirements of the transaction of *murabahah*.

The pillars and conditions are not met including the parties, object, and contract/*sighot* that should be clear. This leads to *riba*, *gharar*, and *maysir*. As a result, transactions conducted unfairly do not provide benefits, but may cause harm instead.

The findings of the data in the implementation of *murabahah* above show the discrepancy between the implementation of *murabahah* financing and *sharia* principles. Mismatches occur in the process of *murabahah* financing at the stage of pre-contract, at the time of the contract, and post-contract or in the implementation phase of *murabahah* financing payment obligations. The findings of lack of conformity are in line with a study conducted by Shofawati (2014, p. 15).

4.2 *Murabahah financing which is consistently applied in accordance with sharia banking principles*

To ensure that *murabahah* financing can be implemented consistently with *sharia* principles, a number of measures need to be taken, among others:

1. Efforts to improve the relevant regulatory instruments, namely filling the regulatory gaps and improving the substantive aspects of the legislation. This needs to revise the regulatory aspect stems from certain problems arising in relation to the regulation of *murabahah* financing, among others:
   - lack of a complete set of BI/OJK regulations that specifically govern *murabahah* financing, notwithstanding the recent issuance of a *murabahah* Financing Guiding Standards by OJK at the end of June 2016;
   - lack of regulatory provisions regarding security compatible with the *sharia* scheme, given that every *murabahah* financing involves collateral in the form of immovable/movable assets;
   - lack of optimal dissemination of legislation regarding *murabahah* financing.
2. Enhancement of the quality as well as the quantity of human resources in *sharia* banking, particularly those that deal with *murabahah* financing.
3. Enhancement of the capacity of internal supervision by the DPS and internal audit mechanisms of *sharia* banks as well as external oversight by banking authorities.

In *murabahah* financing, when the principles of *sharia* are complied with in a consistent manner, it can provide a great benefit to the public/customers, to the *sharia* banks, and the state.

5 CONCLUSION

Murabahah is one of the products of Islamic banking which is a special form of buying and selling on legal basis by referring to the *Qur'an*, *hadith* and *ijtihad* governing the sale and purchase of *murabahah*. Thus in *murabahah*, a sale and purchase agreement of both parties (buyer and seller) is for their mutual benefit. Implementation of *murabahah* financing in buying and selling that is based on *sharia* must fulfil the pillars and terms.

Findings pertaining to the implementation of *murabahah* indicate in certain cases that there are inconsistencies with the principles of *sharia*. Such inconsistencies occur along the process of *murabahah* financing namely before the execution of contract, during the execution of contract, and during the period following the execution of contract. The specific issues of inconsistency involve: 1) the transaction parties; 2) the object being transacted; and 3) stages of the contracting process.

Measures to make *murabahah* financing in line with the *sharia* principles and provide benefit, and appeal to the public can be taken in a number of ways. Firstly, it is by amending various regulations, namely through filling the regulatory gaps and improving the substantive aspect of the instruments. Secondly, it is by enhancing the quality and quantity of human resources of *sharia* banking. Thirdly, it is with the enhancement of internal oversight mechanisms by DPS and the internal audit unit of the *sharia* banks, and external oversight by the appropriate banking authorities. Once the *sharia* principles are implemented consistently, this form of financing will provide the fullest benefit to the stakeholders.

REFERENCES

Ahmad, K. (2000). Islamic finance and banking: The challenge and prospects. *Review of Islamic Economics* 9, 57–82.

Alsoud, G.F.A. & Abdallah, M.K.S. (2013). Customer awareness and satisfaction of Islamic retail products in Kuwait. *Research Journal of Finance and Accounting*, 4(17), 36–52. Retrieved from: http://www.iiste.org/Journals/index.php/RJFA/article/viewFile/8974/9207.

Antonio, M.S. (2004). *Bank syariah: Dari teori ke praktek*. Jakarta, Indonesia: Gema Insani. (Isamic Bank from Theory to practice).

Ash-Shawi, S. & Al-Muslih, A. (2004). *Fikih ekonomi keuangan Islam (Ma la yasa' at-tajira jahluhu)* (A.U. Basyir, Trans.). Jakarta, Indonesia: Darul Haq. (Fiqh of Islamic Economic Finance.

Bank Indonesia. (1999). Surat *Keputusan Direksi BI* [Decree] *No. 32/34/1999*. 1999, Indonesia: Bank Indonesia.

BPS. 2010. Penduduk menurut wilayah dan agama yang dianut.:. Retrieved from: http://sp2010.bps.go.id/index.php/site/tabel?tid = 321&wid = 03 (Statistic Center Bureau—Population by region and religion adhered to)Dewan Syariah Nasional Majelis Ulama Indonesia. (2000). *Fatwa 04/DSN-MUI/IV/2000 tentang murabahah*. Sharia National Board—the Indonesian Council of Ulama.

Dewan Syariah Nasional Majelis Ulama Indonesia (DSN-MUI). (2000). *Putusan Dewan Syariah Nasional Majelis Ulama Indonesia No. 1 Tahun 2000 tentang Pedoman Dasar DSN-MUI* [Decree National *Sharia* Board of Indonesia's Council of Ulama No. 1 2000 regarding Basic Guiding Principles of the National *Sharia* Board of Indonesia's council of Ulema]., Indonesia: DSN-MUI.

Hasan, M.A. (2002). *Berbagai transaksi dalam Islam*. Jakarta, Indonesia: Rajawali. (Various transaction in Islam).

Indonesia. (1992). *Undang-undang No. 7 Tahun 1992 tentang Perbankan* [The Law No. 7 1992 regarding Banking].

Indonesia. (1998). *Undang-undang No. 10 Tahun 1998 tentang Perubahan atas Undang-undang No. 7 Tahun 1992 tentang Perbankan* [The Law No. 10 1988 regarding Amendment of The Law No. 7 1992 regarding Banking].

Indonesia. (2008). *Undang-undang No. 21 Tahun 2008 tentang Perbankan Syariah* [The Law No. 21 2008 regarding Islamic Banking].

Indonesia. (2009). Peraturan Bank Indonesia No. 11/3/PBI/2009 tentang Bank Umum Syariah [Bank Indonesia Regulation No. 11/3/PBI/2009 regarding Sharia Commercial Bank].

Ismal, R. (2011). Islamic banking in Indonesia: Lessons learned. *United Nations Conference on Trade Development (UNCTAD) Multi-Year Expert Meeting on Services, Development and Trade: The Regulatory and Institutional Dimension, April 6–8, 2011, Geneva, Switzerland*. Geneva, Switzerland: United Nations.

Majelis Ulama Indonesia. (2014). *Himpunan fatwa keuangan syariah dewan syariah nasional majelis ulama Indonesia*. Jakarta, Indonesia: Erlangga. The set of Islamic Financial Fatwa—Sharia National Board The Indonesian Council of Ulama.

Muhammad. (2011). *Manajemen bank syariah*. Yogyakarta, Indonesia: UPP-STIM YKPN. (Management of Islamic Bank).

Noor, Z. (2006). *Bank muamalat sebuah mimpi, harapan, dan kenyataan: Fenomena kebangkitan ekonomi Islam*. Jakarta, Indonesia: Bening. (Bank Muamalat a dream of hope and reality: Phenomenon of Islamic economic revival.

Otoritas Jasa Keuangan [Financial Service Authority] Republik Indonesia. (2014). *Statistik perbankan syariah* [Islamic banking statistics] *November 2014*. Jakarta, Indonesia: OJK RI Retrieved from: http://www.bi.go.id/id/statistik/perbankan/syariah/Documents/SPS%20 November%20 2014.pdf.

Otoritas Jasa Keuangan [Financial Service Authority] Republik Indonesia. (2016). *Laporan perkembangan keuangan ayariah 2015*. Jakarta, Indonesia: OJK RI. Retrieved from: http://www.ojk.go.id/id/kanal/syariah/berita-dan-kegiatan/publikasi/Documents/Pages/Laporan-Perkembangan-Perbankan-dan-Keuangan-Syariah-2015/LPKS%202015%20(Indonesia).pdf (Report on the development of Islamic finance).

Sarwer, M.S., Ramzan, M. & Ahmad, W. (2013). Does Islamic banking system contributes to economy development. *Global Journal of Management and Business Research*, 13(2), 61–68.

Shofawati, A. (2014). *Murabahah* financing in Islamic banking: Case study in Indonesia. Proceedings of 5th Asia-Pacific Business Research Conference, February 17–18, 2014, Kuala Lumpur, Malaysia. Melbourne, Australia: World Business Institute Australia. Retrieved from: https://wbiworldconpro.com/pages/previous_confo/malaysia-conference-2014.

Usmani, M.T. (2002). *An introduction to Islamic finance*. Dordrecht, The Netherlands: Kluwer Law International.

Environmental damage and liability in Indonesia: Fancy words under conventional wisdom

A.G. Wibisana & T.A. Dewaranu
Faculty of Law, Universitas Indonesia, Depok, Indonesia

ABSTRACT: In recent years, the Ministry of the Environment has issued several regulations regarding economic valuation of environmental damage that can be used in a civil lawsuit. This article attempts to answer the questions of how economic valuation is implemented in various rulings. In particular, this paper analyzes whether an "objective" damage valuation can be achieved through economic valuation methods. In doing so, before discussing some theoretical foundations of economic valuation methods, this paper will first explore the concept of environmental damage under Indonesian environmental law. In addition, the paper is also interested in comparing the implementation of environmental damage assessment in the US, Europe, and Indonesia particularly based on the Indonesian Supreme Court ruling on Menteri KLH v. PT. Kalista Alam. This paper observes that attempts to provide objective valuation for environmental damage in Indonesia have failed, as indicated by the failure to make a clear distinction between real environmental damage and remedial costs. This paper shows that the government has failed to refer to and apply valuation methods endorsed in various Ministerial Regulations. Surprisingly, the government claimed remedial costs as the basis for calculating environmental damage, although they never took any remedial action and in fact they asked the defendant to conduct the remedial action. To ensure that remedial action is fully implemented, the paper argues that the availability of remedial funds is in order.

1 INTRODUCTION

In recent years, the Ministry of the Environment has issued several regulations regarding economic valuation of environmental damage that can be used in a civil lawsuit. As stated in their consideration, the regulations are issued as guidance for the implementation of government's legal standing in environmental cases. As stated in Article 90, paragraph 1 of the Law No. 32 of 2009 concerning Environmental Protection and Management [hereinafter referred to as the 2009 Environmental Law], the Minister of the Environment and the Local Government have the authority to file a lawsuit related to "environmental damage". The law also mandates the Minister to provide guidance for the implementation of government's standing to sue (art. 90, par. 2).

In order to implement the standing, the government has issued several ministerial regulations. By issuing these regulations, it seems that the government attempts to provide scientific basis for the compensation. In this case, the use of economic valuation is seen as a means to provide justification for the calculation of damage. Hence, the regulations not only provide some "scientific and objective" methods for calculating the damage, but also some administrative requirements, such as certification of competence, for anyone wishing to undertake economic valuation.

The Indonesian approach to the calculation of environmental damage seems to rely on theoretical methods, which might be used regardless of the actual costs incurred by the Government to rehabilitate the damage. Hence, it could be the case that the Government asks a certain amount of money for compensation of remedial costs [*ganti rugi biaya pemulihan*], although in reality they never undertook any remedial measure.

In contrast to the Indonesian approach, the Assembly of the International Oil Pollution Compensation Fund (IOPC Fund) was very critical to any economic valuation method, based almost entirely on theoretical quantification of damage. In their Resolution of 1980, in the aftermath of the Patmos case, the IOPC Fund declares its intention that "the assessment of compensation to be paid by the International Oil Pollution Compensation Fund is not to be made on the basis of an abstract quantification of damage calculated in accordance with theoretical model" (Bianchi, 1997, p. 114).

This paper attempts to answer the question of whether the "objective" of damage calculation can be achieved through economic valuation methods. In doing so, this paper will first explore the concept of environmental damage under the Indonesian environmental law (Section II). Afterwards, Section III discusses some theoretical foundation of economic valuation methods. In addition, the paper is also interested in comparing the implementation of environmental damage assessment in the US and Europe. This comparative perspective will be presented in Section IV, discussing the Indonesian experience in economic valuation as found notably in the Indonesian Supreme Court ruling on Menteri KLH v. PT. Kalista Alam (Mahkamah Agung Republik Indonesia, 2015). Some concluding remarks will be provided in Section V.

2 ENVIRONMENTAL DAMAGE AND INDONESIAN ENVIRONMENTAL LAW

Under civil law tradition, damage is often associated with liability theories, in the sense that before one answers the question of how much a defendant has to pay for compensation and how to calculate the damage, one needs to analyze first who is liable for damage incurred by a plaintiff. It follows, thus, that liability rules can only be effective if the law provides remedy to correct damage (van Dam, 2006, p. 301).

In general, Sappideen and Vines (2011, 268–276) classify damage into pecuniary loss and non-pecuniary loss. A pecuniary loss is a loss/damage that that can actually be measured and monetized by money. However, a non-pecuniary loss is a loss that, by its nature, cannot or is difficult to be measured by money. Even though a non-pecuniary loss is not measurable in monetary value, there is a need to convert such a loss into a nominal amount of money in order to ease the compensation process from the tortfeasors to the victims (Schwartz, Kelly & Partlett, 2010, p. 535).

Under the 2009 Environmental Law, it seems that environmental damage is always related to environmental pollution. This is apparent when one looks at the definition of environmental damage stated in the elucidation of article 90 paragraph (1) of the Law, which reads "[y]ang dimaksud dengan "kerugian lingkungan hidup" adalah kerugian yang timbul akibat pencemaran dan/atau kerusakan lingkungan hidup yang bukan merupakan hak milik privat— [what is referred to as "environmental damage" is damage arising out of pollution and/or losses to the environment that is not a private ownership property]."

In order to implement the government's right to sue, the Minister issued several ministerial regulations providing guidance for the assessment of environmental damage, which is very crucial for the implementation of the standing. These regulations are the Minister of Environment Regulation Number 13 of 2011 regarding Damages Due to Environmental Damage and Pollution, the Minister of Environment Regulation Number 14 of 2012 regarding the Guide of Economic Valuation of the Peat Land Ecosystem, and the Minister of Environment Regulation Number 15 of 2012 regarding the Guide of Economic Valuation of the Forest Ecosystem.

One could observe that these regulations contain similar provisions. First, they explain the concept of economic valuation of environment and natural resources. Second, they describe the available methods and approaches to calculate the total economic value of the environment. In this regard, the regulations embrace not only the use of market prices, but also some methods to indicate non-use value of the environment, including the travel cost method, the hedonic price method, the benefit transfer method, and the contingent valuation method. Third, they recommend procedures to measure the economic values of the environment, including the requirements for assessors. Fourth, they provide an overview and example of the use of economic valuation of certain natural resources in particular sites.

Furthermore, in addition to the definition of damage set forth in art. 90 of the 2009 Environmental Law, the Regulation of the Ministry of Environment No. 13 of 2011 also defines losses to the environment as direct or indirect changes to the physical and chemical and/or biological characters of the environment that surpass either environmental quality or damage standards. From these definitions, one can consider environmental damage as damage resulting from pollution and/or from direct or indirect changes to the physical and chemical and/or biological character of the environmental resources, namely those that are not part of private property, which are indicated by the surpassing of environmental quality or damage standards. Consequently, any damage or harm to environmental and natural resources objects which are part of private property is excluded from the definition of environmental damage. In this case, the damage could be considered damage to private property against which the suffering persons themselves have the standing to sue.

With regard to the issues of compensation and liability, the 2009 Environmental Law formulates several provisions concerning liability rules and alternatives, as well as the mechanism for compensation. As far as the compensation mechanism is concerned, the 2009 Environmental Law urges the Government to implement economic instruments, including environmental funds and incentive or disincentive instruments (art. 42). Meanwhile, art. 43 par. 2 explains environmental funds in the forms of a guarantee fund (dana jaminan pemulihan), a mitigation fund (dana penanggulangan), and environmental trust fund (dana hibah/bantuan). It is, however, unclear how such funds will be collected, organized, and used. It is also far from clear whether the funds, especially the guarantee and mitigation funds, will function like the US superfund or more like a performance bond.

In addition to various possible environmental funds, the 2009 Environmental Law also indicates the possibility of implementing an environmental insurance (*asuransi lingkungan*), namely an insurance that provides protection at the event of pollution and/or damage to the environment (art. 43, par. 3f). However, it is unclear, for example, whether the insurance will function as the third party/liability insurance or as the first party insurance. It is also unclear whether such an insurance scheme will be mandatory or not.

Finally, the 2009 Environmental Law also recognizes liability rules based on fault and strict liability, which are formulated respectively in art. 87 and 88 of the 2009 Environmental Law. Hence, it is possible for a plaintiff to ask for compensation due to the breach of the duty of care owed by the defendants to the plaintiff. It is also possible for those who carry out abnormally dangerous activities to be held liable on the grounds of strict liability.

Since all of implementing regulations regarding funds are still absent, while the third party insurance for environmental damage is also of limited application, one could easily find that the compensation mechanism in Indonesia is heavily dependent on the liability system, without a clear alternative compensation if the liability system fails. For this reason, it could be argued that the compensation mechanism in Indonesia is ill-structured, leaving the victims, including the government, with liability rules as the only possible way to get compensation.

3 IN SEARCH OF THEORETICAL GROUNDS FOR ECONOMIC VALUATION

Economic valuation is a process of assigning monetary value to environmental objects. Such values are useful when it comes to making economic choices appertain to environment and natural resources. Economic valuation of ecosystem and natural resources is expected to make explicit, to society in general and policy making in particular, and that biodiversity and ecosystem are scarce and that their depreciation or degradation has associated costs to society (Pascual *et al.*, 2010). According to Cross (1989, p. 280), in conducting an economic valuation, one needs first to determine what kind of value that is contained in such an environmental object, how many of the value is lost, and how much it is worth. Afterwards, one will have to proceed with particular methods to find out how much monetary value attaches to the object in question.

3.1 *Which values?*

Following Cross (1989, pp. 280–297), there are at least, three types of value that can be attributed to environmental and natural resources, namely use value, existence value, and intrinsic value. These values are briefly discussed below.

Use value is simply the worth of natural resources to the people who use them. This value can be divided into consumptive use value, non consumptive use value, direct use value, and indirect use value (Pearce, 1993, pp.16–22). Consumptive use value is a value whose utilization by humans can only be done by directly reducing the environmental functions. Such utilization includes fishing, hunting, wood cutting. For a non-consumptive use value, a utilization of a resource does not reduce any function of the resource. Included in this value is value derived from enjoying scenery, bird watching, hydroelectric power plant, etc. A direct use value is attributed to direct utilization of the environment and natural resources by humans. Such utilization includes hiking, surfing, processing wood into furniture, etc., while an indirect use value is derived from indirect utilization of ecosystem services. This value is gained from the ecological value of the environmental object, such as breathing fresh air from around a forest area.

People may also attach values to natural resources because of their existence. That is because they simply exist. According to Cross (1989, pp. 286–288), existence value can be divided into option value, vicarious value, and intertemporal value. An option value indicates a value of an option, namely to choose when and how to use the available natural resources. Hence, an option value is only relevant if the environment or natural resources are still available and have not been exploited yet. Option value is often related to an investment activity where some people tend to wait for potential future that they may get. It appears that the same thing applies to the environment and natural resources management (Brookshire, Eubanks, & Randall, 1983, pp. 1–3, Weisbord, 1964, p. 472). One variant of option value concerns with uncertainty in future benefit that may be provided by natural resources. Such potential future merit is called as quasi-option value, which certainly has a great deal to do with uncertainty and unpredictable condition (Traeger, 2014, pp. 243–244). One is considering a vicarious value if one thinks of the important of a particular natural resource without necessarily relating such importance to one's future use. Hence, one could state that a tiger needs to be protected, although it is unlikely that one is going to use (or derive benefits from) the beast in the future. An intertemporal value of natural resources indicates a value aimed to benefit future generations. If society values these resources today, future generations almost certainly will value the preservation of these resources. In this case, one takes a preservation action for a species not to meet one's interest, but to ensure that future generations will also experience the environment we are living in today.

Nevertheless, some would argue that nature has an intrinsic value, namely a value in itself, regardless of whether it has any benefit or use for humans. This opinion could be traced back to the thought of "deep ecology", which argues that every living thing has inherent value and moral significance independent of their use by human beings, or even of human existence (Pollack, 1985, p. 401). This perspective considers every other living thing has every right to live as much and as important as human beings (Naess, 1973). One might, however, argue that the idea of monetizing intrinsic value of environmental resources seems to run counter the belief that nature has intrinsic value, since monetary value is a concept that specifically emerges from an anthropocentric point of view. If one truly believes that environment and natural resources have the same right as humans, monetizing this value is considered to be degrading the value itself. In this regard, dollarization of nature's intrinsic value is considered a cruel and meaningless procedure (Halter & Thomas, 1982, p. 8). Despite such critiques, others could still argue that giving monetary value to nature is an inevitable and the most concrete instrument in order to assess the nature's value and the extent it has been damaged, which otherwise would not have been taken into account in the decision making process and, hence, would not also be compensated.

3.2 *Valuation methods*

There are several methods commonly used in economic valuation. This paper will discuss three of them, namely market valuation, behavioral use valuation, and contingent valuation method (CVM).

The most common method of valuation is the market valuation method, which refers to the use value as it appears in the market price. Market helps to set the exact worth of environmental use value. An appraisal of environmental value through the market system is probably the easiest and simplest method, in which the value of certain environmental object is assessed simply by observing the market price of such an object. Despite its advantages, critics might point to the most common weakness of the method, namely the failure of market to provide prices for the most of environmental values. Hence, valuing the forest simply based on the price of its trees will doom to fail as it ignores most of the values of the forest. One needs other methods to capture the values that have no market price.

The behavioral valuation appraises the value of environment and natural resources based on human behavior. This method is divided into travel cost valuation and hedonic price valuation. A travel cost valuation is a method to monetize environmental value based from the costs incurred by those who want to visit certain environmental objects. This procedure is motivated by the intuitive assumption that says that the environmental value of an object can be seen of the expenditure incurred by those who are willing to visit the object. These expenses typically include the cost of transport, the cost of the sacrifice of time, entrance fees (if any), and any other expenses.

A hedonic price valuation is used to calculate the economic value of the environment, which directly affects the market price of a particular item that is available in the market. This method will reveal the value from the market price of another object. For example, to measure the price of the air in the open space that has no market price, one can see it from the market price of the property around the open space. How much would people pay for the property in a place with such air quality? If the air in the place is clean and fresh from pollution, the price of the property there will also be higher. On the contrary, if the air in the surrounded area is polluted, the property prices in the area most likely will decrease. This indicates how air quality affects the market price of the property in vicinity. Hence, the differences in price of the otherwise identical objects actually reveal the differences in the economic value of environmental quality.

The CVM approach simplifies the process of valuing natural resources by asking people directly to state what monetary value they place on identified resources. The valuation usually employs either people's willingness to pay (WTP), namely by asking people how much they are willing to pay to enjoy or "use" certain environmental and natural resources objects, or people's willingness to accept (WTA), namely by asking people how much money they are willing to accept to replace some environmental or natural resources objects. Since the result of both approaches may come up differently, it becomes a topic on which one should be used to estimate the economic value of natural resources. Despite the critics and doubts addressed to the CVM, this method is considered the most effective assessment which covers all values of natural resources and environment. Even intrinsic and existence values that are rarely covered by other methods can be accommodated properly by the CVM.

4 ECONOMIC VALUATION IN ACTION

In 2012, the Minister of Environment of the Republic of Indonesia sued a company named PT Kalista Alam for forest fires in Aceh Province. In its verdict, the District Court [*Pengadilan Negeri*] of Meulaboh ordered the defendant to pay a compensation of IDR 114.3 billion and to undertake rehabilitation measures in 1000 acres of burned land and rehabilitation costs of at least IDR 251.8 billion in Meulaboh District Court, Decision No. 12/PDT.G/2012/PN.MBO (Pengadilan Negeri Meulaboh, 2013, p. 231). The ruling was upheld by the Court of Appeal High Court [*Pengadilan Tinggi*] of Banda Aceh in Decision No. 50/PDT/2014/PT.BNA (Pengadilan Tinggi Banda Aceh, 2014) and finally by the Supreme Court [*Mahkamah Agung*] Decision No. 651 K/Pdt/2015 (Mahkamah Agung RI, 2015). In this case, the plaintiff maintained two types of claims, namely a compensation for ecological damage of degraded soil on the one hand, and rehabilitation measures on degraded soil on the other hand.

It appears, thus, that since the outset, the plaintiff was faced with difficulties in defining what constitutes ecological damage, and to what extent the defendant needs to carry out rehabilitation measures and bear the corresponding restoration/replacement costs. Discussion on this matter is inevitable because an unclear definition about such claims runs the risk of double counting, in which the plaintiff asks a similar thing twice namely through compensation and remedial costs.

For ecological damage, one of the losses claimed by the plaintiff is the degradation of water storage functions of peatland due to forest fires. Ecological damage from this loss reached IDR 71.33 billion. To come to this number, the plaintiff based its calculation on the rebuilding and maintenance costs of the so-called reservoir as follows: IDR 63.5 billion for the building of artificial reservoir costed; IDR 1.5 billion for the maintenance of the reservoir; IDR 30 million for water management of the reservoir; IDR 1.2 billion for erosion prevention land; IDR 50 million for soil reformation; IDR 4.6 billion for the recycle of nutrients; and IDR 435 million for waste decomposition in the Supreme Court [*Mahkamah Agung*] Decision No. 651 K/Pdt/2015 (Mahkamah Agung RI, 2015, pp. 21–25). The plaintiff also asked compensation for losses of biodiversity, carbon release, and economic losses due to fires. The total ecological damage, including losses due to the degradation of water storage, was IDR 114.3 billion (Mahkamah Agung RI, 2015, p. 28).

In addition to such claim of the compensation for ecological damage, the plaintiff also asked for the defendant to take rehabilitation measures on the degraded peat land. The costs for these measures are IDR 251.8 billion, consisting of costs for compost formation and spread, and restoration costs in the Supreme Court Decision No. Decision No. 651 K/Pdt/2015 (Mahkamah Agung RI, 2015, pp. 28–29).

Apparently, the courts never took a hard look at the way the plaintiff presented its claim. It seems that the court fully agreed with the plaintiff's claim for the damage and costs, and hence no consideration can be found in the ruling on how the plaintiff assessed the ecological (environmental) damage and the restoration costs.

This is an unfortunate situation for several reasons. First, if the plaintiff intended to seek for ecological damage, they should calculate the claimed losses based on economic valuation and on actual remedial efforts, and to make a clear distinction between the losses. Unfortunately, although the plaintiff based its claim of ecological damage on the installation of "reservoir", it is clear from the evidence that such a "reservoir" was never installed or rebuilt by any party. Hence, one might challenge the use of hypothetical rebuilding of reservoir as the basis of the plaintiff's claim of ecological damage. Second, the use of such a hypothetical reinstallation of a water reservoir for as the basis for calculating ecological damage seems to give impression that it was the plaintiff who has restored and rebuilt the reservoir, while in fact it was the defendant who was asked to do the restoration. In this regard, one is likely to consider this way of calculation as a double-counting. The defendant was asked to pay compensation on the grounds of restoration costs, yet the defendant was also asked to conduct the restoration. Third, it is unclear whether the monetary values presented in the case come from economic valuation methods, such as the CVM. No discussion can also be found on the merits of the methods, if any, used to assign monetary value to ecological damage.

This approach is clearly different from the approach used in the Europe and US. According to the Environmental Liability Directive of 2004 (Directive of 2004/35/CE), the operator of a damaging activity has the primary responsibility to conduct preventive and remedial actions. Only when an operator does not comply with the instruction and obligations, or when the operator cannot be identified or is not liable for the occurring damage, will the government authority take the responsibility to conduct the preventive and remedial actions (Jans & Vedder, 2008, p. 343). For these reasons, it is the operator who has the responsibility to bear the costs for preventive and remedial actions. Consequently, the competent government authority who has conducted the actions shall recover the costs through, inter alia, security over property or other appropriate guarantees from the operator who has caused the damage or the imminent threat of damage (Art. VIII). Clearly, the European Directive mandates the government to recover all costs that have incurred in conducting preventive and remedial actions. Hence, the Directive that seems to hold that recoverable costs should correspond to actual actions conducted by the government. It also appears that before taking the actions, the government should first resort to the operator to conduct prevention and remediation.

For the US approach, one could refer to CERCLA (Comprehensive Environmental Response, Compensation, and Liability Act) stating that the potentially responsible party (PRP) is liable for, on the one hand, "all costs of removal or remedial action incurred by the United States Government or a State or an Indian Tribe not inconsistent with the National

Contingency Plan." (CERCLA § 107(a)(4)(A)), and on the other hand for "damages...to... natural resources" (CERCLA § 107(a)(4)(C)).

When it comes to the costs for removal or remedial actions, those who undertake the actions are entitled for cost recovery from the PRP, so long as the actions conducted in accordance with the Plan previously approved by the government. These costs include payments for investigation, monitoring, assessing, and reimbursement for contractor conducting the remedial action (Applegate & Laitos, 2006, pp. 222–226). Again, the costs are recoverable if they are arising out of actual actions or actions that have been planned to occur (Grady, 1980).

With respect to natural resources damage, it is important to note the tendency to base the calculation on economic valuation. In this regard, it is important to consider that CERCLA prescribed two procedures for conducting natural resources damages assessments, namely the Type A rule, which is "standard procedures for simplified assessments requiring minimal field observation", and the Type B rule, which is "alternative protocols for conducting assessments in individual cases" (CERCLA § 301(c)(2)). This provision is further explained by the Regulation of the Interior Department, which provides that recoverable costs are either restoration/replacement costs or the diminution of use value pending the restoration, whichever was the less. In addition, the Regulation also relies heavily on use value for the calculation, and allows the CVM only if no market price is found for a particular object (Applegate & Laitos, 2006, p. 231). Finding that the Regulation was not consistent with the CERCLA, the State of Ohio filed a petition against the Regulation. The DC Circuit ruled in favor of the petitioners and invalidated most part of the Type B Rule under the Regulation. In this regard the Court held that the "lesser of rule" under the Regulation was contrary to the intention of Congress. The Court also was of the opinion that the preference of use values and market price over non-use values and CVM was unreasonable [Ohio v. United States Dep't of Interior, 880 F.2d 432, 438] (D.C. Cir., 1989).

5 CONCLUSION

This paper observes that attempts to provide objective valuation, which can be justified on the grounds of economic valuation, for environmental damage in Indonesia have failed. The failure should not come as surprise since in practice the government often failed to make a clear distinction between real environmental damage and remedial costs. This paper shows that not only the government fails to refer to economic valuation methods endorsed in various Ministerial Regulations, but also they use remedial costs as the basis for calculating environmental damage, while they never took any remedial action and in fact they asked the defendant to conduct the action.

As this paper indicates, there is an urgent need to ensure that the government will seek for remedial costs only if they have conducted the actual remedial or if this action has been planned. For this purpose, the availability of remedial funds mandated by the Law is in order. In addition, a claim for environmental damage should really be about the damage pending the remedial action. This damage can be assessed through various methods of economic valuation.

REFERENCES

Applegate, J.S & Laitos, J.G. (2006) *Environmental Law: RCRA, CERCLA, and the Management of Hazardous Waste*. New York, Foundation Press.

Bianchi, A. (1997) Harm to the environment in Italian practice: The interaction of international law and domestic law. In: Wetterstein, P. (ed.) *Harm to the Environment: The Right to Compensation and the Assessment of Damages*. Oxford, Clarendon Press. pp. 104–129.

Brookshire, D., Eubanks, L.S. & Randall, A. (1983) Estimating option prices and existence values for wildlife resources. *Land Economics*, 59 (1), 1–15.

Cross, F.B. (1989) Natural Resource Damage Valuation. *Vanderbilt Law Review*, 42 (2), 270–341.

van Dam, C. (2006) *European Tort Law*. Oxford, Oxford University Press.

Grady, K.T. (1980) Commonwealth of Puerto Rico v. SS Zoe Colocotroni: State actions for damage to non-commercial living natural resources. *Boston College Environmental Affairs Law Review*, 9 (2), 397–429.

Halter, F. & Thomas, J.T. (1982) Recovery of damages by states for fish and wildlife losses caused by pollution. *Ecology Law Quarterly*, 10 (1), 5–35.

Indonesia (2009) *Undang-undang No. 32 Tahun 2009 tentang Perlindungan dan Pengelolaan Lingkungan Hidup* [*The Law Number 32 of 2009 concerning Environmental Protection and Management*].

Indonesia (2011) *Peraturan Menteri Lingkungan Hidup No. 13 Tahun 2011 tentang Ganti Kerugian Akibat Pencemaran Dan/Atau Kerusakan Lingkungan Hidup* [*Ministry of Environment Regulation Number 13 of 2011 regarding Compensation of Environment Contamination or Damage*].

Indonesia (2012) *Peraturan Menteri Lingkungan Hidup No. 14 Tahun 2012 tentang Panduan Valuasi Ekonomi Ekosistem Gambut* [*Ministry of Environment Regulation Number 14 of 2012 regarding Economic Valuation on Peat land Ecosystem*].

Indonesia (2011) *Peraturan Menteri Lingkungan Hidup No. 15 Tahun 2012 tentang Panduan Valuasi Ekonomi Ekosistem Hutan* [*Ministry of Environment Regulation Number 15 of 2012 regarding Economic Valuation Guidelines on Forest Ecosystem*].

Jans, J.H. & Vedder, H.B. (2008) *European Environmental Law*. 3rd edition. Groningen, Europa Law Publishing.

Mahkamah Agung [Supreme Court] Republik Indonesia (2015) Putusan [Decision] No. 651 K/Pdt/2015.

Naess, A. (1973) The shallow and the deep, long-range ecology movement. A Summary. *Inquiry*, 16 (1), 95–100.

Pascual, U. et al. (2010) The economics of valuing ecosystem services and biodiversity. In: Kumar, P. (ed.) The Economics of Ecosystems and Biodiversity Ecological and Economic Foundations. London, Eartscan.

Pearce, D. (1993) *Economic Values and the Natural World*. London, Earthscan Publications.

Pengadilan Negeri [District Court] Meulaboh (2013) Putusan [Decision] No. 12/PDT.G/2012/PN.MBO.

Pengadilan Tinggi [High Court] Banda Aceh (2014) Putusan [Decree] No. 50/PDT/2014/PT.BNA.

Pollack, S. (1985) Reimagining NEPA: Choices for Environmentalist, *Harvard Environmental Law Review*, 9, 359–418.

Sappideen, C. & Vines, P. (2011) *Fleming's The Law of Torts*. 10th edition. Pyrmont, Thomson Reuters Australia/Lawbook Co.

Schwartz, V.E., Kelly, K. & Partlett, D.F. (2010) *Prosser, Wade and Schwartz's Torts: Cases and Materials*. 12th edition. New York, Foundation Press.

Traeger, C.P. (2014) On option values in environmental and resource economics. *Resource and Energy Economics*, 37, 242–252.

Weisbrod, B.A. (1964) Collective-consumption services of individual consumption goods. *The Quarterly Journal of Economics*, 78 (3), 471–477.

Review of the principles of the criminal law on the ancient archipelago law book: The study of the Book of Kutara Manawa Darma Sastra of the Majapahit Kingdom and the Laws on Simbur Tjahaja of the Sultanate of Palembang in the context of criminal law development and condemnation in Indonesia

I. Darmawan & H. Harkrisnowo
Faculty of Law, Universitas Indonesia, Depok, Indonesia

ABSTRACT: Seeking the ancient criminal principle law in ancient archipelago law books is an important quest in the development of the criminal law in Indonesia, especially on the occasion of the enactment of the Bill of the Indonesian Criminal Code. There are at least two relevant ancient archipelago law books, namely the Book of Kutara Manawa Darma Sastra of the Majapahit Kingdom and the Laws on Simbur Tjahaja of the Sultanate of Palembang. Those books, which contain the important principles of the criminal laws, are considered very important and can contribute to the formulation of the Bill of the Indonesian Criminal Codes. Both of the ancient books of the law of archipelago should be reviewed and researched in-depth, because they contain very high local wisdom values and historical values of the development of criminal law and condemnation, especially in the era of the Majapahit Kingdom and the Sultanate of Palembang. With the exploration and the study of the principles of the criminal law in both books, it is expected that an understanding of the historical development of the criminal law in Indonesia can be gained, instead of only referring to western criminal law. This will lead to an enrichment of the current Indonesian criminal law and in particular, in its potential reform.

1 INTRODUCTION

At present, the government is formulating and codifying the Bill of the Indonesian Criminal Code, which is a strategic policy in the development of national law, in the context of criminal law reform in Indonesia. This research aims to study the local wisdoms of Indonesian people found in the books of ancient law and which have philosophical content in relation to criminal law. The ancient law system that once existed in the territory of Indonesia, had regulated many criminal cases, punishment processes, as well as related philosophy. Unfortunately, after the Indonesian independence, legal experts were not very attentive about this. Delving into the two ancient law books that formally prevailed in this archipelago, it was discovered that various principles developed under western criminal law were also formulated in both books. The issue of criminal penalty then shall remains in line with human rights norms, takes into account local wisdom, while still considers current development in the society.

Harkristuti Harkrisnowo (2003) states that:

> 'Punishment is always a problematic issue, because punishment is always related to act which, unless performed by the state in accordance with the law, constitutes an t that violates public morale. State's act to penalise raise fundamental question, whether or not such act is justified and whether or not the issue of human rights has been considered. Hence the philosophy of punishment is necessary to justify this act'.

Regarding condemnation, Remmelink (2003) explains that we would need a justification for the imposition of pain by criminal law, for it should be done with awareness against the perpetrators of crimes.

Furthermore, Alice Ristroph was of the opinion that:

> 'A theory of crime, or criminal law, thus needs to be attentive to the collective entity that makes crimes. Criminal law theory needs political theory. This point has been made before, most often with respect to justification of punishment, but it bears re-emphasis. There is a difference between the claim that wrongdoers should be punished and the claim that the state should punish them, so a properly political theory must address the latter crime' (2013).

Based on the rationale above, this study was completed because of the limited exploration of condemnation philosophical values in the Indonesian ancient law books. Two books of law which become the focus of this research are the Book of Kutara Manawa Dharma Sastra of the Majapahit Kingdom and the book of Laws on Simbur Tjahaja of the Sultanate of Palembang. The contents of these two books should receive serious attention from the Indonesian Government, practitioners, and academicians, since they will be useful in the development of criminal law reform in Indonesia. Hence, this research will attempt to find the answer to the following questions by investigating both the Book of Kutara Manawa Dharma Sastra of the Majapahit Kingdom and the Laws on Simbur Tjahaja of the Sultanate of Palembang: (1) What are the principles of criminal law? (2) Are the condemnation principles in both books still relevant to the current circumstances and law development in Indonesia? (3) How do we compare the concept of condemnation with western law and criminal code, especially in terms of codifying the Bill of the Indonesian Criminal Code?

2 RESEARCH METHODOLOGY

The method used in this study is the document study methodology which focuses on two main documents, namely the Book of Kutara Manawa Dharma Sastra of the Majapahit Kingdom and the Laws on Simbur Tjahaja of the Sultanate of Palembang, and is supported by a wide range of literature concerning the subject of research.

3 RESULTS AND ANALYSIS

The main source of this research was the Book of Kutara Manawa Dharma Sastra of the Majapahit Kingdom translated by Slamet Muljana (1967) who states:

> 'that the original script of this book consists of 275 Chapters, but apparently there are several articles which are the same or similar. The translation only consists of 272 Chapters. It is noteworthy that as the other ancient law books, the Book of Kutara Manawa Dharma Sastra, does not distinguish between criminal and civil regulations, although there are more regulations related to criminal cases than the ones related to civil cases. The systematic content of the Book of Kutara Manawa Dharma Sastra is as follows: 1) General provisions regarding fines; 2) Murder (Astadusta); 3) The treatment of servants (Kawula); 4) Theft (Astacorah); 5) Compulsion (Sahasa); 6) Purchase and Sell (Adol Atuku); 7) Pawning (Sanda); 8) Debts and receivables (Ahutang apiutang); 9) Deposit; 10) Brideprice (Tukon); 11) Marriage (Kawarangan); 12) Nasty deed (ParaDara); 13) Heritage (Drewe Kaliliran); 14) Mock (Wakparusya); 15) Hurting (Dandaparusya); 16) Omission (Kagelehan); 17) Fights (Atukaran); 18) Land (Bhumi); and 19) Defamation (Duwilatek)'.

The criminal regulations in Kutara Manawa Dharma Sastra include: 1) Murder (Astadusta); 2) Theft (Astacorah); 3) Compulsion (Sahasa); 4) Nasty deeds (Para Dara); 5) Mock (Wakparusya); 6) Hurting (Dandaparusya); 7) Omission (Kagelehan); 8) Fights (Atukaran); 9) Defamation (Duwi Latex); 10) Six kinds of crime (burning a house of amonarch, poisoning human beings, bewitching human beings, rampaging, slandering the King, and undermining the respect for women (*tatayi*) (Muljana, 1967). The Book of Kutara Manawa Dharma Sastra of the Majapahit Kingdom also contains a description of the criminal acts which are subject to a fine or a penalty in the form of money, goods, or death sentence (Kawuryan, 2006).

In total, there are 117 Articles on criminal stipulation. For example, Article 3 concerning *astadusta* states:

> 'Astadusta consists of 1) Murdering innocent people; 2) Ordering to murder innocent people; 3) Injuring innocent people; 4) Eating together with a murderer; 5) Following the footsteps of the murderer; 6) Being friendly with the murderer; 7) Giving place to a murderer; 8) Giving help to a murderer. From those astadusta, the ransom of the first three is starch, and the other five is money. In the chapter on Astacorah, article 55 states: "If a thief is caught in the action, he will be imposed to the death penalty, his wife, children, and his land will be taken over by the reigning monarch. If that thief has male and female servants, they will not be taken over by the reigning monarch, but exempted from all debts to the thief' (Muljana, 2006).

The penalties that could be imposed based on the Book of Kutara Manawa Dharma Sastra of the Majapahit Kingdom, were divided into primary punishment and additional punishment. Primary punishment includes the death penalty, corporal punishment, fines, and redress or *panglicawa* or *patukucawa*. The additional punishment, or those punishment added to the primary punishment, includes payment of ransom, confiscation of goods or property, and obligation to pay for the medicine needed by the victim. Corporal punishment was applied to the thief in certain matters (Article 69). As for Laws on Simbur Tjahaja of the Sultanate of Palembang, it was created during the reign of *Sido Ing Kenayan* (1629–1636). This law was made by *Queen Sinuhun*, the consort of *Sido Ing Kenayan*, with the help of her own prince, royal officials, and the *mufti (religious leader)*. The original form was written in an Arabic script (Hanifah, 1994). This law on Simbur Tjahaja of the Sultanate of Palembang consists of six Chapters and 179 Articles: Chapter I Indigenous Single Girls and Marriage consists of 32 Articles; Chapter II Rules of Clan consists of 29 Articles; Chapter III Rules of Hamlet and Farming consists of 34 Articles; Chapter IV Rules of Tribe consists of 19 Articles; Chapter V Indigenous Punishment consists of 58A Articles; and Chapter VI Rules of Bahagi Money Penalty consists of 6 Articles.

The Laws on Simbur Tjahaja of the Sultanate of Palembang does not separate criminal and civil regulations; however regulations concerning criminal acts could be specified in several categories, among others: 1) Adultery (pregnancy with no definite father–Article 11, Chapter I); 2) Nasty deed (*meregang gawe*–Article 21 of Chapter I); 3) Escaping with another man's wife (Article 25 of Chapter I); 4) Leaving the King's *gawe* or event (Article 14 of Chapter II); 5) Prohibition on going into another clan without carrying *pas* or seal from *Pasirah or head of the clan* (Article 18 of Chapter II); 6) Accepting foreigners without *Proatin's (village officer)* permission (Article 9 of Chapter III); 7) Omission that causes house fire (Articles 13 and 14 of Chapter III); 8) Prohibition of wasting logs (Article 19 of Chapter 3), omission resulting in the burning of a field (Article 20 of Chapter 3); 9) Omission to keep buffalo (Article 22 of Chapter 3); 10) Obligation to ask permission from the *Pasirahor head of the clan* for going to the farm area of another clan (Article 27 of Chapter III); 11). Theft of a beehive tree (Article 32 of Chapter III); 12). Prohibition of gambling and cockpit without permission from the authority (Article 34 of Chapter III); 13). Giving perjury in the case/witness (Article 12 of Chapter V), fighting (Articles 13 –Article 17 of Chapter V); 14). Insult (Article 18 and Article 19 of Chapter V); 15). Theft (Articles 21–35); 16). Omission that causes death (Article 38 of Chapter V); 17). Omission that causes injury (Article 39 of Chapter V); 18). Hiding in people's houses at night with no good intention (Article 42 of Chapter V); 19). Stealing or desolating the King's goods (*Ketujuwalah*–Article 43 of Chapter V); 20). Abuse towards animals (Article 45 of Chapter V), prohibition on installing *tukos*, strainer, and broker near field/road (Article 46 of Chapter V); and 21). Prohibition on cutting down the trunk of a beehive (Article 48 of Chapter V).

Article 21 concerns itself with nasty deeds (*meregang gawe*) and states:

> 'If a man holds a girl or rangdo then hugs her body, it is called 'meregang gawe', and he will be fined 12 Ringgit, if the woman files any complaint. 6 Ringgit shall be given to her, and the other 6 Ringgit shall be for Pasirah if it happens in the hamlet of Pasirah. If it happens in the hamlet of Pengandang, 3 Ringgit shall be paid to Pasirah, and the other 3 Ringgit shall be for the Chief of hamlet and his penggawa'.

The penalties imposed based on the Laws on Simbur Tjahaja of the Sultanate of Palembang includes the King's penalty, the monetary fine penalty, bondage, *panjing* (servant of *Pasirah*), and *tepung satu kambing or payment of a goat* (Hanifah, 1994). In addition to a fine as a penalty, there were other penalties and bondage, especially if it involved the safety of the King and officials. Offences against the *Proatin* and *Pasirah or village officer and head of the clan* were subject to imprisonment for three to six months (Harun, 2008).

From an in-depth assessment, examination, and analysis of the Book of Kutara Manawa Dharma Sastra of the Majapahit Kingdom and the Laws on Simbur Tjahaja of the Sultanate of Palembang, the findings are as follows:

1. In the Book of Kutara Manawa Dharma Sastra of the Majapahit Kingdom, there is no penalty of liberty deprivation in the form of imprisonment and detention and this could be seen in all of the Articles. For example, Article 3 and Article 4 regarding *astadusta*, only includes the death penalty and a fine. This is reinforced by Slamet Muljana's opinion which states that the *Majapahit's* Code did not recognise imprisonment and detention (1967 or 2006?). Death penalty and fines were found to be imposed for many crimes. In the Laws on Simbur Tjahaja of the Sultanate of Palemban, a penalty fine is more dominant, but the death penalty and imprisonment are not recognised. Various types of sanctions recognised among others are a fine in the form of money, a penalty in the form of bondage and a penalty in the form of *tepung satu kambing or payment of a goat*: 1) |The King's penalty (Article 3, Article 27 of Chapter I, Article 34 of Chapter III, Article 12, Article 16 and Article 27 of Chapter V; 2) Corporal punishment (Article 2 of Chapter V); 3) *panjing* (servant of *Pasirah*); 4) The prisoner's stocks (Article 6 of Chapter II); 5) *tepung* (to redress) (Article 14 of Chapter V); 6) Drugs money (Article 16 of Chapter V); and 7) Penalty in the form of a bushel of rice, one coconut, one chicken, and one boiled betel vine (Article 18 of Chapter V).
2. In the Laws on Simbur Tjahaja of the Sultanate of Palembang, there are some regulations relating to criminal acts which are not followed by criminal sanctions, among others, Article 24, Article 25, Article 26, Article 27, Article 28 of Chapter II Rules of the Clan, and Article 10 of Chapter III Rules of Hamlet and Farming. Article 24 states: 'No person may save any gun or cannon, without the permission from the authority. And regarding the local rural people who get mad and get crazy, let the society take care of them'.
3. The Book of Kutara Manawa Dharma Sastra of the Majapahit Kingdom has already recognised the principle of equality before the law, as it can be seen in Article 6 and Article 11. Article 11 states: 'If any person, be they teachers, children, aged people, scholars, and everyone who are viewed by pastors, commits Tatayi (crime), and such offence is proven afterwards, they shall be sentenced to death'. Similarly, the Law on Simbur Tjahaja of the Sultanate of Palemban has also already recognised the principle of equality before the law, as can be seen in Article 8 of Chapter III, which states: 'If someone's house is entered by a bad guy, or if a thief enters a hamlet, unbeknown to the hamlet's guards, the guards shall get a penalty in the form of Panjing or used as slave (servant in Pasirah) for 1 to 3 months'.
4. The Book of Kutara Manawa Dharma Sastra of the Majapahit Kingdom has already recognised the principle of the enactment of criminal law by time, the principle of criminal liability and the principle of exemptions from liability (*strafuitsluiting gronden*), either as a legal excuse or justification. This could be seen in several Articles, including Article 1 containing the principle of exemptions from liability (legal excuse), which states: 'Children who cannot discern good from evil, and he is under ten years, if he misbehaves, he is not proper and worth to be imposed any fine by the King'. The legal justification could be seen in Article 20 which states:

'For liars, witches, and thieves, if their deeds are proven, they shall be sentenced to death by the reigning monarch. Their children and grandchildren shall be announced to the society, and the reigning monarch shall not allow them to live. If they escape from the executor's supervision, they shall not be tolerated by the reigning monarch. If they flee to another hamlet, they shall be killed. They shall be called as fugitive'. In the Laws on Simbur Tjahaja, the principle of exemptions from liability could be seen in several articles including Article 15 of Chapter II, Article 40, and Article 41 of Chapter V. Article 40 of Chapter V states: 'If someone kills a burglar who is at his house, he shall not be fined (money)'.

5. Both the Book of Kutara Manawa Dharma Sastra of the Majapahit Kingdom and the Laws on Simbur Tjahaja of the Sultanate of Palemban have already recognised the teachings about omission (*culpa*). Article 247 of the Book of Kutara Manawa Dharma Sastra of the Majapahit Kingdom states:
6. 'If someone cuts down a tree, and it falls on a person, or someone throws something unintentionally and hits a person or an animal, and afterwards he/she says that it is the fault of another person who gets into the accident, or a buffalo or a cow or any animal dies as a result thereof, then he/she should indemnify half of the estimated price'. Article 38 of Chapter V of the Laws on Simbur Tjahaja states: 'If someone commits manslaughter, he or she is obliged to pay 'bangun' or restitution according to article 36 mentioned above and also pay 4 Ringgit for 'belanja mati', or expenses due to death, and for the sin to the Pasirah, and 12 Ringgit to the Proatin. However, this case should be decided before the monarch'.
7. In the Book of Kutara Manawa Dharma Sastra of the Majapahit Kingdom, the principles of criminal law that could be found included the principle of the enactment of the criminal law by time (Article 1), the principle of exemptions from liability (Article 1 and Article 20), the principle of equality before the law (Article 6 and Article 11), the principle of complicity (*deelneming*) (Article 54 and Article 58), and the principle of fault (*opzet and culpa*) (Article 87 and Article 241). In the Laws on Simbur Tjahaja of the Sultanate of Palemban, the principles of criminal law that could be found among other things were the principle of equality before the law (Article 8 of Chapter III), the principle of exemptions from liability (Article 15 of Chapter II, Article 40 and Article 41 of Chapter V), the principle of fault (*schuld, opzet* and *culpa*) (Article 38 and Article 41 of Chapter V), the principle of complicity (*delneming*) (Article 11 of Chapter I) and the principle of special minimum and maximum sentence (Article 13–Article 16 of Chapter V). It recognised the difference between normal offences (*gewone delic*) and complaint offences (*klach delic*), where complaint offences can be found in Articles 18–21 and Articles 23–24, while normal offences can be found in Article 22, Article 25 of Chapter I. This same book recognised the system of sentences alternatively and cumulatively, where an alternative system can be found in Article 20 of Chapter V and a cumulative system can be found in Article 20 of Chapter III, Article 17–18 and Article 45 of Chapter V. The offence of murder and the offence against the government by armed weapons, should not be decided by *Pasirah* and *Proatin or head of the clan and village afficer* because it is within the King's authority (Article 10 of Chapter V). In the Bill of the Indonesian Criminal Code, the principles mentioned above are scattered in several Chapters, namely Chapter I concerning the extent of operation of the statutory penal provisions, Chapter II concerning the penal code, Chapter III concerning the things that exempt the liability, mitigating, or aggravating penalty, and Chapter IV concerning the complicity of the offence. In the Bill of the Indonesian Criminal Code, the formulation of those principles is contained in Chapter I concerning the scope of extent of operation of the statutory penal provisions, Chapter II concerning crime and criminal liability, and Chapter III concerning condemnation, criminal, and action.

4 CONCLUSION

The ancient archipelago law books scattered in several parts of Indonesia, which was formerly called the Archipelago (*Nusantara*), have historical values and Indonesian local wisdoms of very high value, and they provide reinforcement to the existence of great and qualified works of the Indonesians in the past. This work will be a reflection as well as a future projection for the creation of the national criminal law, which should not only rely on the western criminal law, but also on the Indonesian local wisdoms that are still appropriate and relevant to the present condition, and incontext of the present life.

The Book of Kutara Manawa Dharma Sastra of the Majapahit Kingdom and the Laws on Simbur Tjahaja of the Sultanate of Palembang are two of the most ancient of law books in the archipelago of Indonesia, and contain the deepest values and philosophy of condemnation. This is because both of these books were born in the *Majapahit* Kingdom and the *Sultanate* of *Palembang*, which had enormous influences in the law enforcement and the protection for the society at that time.

From the assessment of both these books, it is expected that this work can positively contribute to the codifying process of the Bill of the Indonesian Criminal Code, so the planned national criminal law reform can fully satisfy the aspirations of the Indonesian people. This will have good implications for future law protection and enforcement, and contribute towards a more equitable enforcement of criminal law.

REFERENCES

Hanifah, A. (1994). *Undang-undang simbur tjahaja.* (The Laws on Simbur Tjahaja) Jakarta, Indonesia: Pusat Pembinaan dan Pengembangan Bahasa Departemen Pendidikan dan Kebudayaan Republik Indonesia (The Language development and Fostering Agency Department of Education and Culture of the Republic of Indonesia).

Harun, J. (2008). *Undang-undang kesultanan melayu dalam perbandingan* (The laws on the Malay Sultanate in comparison). Pulau Pinang, Malaysia: Penerbit Universiti Sains Malaysia.

Harkrisnowo, H. (2003). *Rekonstruksi konsep pemidanaan: Suatu gagasan terhadap proses legislasi dan pemidanaan di Indonesia. Depok (Reconstruction of the concept of punishment: some thoughts on the legislation and sentencing processes in Indonesia)*, Indonesia: Faculty of Law Universitas Indonesia.

Kawuryan, M.W. (2006). *Tata pemerintahan negara kertagama keraton majapahit (State Governance of Kertagama of Keraton Majapahit)*. Jakarta, Indonesia: Panji Pustaka.

Muljana, S. (1967). *Perudang-undangan majapahit* (The Laws of Majapahit). Jakarta, Indonesia: Bhratara.

Muljana, S. (2006). *Tafsir sejarah negara kretagama* (Historical interpretation of Negara Kretagama). Yogyakarta, Indonesia: Lembaga Kajian Islam dan Sosial Yogyakarta (Yogyakarta institute for Islamic and social Studies).

Remmelink, J. (2003). *Hukum pidana: Komentar atas pasal-pasal terpenting dari kitab undang-undang hukum pidana belanda dan padanannya dalam kitab undang-undang hukum pidana Indonesia. (Criminal Law: Commentary on the most important articles of the Dutch Criminal Code and their equivalence in the Indonesian Criminal Code)*. (translated by T.P. Moeliono, Trans.). Jakarta, Indonesia: Gramedia Pustaka Utama.

Ristroph, A. (2013). Responsibility for the Criminal. In R.A. Duff & S. Green (Eds.), *Philosophical Foundations of Criminal Law* (pp. 107–124). New York, United States: Oxford University Press.

The existence of arbitration principles in commercial agreements: Lessons learned from an Indonesian court

Y.K. Dewi
Faculty of Law, Universitas Indonesia, Depok, Indonesia

ABSTRACT: Arbitration is seen as a method to avoid going to the courts when settling parties' commercial disputes. However, in many cases, parties still seek to invoke the jurisdiction of the national courts even when there is a valid arbitration clause. There are two principles to ensure the effectiveness of arbitral awards: autonomy and competence-competence principles. The "competence-competence" principle provides an arbitration tribunal with the power to decide its own jurisdiction and, at the same time, requires national courts to refer to the arbitration proceedings or preclude a court from addressing the same issue. In Indonesia, this principle has already been embodied in Article 11 (2) of Law No. 30 of 1999 regarding arbitration and alternative dispute resolution. This work will focus on this principle. To what extent does an Indonesian court preclude itself from exercising jurisdiction over matters subject to arbitration? To provide a more comprehensive view, this article will scrutinise relevant cases in an Indonesian court. This work will show that the court attitude creates uncertainty on a jurisdictional issue and reduces the simple efficiency of the arbitral process as one of arbitration's basic virtues.

1 INTRODUCTION

Lately, arbitration has been a prevalent dispute resolution approach exhausted to settle (international) commercial disputes. However, this type of dispute settlement mechanism must be explicitly chosen by the contracting parties in one of two forms: an arbitration clause in the main contract or an independent submission once the dispute has arisen. Both of them are referred to as "arbitration agreement" (Ferrari & Kroll, 2011). In the absence of an arbitration agreement, any dispute shall be brought before the courts (Born, 2010). Consequently, a valid arbitration agreement will provide the absolute jurisdiction for the arbitration tribunal to hear the dispute and rule the award. Interference by the courts, if not limited, can destroy the whole arbitration process, eliminating its benefits and, in the long term, leading to a diminution of the arbitration mechanism as a dispute resolution mechanism for the country. The autonomy and the competence-competence principles are the cornerstones for successfully building the whole process of arbitration and are intended to give primary responsibility to arbitration to determine whether it has jurisdiction (Lew et al., 2003).

This paper highlights the implementation of the competence-competence principle which is embodied in Law No. 30 of 1999 regarding arbitration and alternative dispute resolution ("Indonesian Arbitration Act") (Republic of Indonesia, 1999) by examining several cases. The issue concerning the power of the courts to intervene in the arbitration process when there is a valid arbitration clause has become problematic. If there is a valid arbitration clause and a party tries to overrule it by also submitting a dispute before the court, the question would be whether the court should refuse the dispute on the basis of a lack of jurisdiction.

The paper begins by explaining the two foundation principles in establishing an arbitration jurisdiction emphasising the competence-competence principle under the international law framework such as The United Nations Commission on International Trade Law (UNCITRAL) Model Law, International Chamber of Commerce (ICC) 1995, and Indonesian arbitration law. It then continues to examine several relevant cases, which have a valid agreement to arbitrate, to describe the attitude of the Indonesian courts when dealing with the competence-competence principle. It then concludes that the right attitude of the parties and the courts towards the arbitration is the key to the success of the arbitration. Thus, in order to maintain certainty on a jurisdictional issue and preserve the efficiency of the arbitral process, the court must avoid litigating cases that have a valid arbitration agreement.

2 METHODOLOGY

The method followed for this work is normative legal research using two approaches: statutory and case law (McConville & Chui, 2007). The statutory approach is used because this work examines relevant provisions on the existence of an arbitration clause in a commercial or business contract. The case law approach is used because this work examines several court decisions that are not in line with the provisions stipulated in the Indonesian Arbitration Act.

Data used in this work are secondary data, consisting mainly of textbooks, academic papers published both on a national and international level, and court decisions on the matters of arbitration principles contained in an arbitration clause (Crowther & Lancaster, 2014). Data were obtained through library research including sources from relevant websites. Once the data were obtained, they were analysed qualitatively by linking principles on commercial arbitration with several verdicts that have tended to ignore well-defined provisions stipulated in the Indonesian Arbitration Act.

3 DISCUSSION

3.1 *The arbitration principles: Principle of autonomy/separability/severability and principle of competence-competence*

To submit a dispute before an arbitration tribunal, it is critical that a disputing party really comprehend the autonomy principle, or sometimes is referred to as principle of separability or principle of severability (Born, 2001), and the competence-competence principle (Lew et al., 2003).

The principle of autonomy. The autonomy principle refers to the detachment of an arbitration clause from the underlying contract. Based on this principle, the arbitration clause is viewed as a separate agreement from the remaining contract between the parties. Consequently, the arbitration agreement is still intact when the purpose of the contract itself is already completed. The principle also refers to the autonomy of the arbitration agreement from all national laws (Gaillard & Savage, 1999). By its legal nature, the autonomy principle creates a question of substance. International Chamber of Commerce (ICC) 1995 Article 6 (4) and its revision, the ICC Rules 1998, confirm this principle by stating that an arbitration tribunal shall continue to have jurisdiction to determine the respective rights of the parties and to adjudicate their claims and pleas, even when the contract itself may be non-existent or null and void.

Likewise, UNCITRAL Arbitration Rules 1976 Article 21 (2) states that an arbitration clause which forms part of the contract and which provides for arbitration under these rules shall be treated as an agreement independent of the other terms of the contract. A decision by the arbitration tribunal that the contract is null and void shall not entail *ipso jure* the invalidity of the arbitration clause.

Many countries, including Indonesia, have implemented such a principle, which will be elaborated in the next part. The autonomy is based on the choice of law in a contract (Collins et al., 2010). Since the parties have surrendered their right to bring a dispute before the court and instead chose the arbitration route, they have the freedom to determine their own rules of the game in settling their disputes (Moses, 2008). They are free to select the seat of the arbitration tribunal, determine the composition of the tribunal, the law applicable, the arbitration procedure and the merit of the dispute (Briggs, 2008). The arbitration agreement constitutes a predominant/primary resource of the whole arbitration process. In this case, the arbitration agreement serves as a guidance to the disputing parties and arbitration tribunal during the whole arbitration course of action. Moreover, the arbitration agreement is a compelling proof of the existence of party autonomy by stipulating the chosen law and carrying out the arbitration process autonomously on the basis of the arbitration agreement. Thus, it is clear that the principle/doctrine of autonomy's function is to extricate the arbitration clause from the main contract and provide it with a sovereign existence. This principle improves the integrity of the arbitration process.

The principle of competence-competence (jurisdiction to determine jurisdiction). The competence-competence principle refers to the power of an arbitrator to make jurisdictional decisions over its own jurisdiction. In other words, it provides arbitration with the power to rule on its own jurisdiction without any obligation to adjourn proceedings should a court be concomitantly held. In this context, the tribunal can establish the existence of the arbitration clause, its legality and scope, without invoking the jurisdiction of a court. By its legal nature, this principle creates question of procedure. This principle is also acknowledged by primary international treaties on arbitration. Article 16 (3) of the UNCITRAL Model Law states that the arbitration tribunal may rule on a plea that the arbitration tribunal does not have jurisdiction either as a preliminary question or in an award on the merits, and that in the event of an action to set aside a partial award concerning jurisdiction, the arbitration tribunal may continue the arbitral proceedings and make an award.

The UNCITRAL arbitration rules provide that the arbitration tribunal shall have the power to rule on objections that the tribunal has no jurisdiction, including any objections with respect to the existence of validity of the arbitration clause or of the separate arbitration agreement.

The ICC rules also state that in all matters decided by the court under Article 6 (4), any decision as to the jurisdiction of the arbitration tribunal, except in relation to parties or claims for which the court decides that the arbitration cannot proceed, shall then be taken by the arbitration tribunal itself (see Art. 6.5).

The competence-competence principle has both positive and negative effects. The positive result of the principle is to enable the arbitrators to decide a binding award on their jurisdiction preceding to any court and restrict the function of court to conduct the examination of the arbitration award. This principle can be described as the rule stating that the arbitrators must have the first chance to hear objection on the jurisdictional challenge, expose to subsequent review by the court. The rule is aimed at ensuring that a disputing party cannot thrive in postponing the arbitral proceedings by claiming that the arbitration agreement is invalid or non-existent. However, it does not mean that a disputing party, having a lawful course of action on the illegality of the arbitration agreement, is less protected because such party will be able to bring the claim before the court if the arbitrator opts to decline the request to arbitrate.

The negative effect of the principle refers to the function of the court to then assess the tribunal's decision (Gaillard & Savage, 1999). It is understood that claiming parties pursue tactical advantages (Park, 1995) and that contesting jurisdiction is an effective approach to delay an arbitration for a strategic reason (Lew et al., 2003).

In some cases, when a dispute arises, a party, who is bound by an arbitration agreement, will question the tribunal's jurisdiction for a number of reasons. For example, such party now discovers that the arbitration proceedings are undesirable for some reason or merely needs to hold back the progress of the proceedings (Kawharu, 2008). Impeding opportunities to misuse legal proceedings will improve the efficacy of arbitration and reduce the exploitation of legal rights.

3.2 *The Indonesian Arbitration Act and its fundamental principles*

Arbitration in Indonesia has been practised since the Dutch colonial period (Gautama, 2006). During this time, exporters, importers, and businessmen used arbitration as the preferred mechanism to settle commercial disputes (Usman, 2013). An arbitration process is governed by the Indonesia Code of Civil Procedure, Article 615–651 (R. Subekti, 1977) and, in 1999, Indonesia enacted Law No. 30 Year 1999 regarding arbitration and alternative dispute resolution ("Indonesian Arbitration Act") (Republic of Indonesia, 1999).

The Indonesian Arbitration Act defines arbitration as a method of settling civil disputes outside the District Court based on the arbitration agreement in writing made by the disputed parties (Republic of Indonesia, 1999, art. 1, ¶ 1). The arbitration agreement is an agreement in the form of the arbitration clause contained in a written agreement made by the parties before a dispute arises, or an arbitration agreement made by the parties after a dispute arises (Republic of Indonesia, 1999, art. 1, ¶ 3).

The Indonesian Arbitration Act applies the two fundamental principles discussed above. The autonomy principle set out in Article 10 (Republic of Indonesia, 1999) states that an arbitration agreement shall not be cancelled due to the following circumstances: death; bankruptcy; novation; insolvency (a state of one's inability to pay); inheritance; abolishment of the main contract terms; when the performance of the agreement is transferred; or termination or cancellation of the main contract.

The competence-competence principle is governed by Article 3 (Republic of Indonesia, 1999), which states that: "A District Court is not authorised to adjudicate disputes that have bound the parties in the arbitration agreement."

The principle, also laid out in Article 11, makes the following provisions:

1. The existence of an arbitration agreement in writing eliminates the right of the parties to submit the dispute or difference of opinions contained in the agreement to the District Court;
2. The District Court shall refuse and shall not interfere with a settlement of dispute established by arbitration tribunal, except in certain cases stipulated in the Indonesian Arbitration Act.

Despite its imperfections, the Indonesian Arbitration Act has followed fundamental principles that have been recognised under international auspices. As elaborated, the competence-competence principle set out in Articles 3 and 11 of the Indonesian Arbitration Act empowers an arbitration tribunal to rule on its jurisdiction. The arbitration tribunal power is also espoused by the autonomy principle established in Article 10 that indulges an arbitration clause in a main contract as a separation from the main contract, consenting the arbitration clause, and consequently jurisdiction, to prevail from the contract invalidity or contract termination. It is then interesting to observe how the Indonesian court treats those two principles when there is a claim brought before it with a valid arbitration agreement.

3.3 *Case studies in the Indonesian courts: Lessons learned*

The following describes four cases related to the existence of an arbitration clause and the attitude of the courts.

The first case is PT Zhongjuan South East Asia vs. PT Pertamina & PT Golden Spike Energy Indonesia (South Jakarta District Court Decision No. 342/Pdt.G/2012/PN.Jkt.Sel.). The dispute involved PT Zhongjuan South East Asia as claimant, which is an Indonesian legal entity duly established under Indonesian company law, and a Joint Operation Body (JOB) between PT Pertamina and PT Golden Spike Energy Indonesia as respondent. The co-respondent is BP Migas RI (Indonesia Upstream Oil and Gas Agency). The dispute arose from an unlawful act (tort) committed by the respondent and co-respondent resulting in losses to the claimant. Although the parties had a valid arbitration clause in their contract, the District Court agreed to hear the dispute by reason that, unless it is a breach of contract, the District Court has the jurisdiction to adjudicate a dispute concerning tort.

The second case is PT Ciputra Surya Tbk vs. PT Taman Dayu (Central Jakarta District Court No. 517/PDT.G/1999/PN.JKT.Pst.). In summary, in December 2004, PT Taman Dayu concluded a Joint Development Contract (JDC), which contained a validity period of 20 years with PT Ciputra. The parties agreed to share properties in three categories, namely, developed area, semi-developed area, and undeveloped area (this area was excluded within the written agreement). The JDC contained an arbitration clause stating that in the event of a dispute, they would refer to Indonesia's National Arbitration Centre (*Badan Arbitrase Nasional Indonesia* (BANI)). In December 2012, PT Taman Dayu sold land covering around 436,320 m^2 from the undeveloped area to PT Lim Seng Tee, a company affiliated with PT Sampoerna. Instead of submitting the dispute to BANI, PT Ciputra brought it before the District Court. PT Taman Dayu argued that the dispute should have been settled by arbitration. However, the District Court ignored this argument and decided to hear the dispute. Unfortunately, the District Court did not provide any explanation nor elaboration addressing the argument made by the claimant.

The third case is PT Berkah Karya Bersama vs. Siti Hardiyanti Rukmana (Central Jakarta District Court No. 10/Pdt.G/2010/PN Jkt. Pst.). The case was triggered by an investment agreement to establish national education television between the claimant, PT Berkah Karya Bersama, and Siti Hardiyanti Rukmana, the owner of a television station, *Televisi Pendidikan Indonesia* (TPI) in 1991, as the respondent. The dispute aroused because Rukmana, as owner of TPI, claimed that PT Berkah has not fulfilled its obligation to make equity capital into TPI to entitled it as PT TPI's shareholder. However, PT Berkah denied it and declared that it has performed its obligations. Thus, PT Berkah is entitled to a 75% share in PT TPI.

The investment agreement contained a valid arbitration clause stating that if there was any dispute concerning interpretation, execution, validity, or decisions on rights and obligations of the parties, it would be settled by arbitration in Indonesia's National Arbitration Centre (BANI). However, although the proceedings in the arbitration tribunal were still ongoing, the respondent filed a lawsuit in the Central Jakarta District Court on the same grounds. The Central Jakarta District Court decided to adjudicate the dispute and later on rendered its decision.

The last case is PT Environmental Network Indonesia vs. PT Putra Putri Fortuna Windu (Supreme Court Decision No.12 K/N/1999). This case involved PT Environmental Network Indonesia (PT Enindo) as the claimant and PT Putri Putra Fortuna Windu (PT PPFW) as the respondent. The dispute was about a turnkey management and construction agreement for a shrimp farm project in which PT Enindo was obliged to provide management as well as construction services to PT PPFW for ten years. The agreement incorporated an arbitration clause. However, two years after signing the agreement, PT PPFW unilaterally terminated the agreement. PT Enindo soon filed a bankruptcy petition for costs incurred during its performance of the agreement. In the first stage, the Commercial Court denied the case by saying that it had no jurisdiction to hear and settle the disputed matter because of the valid arbitration clause. However, in cassation stage, the Supreme Court annulled a the Commercial Court's refusal to examine bankruptcy petitions due to the existence of an arbitration clause between the disputing parties. In the case of bankruptcy, an arbitration clause cannot set aside the jurisdiction of the Commercial Court. This is because the power to grant or refuse the request for bankruptcy is vested by the State to the Commercial Court, not to the contracting parties.

As elaborated in the previous sections, the crucial requirement that the parties to an arbitration agreement pledge their undertaking to refer to arbitration any disputes covered by their agreement gives rise to the effect that the courts of a given country are barred from giving a ruling in such disputes. If the dispute falls within the matter covered by an arbitration agreement, the courts will be compelled to direct the parties to an arbitration mechanism. When a dispute is subject to arbitration depends on a two-part question: First, did the parties have a valid agreement to arbitrate? Second, is the dispute at issue within the substantive coverage of that agreement? If the answers to both questions are "yes", then arbitration applies and a court has no jurisdiction to resolve the dispute. This means that court should look at any dispute in which the agreement contains an arbitration clause as a way to resolve a dispute that has been selected by the parties when signing the agreement.

Based on the first three cases described above, it is clear that the main contract contained a valid arbitration clause, thus creating an arbitration jurisdiction. Unfortunately, the district courts shared the same attitude in dealing with a dispute that has a valid arbitration clause, that is to remain accepting to adjudicate the dispute, rather than to declare that they were not have the power to adjudicate. In other words, those courts did not adjourn themselves from resolving the disputes; they still accepted a lawsuit filed by one of the parties when it was clear there was an arbitration agreement. If we apply the provisions as stipulated in Law 30/1999 (Republic of Indonesia, 1999) in relation to the autonomy and competence-competence principles, it can be predicted that the court will reject or refuse the case when it is clear that the dispute has an agreement to arbitrate. The court shall provide the opportunity for arbitration tribunal to first examine the case.

4 CONCLUSION

The reason to establish an independent dispute settlement method for commercial or business issues is based on the excessive adjournments and technicalities that afflict legal procedures. Arbitration dispute settlement is based on the free will of the contracting parties to determine their own ways of solving their disputes. This is done by selecting the arbitrators, the procedures, and the subject matter to be presented before the arbitration tribunal.

The autonomy principle is coupled with the competence-competence principle. The autonomy conferred by the arbitration clause assists in inhibiting disputing parties from struggling to postpone arbitral proceedings by striving to negate the authority of arbitration tribunals instituted by the contract. The success of arbitration as an alternative dispute resolution mechanism is reinforced by the acknowledgement of its jurisdiction within the arbitration agreement.

If Indonesia wishes to improve its arbitration dispute mechanism as one of the alternative dispute resolution routes, its existence must be supported not only by the contracting parties but also by the courts. This article does not undermine the importance of court involvement, which may be vital and indispensable; it only emphasises the need to strike a balance, so the efficiency of the arbitration process is not adversely affected. It also serves as a reminder of the importance of the preservation of the fundamental pillars of arbitration as expressed in the autonomy principle and the competence-competence principle.

REFERENCES

Bergsten, Eric. *Dispute Settlement: International Commercial Arbitration*, Module, (New York and Geneva: UNCTAD), 2005, can be accessed at: http://unctad.org/en/Docs/edmmisc232add38_en.pdf.

Born, G. (2010). *International arbitration and forum selection agreements: Drafting and enforcing* (3rd ed.). Alphen aan den Rijn, The Netherlands: Wolters Kluwer.

Briggs, A. (2008). *The conflict of laws*. Oxford, UK: Oxford University Press.

Collins, J.L., Morse, C.G.J., McClean, D., Briggs, A., Harris, J. & McLachlan, C. (2010). *Dicey, Morris & Collins: The conflict of laws, 4th cumulative supplement* (14th ed.). London, UK: Sweet & Maxwell.

Crowther, D. & Lancaster, G. (2014). *Research methods: A concise introduction to research in management and business consultancy* (2nd ed.). New York, NY: Routledge.

Ferrari, Franco & Stefan Kroll (Eds). (2011). *Conflict of Laws in International Arbitration*. Munich, Germany: Sellier, European Law Publishers GmBH, p. 96–98.

Gaillard, E. & Savage, J. (Eds.) (1999). *Fouchard Gaillard Goldman on international commercial arbitration*. The Hague, The Netherlands: Kluwer Law International.

Gautama, S. (2006). *Indonesian business law*. Bandung, Indonesia: Citra Aditya Bakti.

Kawharu, A. (2008). Arbitral jurisdiction. *New Zealand Universities Law Review*, 23(2), 238–264.

Lew, J.D.M., Mistelis, L.A. & Kroll, S.M. (2003). *Comparative international commercial arbitration*. New York, NY: Kluwer International Law.

Mahkamah Agung [Supreme Court] Republik Indonesia. (2011). Putusan [Decree] No.12 K/N/1999.

McConville, Mike & Wing Hong Chui (Eds). (2007). *Research Methods for Law*. UK: Edinburgh University Press.

Moses, M.L. (2008). *The principles and practice of international commercial arbitration*. New York, NY: Cambridge University Press.

Park, W.W. (1995). *International forum selection*. London, UK: Kluwer Law International.

Pengadilan Negeri [District Court] Jakarta Pusat [Central Jakarta]. (1999). Putusan [Decree] No. 517/PDT.G/1999/PN.Jkt.Pst.

Pengadilan Negeri [District Court] Jakarta Pusat [Central Jakarta]. (2010). Putusan [Decree] No. 10/Pdt.G/2010/PN Jkt.Pst.

Pengadilan Negeri [District Court] Jakarta Selatan [South Jakarta]. (2012). Putusan [Decree] No. 342/Pdt.G/2012/PN.Jkt.Sel.

Republic of Indonesia. (1999). *Undang-undang No. 30 Tahun 1999 tentang Arbitrase dan Alternatif Penyelesaian Sengketa [Law No. 30 of 1999 regarding Arbitration and Alternative Dispute Resolution]* can be accessed at: http://en.hukumonline.com.

Subekti, R. (1977). *Hukum Acara Perdata*. Bandung, Indonesia: BPHN & Bina Cipta [Subekti R. (1977). *Civil Procedural Law*. Yogyakarta, Indonesia: BPHN & Bina Cipta].

Usman, R. (2013). *Pilihan Penyelesaian Sengketa di Luar Pengadilan*. Yogyakarta, Indonesia: Citra Aditya Bakti. [Usman, R, (2013), *Out of Court Alternatives Dispute Settlement*, Yogyakarta, Indonesia: Citra Aditya Bakti].

Cross-border child adoption: Protection and challenges in Indonesia

A.B. Cahyono
Civil Law Department, Faculty of Law, Universitas Indonesia, Depok, Indonesia

ABSTRACT: The adoption process can be conducted within the national jurisdiction of a state, or by request of foreign citizens outside the territory of the state. The first type is referred to as local adoption, while the second type is called Cross-Border Child Adoption or Inter country Child Adoption. Although inter country adoption can offer a better life for the adopted children, the high number of requests for children to be adopted across country borders can give rise to the commercialization of adoption or even trafficking in persons that can greatly endanger the lives of the adopted children. This raises the question regarding what protection is given by the Indonesian government to avoid the negative impacts of inter country adoption and what challenges are currently faced in providing protection to children who are put up for inter country adoption. Although the legislation regarding adoption, particularly inter country adoption, has undergone a number of improvements, the implementation of such legislation to ensure that the best interests of the children are observed still encounter challenges.

1 INTRODUCTION

Cross-border or inter country child adoption is a global phenomenon that involves cross-border movement of vulnerable children, particularly those with low economic status, from developing countries to more developed nations (Misca, 2014). There has been a sharp increase in the number of inter country adoption cases over the period between 1998 and 2004. However, from 2004 to 2007 the number dropped, although it remained higher when compared to the 1990s. The United States is the country that received the highest number of adoption of all other countries, while China was the country that sent the most children for adoption to other countries (Selman, 2009). The first decade of inter country adoption was prompted by the need for care to be provided to children who were neglected or lost their parents due to the Second World War in countries which lost the War such as Germany, Greece and the Baltic states. Other European countries and the United Sates were the receivers of most of these children (as cited in Graff, 2000). The second wave of inter country adoption occurred after the end of the Korean War (Graff, 2000).

In Indonesia, adoption is specifically regulated through the Government Regulation No. 54 of 2007. Under Article 2 sub-article 2 of such regulation, adoption is defined as the legal act of transferring the custody of a child from his/her parents, legal guardian, or another person who has the responsibility for the care, education, and rearing of the child to the care of adoptive parents. Adoption can occur between Indonesian citizens, or between Indonesian citizens and foreign nationals. The latter is known as cross-border adoption or inter country adoption.

Although inter country adoption can offer a better life for the adopted children, the high number of requests for children to be adopted across country borders can give rise to the commercialization of adoption or even trafficking in persons that can greatly endanger the lives of the adopted children. This raises the question regarding what protection is given by the Indonesian government to avoid the negative impacts of inter country adoption, and what challenges are currently faced in providing protection to children who are put up for inter country adoption. This paper will explain the various protections provided in Indonesia with respect to inter country adoption, and the challenges faced by the national government in providing adequate protection to Indonesian children who are to be adopted by foreign nationals.

2 METHODS

This is a normative or doctrinal qualitative research to analyze the protection provided in Indonesia with respect to inter country adoption and the challenges faced by the national government in providing adequate protection to Indonesia children who are to be adopted by foreign nationals.

The research includes a historical study of regulations regarding the protection of inter country adoption prior and after Indonesia's declaration of independence in 1945. Prior to 1945, the regulation on the adoption of children had been enacted. The provisions were embodied in *Staatsblad* No 129 of 1917. After Indonesia's declaration of its independence, adoption has been regulated in several number of regulations. The regulations consist of law, government regulation, circular of the Indonesian Supreme Court, and regulation of the Minister of Social Affairs. Indonesia has also ratified some conventions regarding child adoption. The development of the concept of adoption and the protection of inter country adoption provided in some regulations are important to be examined or analyzed.

The researcher collected the data from relevant legislation and court decisions as primary resources. Secondary sources such as journal articles or written commentaries regarding child adoption were also collected to support the primary resources. Court decisions are important to be examined or analyzed to describe the implementation of legislation regarding inter country adoption. The role of judges is very important to ensure that adoption process is followed for the best interest of the child rather than to fulfill the interest of the child's prospective parents.

3 RESULTS AND ANALYSIS

3.1 *Adoption regulatory framework prior to Indonesia's independence*

The regulation on the adoption of children in Indonesia had existed well before Indonesia's independence. The regulatory instruments, however, only applied to persons of Chinese descent who wish to continue their family lineage (clans). The provisions were embodied in *Staatsblad* No 129 of 1917. In other words, the regulation on adoption in Indonesia initially was not for the best interest of the child, which is the objective of modern adoption legislation. The early regulation did not specifically provide for the protection of Indonesian children who were adopted by foreign nationals.

Since the purpose of adoption was to continue family lineage (clans) *Staatsblad* No. 129 of 1917 determined that only male children are allowed to be adopted. The legislation also stipulates that if the adopted boy has a family name difference from that of the adoptive parents, then the child will take the family name of the adoptive parents.

The mechanism for adoption under *Staatsblad* No. 129 of 1917 requires the consent of the parents if the adopted child is younger than 15 years old. If the adopted child has reached 15 years old, the consent of the child is also needed. Adoption is effected through a notarial deed, and noted in the margin of the child's birth certificate. Adoption can be undertaken by a husband and wife as a couple, or by a widow or widower. An adopted child must be at least eighteen years younger than the male adoptive parent and fifteen years younger than the female adoptive parent or the widower.

Since adoption at that time was intended to allow for the continuation of family lineage, adoption would sever the legal relationship between the biological parents and the adopted child. This loss of legal relationship causes the child to no longer be entitled to receive inheritance from his or her biological parents, but it will entitle him or her to the inheritance from the adoptive parents. An adoptive child will continue the family lineage of the adoptive parents.

In addition to the written law that applied to the residents of Chinese descent, the indigenous communities also have an adoption mechanism embedded in their traditional laws. Under these traditional laws, if a child is adopted, he or she enjoys the same legal status as their adoptive siblings, in some jurisdictions, such as in Bali, the legal relationship with their biological parents will cease to exist (Zaini, 1995). In indigenous communities, a child who has been adopted has legal and social rights in terms of inheritance law, custody rights, child protection, marriage and civil rights (Kamil & Fauzan, 2008).

3.2 *Regulation on adoption in post-independence era*

After Indonesia gained its independence, adoption is regulated in a limited manner under the Law No. 4 of 1979 on Child Welfare. With the regulation on adoption that is focused on the child's well-being, adoption is no longer restricted to male children for the purpose of continuing family lineage. Child welfare as defined under the Law No. 4 of 1979 is a life and living of a child that can ensure adequate spiritual, physical and social growth and development of the child. However, the law does not provide a formal adoption procedure, causing the process to be undertaken based on the previous legislation.

As regards the procedure for adoption, the Circular of the Indonesian Supreme Court No. 2 of 1979 dated 7 April 1979 regarding Child Adoption stipulates that adoption of a child of Indonesian nationality can only be undertaken through a court ruling. The Circular improves the mechanism for adoption that previously may only be effected by a notarial deed as provided under *Staatsblad* No 129 of 1917. As such, adoption could no longer be sought through private means, but it must be conducted under the state's auspice through a court ruling, and the court determines the legality of the adoption. In 1983 the 1979 Circular was further refined by the Circular of the Indonesian Supreme Court No. 6 of 1983. The new circular lays down more detailed requirements for child adoption, particularly for inter country adoption. The circular is also augmented by the Decree of the Minister of Social Affairs No. 41/HUK/KEP/VII/1984 setting forth the operational procedural guide to effect adoption.

In order to provide stronger protection for the child, the Law 23 of 2002 on Child Protection, which contains provisions on adoption, was enacted in 2002. Pursuant to this law, adoption must be for the best interests of the child. The term 'best interests of the child' is a widely used term all over the world in the context of adoption. This was particularly so following its introduction in the Convention on the Rights of the Child. Indonesia ratified the Convention through the Presidential Decree No. 36 of 1990.

Other provisions in the Law No. 23 of 2002 are that child adoption does not sever blood lineage of the adopted child with the biological parents, the potential adoptive parents must practice the same religion as the adopted child, and adoption by a foreign national can only be granted as the last possible resort.

On 26 December 2004 a severe earthquake and tsunami hit the Indonesian provinces of Nanggroe Aceh Darussalam and North Sumatera. The disaster took many lives and destroyed vast properties. The disaster also caused many children to lose their homes and parents. The condition prompted people in the country as well as overseas to wish to adopt these children who survived the disaster. In order to enforce a stricter control on adoption through the courts, the Supreme Court issued Circular No. 3 of 2005 on Child Adoption, which called upon all district courts in the country to seriously take into account the applicable laws and the Supreme Court Circular No. 6 of 1983 on Improvements to Circular No. 2 of 1979 that provides guidelines for the filing of adoption petitions and their due review. The Supreme Court strongly emphasizes that there should not be any deviation from the court-led process.

Moreover, the government issued the Government Regulation No. 54 of 2007 on the Implementation of Child Adoption. The Government Regulation supersedes all legislation insofar as they relate to the subject matter of the regulation. Pursuant to the Government Regulation No. 54 of 2007, foreign persons who wish to adopt in Indonesia must observe the following requirements:

a. approval must be obtained from the petitioners' country of origin;
b. approval must be obtained from the Indonesian Minister of Social Affairs;
c. the adoption process must done through a child services institution appointed by the Ministry of Social Affairs;
d. the prospective adoptive parents must not be a single parent;

e. the prospective adoptive parents have been legally domiciled in Indonesia for at least two years;
f. the prospective adoptive parents must submit a written statement that they will report the development of the child to the Indonesian Ministry of Foreign Affairs through the Indonesian government representative office in the country.

In addition to the specific regulatory provisions above, foreign nationals who wish to adopt must meet the following general conditions:

1. Regarding the child to be adopted:
   a. The child to be adopted is below 18 years of age. Children below 6 years must be prioritized for adoption. Children between the ages of 6 years and 12 years are allowed to be adopted under exigent circumstances. Exigent circumstances include those experienced by children who are victims of natural disasters or who are displaced persons. Adoption of children aged between 12 to 18 years old is only allowed on children who require special protection including those who have been neglected.

2. Regarding the prospective adoptive parents:
   a. having good physical and mental health,
   b. not younger than 30 years old and not older than 55 years old,
   c. having the same religious denomination as the child to be adopted,
   d. showing good conduct and having never been sentenced for any crime,
   e. being married for at least 5 years,
   f. not in a same-sex marriage,
   g. having no children, or only having one child
   h. possessing economic and social means,
   i. having obtained a consent from the child and a written consent from the child's parents or guardian,
   j. having made a written statement that the adoption is for the best interests, well-being and protection of the child,
   k. having obtained a report from a local social worker,
   l. having cared for the child to be adopted for at least six months following issuance of a child care permit,
   m. having received an approval from the Minister and/or head of social services office.

Furthermore, in order to further elaborate provisions on child adoption as provided under Article 18 of the Government Regulation No. 54 of 2007, the government issued the Regulation of the Minister of Social Affairs No. 110/HUK/2009 regarding Conditions for Child Adoption. Moreover, the government also issued the Regulation of the Minister of Social Affairs No. 37/HUK/2010 on National Deliberation Team for Issuance of Child Care Permit, which is an implementation of Article 25 of the Government Regulation No. 54 of 2007.

3.3 *Indonesia not ratifying the convention on the protection of children and co-operation in respect of inter country adoption (the Hague Convention)*

Although Indonesia has ratified a number of conventions relating to human rights including the Convention on the Rights of the Child or CRC, it has yet to ratify the Convention of 29 May 1993 on the Protection of Children and Cooperation in Respect of Inter country Adoption (the Hague Convention). According to the opinion of the author, the differing concepts of child adoption as embodied in the Hague Convention and with that which is adopted in the Law No. 23 of 2002 pose as the biggest impediment for Indonesia to ratify the convention. According to Article 26 of the Hague Convention, adoption creates a legal relationship between the adopted child and the adoptive parents. The adoption process also severs the legal relationship between the adopted child and the biological parents. This is different from the perspective adopted by the Law No. 23 of 2002 on Child Protection. According to the latter, adoption does not eliminate the legal relationship between the child and the biological parents.

Indonesia's concept of adoption, which does not sever the legal relationship between the adopted child and the biological parents, is influenced by Islamic law. According to Islamic law, adoption does not eliminate the legal relationship between the adopted child and the biological parents. An adopted child is not entitled to inheritance from the adoptive parents but inherits from the biological parents (Alam & Fauzan, 2008). As such, adoption under Islamic law only creates a custodial relationship between the child and the adoptive parents.

Another aspect that differentiates adoption as defined under the Hague Convention and which is referred to in the Indonesian child protection law relates to the prerequisites for adoption. The child protection law stipulates that the child to be adopted must belong to the same religious denomination as the prospective adoptive parents. Additionally, adoptive parents cannot be in a same-sex marriage. These provisions are not included in the Hague Convention.

According to Article 6 of the Hague Convention, a Contracting State shall designate a Central Authority to discharge the duties which are imposed by the Convention upon such authorities. Furthermore, According to Article 7 of the Hague Convention, Central Authorities shall cooperate with each other and promote cooperation amongst the competent authorities in their States to protect children and to achieve the other objects of the Convention. By not ratifying the Hague Convention, Indonesia has no formal avenue by which it collaborates with other countries in providing protection to Indonesian citizens who are adopted by foreign nationals. Without the cooperation of other countries, Indonesia would encounter difficulties in overseeing or monitoring the condition of the children that have been adopted by non-Indonesians, particularly once the adoption petition has been awarded by the court and the child has been taken out of the country by the adoptive parents. An exception to this condition is when a bilateral agreement has been entered into by Indonesia and the other country and intended to ensure that adoptions are awarded for the best interests of the child.

3.4 *Inter country adoption causes forfeiture of Indonesian citizenship*

One of the state's duties with regard to adoption is to supervise the adoption process to ensure that it complies with the national legislation. However, inter country adoption can take away the Indonesian citizenship of the child. When an adopted Indonesian child loses his or her Indonesian citizenship, the situation would make supervision difficult to be effected by the Indonesian government. By renouncing Indonesian citizenship, the adopted person would also lose his or her civil rights and the protection he or she is entitled to.

According to Article 5 paragraph (2) of the Law No. 12 of 2006 on Citizenship, an Indonesian child adopted by a foreign national pursuant to a court ruling prior to the child reaching five years of age shall be considered as an Indonesian citizen. If the child is entitled to be granted citizenship of the adoptive parents, the child will hold dual citizenship until he or she is 18 years old. A problem arises when the adopted child lives overseas with the adoptive parents, while the parents fail to provide regular reports to the Indonesian embassy in the country. This can cause the adopted child to lose his or her citizenship as provided under Article 23 of sub-article (1) of the Law No. 12 of 2006 on Citizenship.

3.5 *Adoption for the best interests of the child not easy to implement effectively*

The term 'for the best interests of the child' is used almost in all countries. The term is also referenced in the Hague Convention and the Indonesian child protection law. Olsen interprets the term as meaning that every child has the right to grow and live in a condition that provides security, stability and love (Olsen, 2004). However, the concept of the best interests of the child in an inter country adoption is difficult to implement. Not all inter country adoption can ensure that the best interests of the child are pursued. This may be for several reasons:

Firstly, adoption is not always driven by the desire on the prospective parents' side to provide assistance or protection for the child and to provide a better life and future. Based on a 1993 report issued by the U.S. government, it was found that many people seek inter country adoption because of the fact that the process involved is much quicker and easier than domestic adoption (Olsen, 2004). There are also parents who choose inter country adoption for commercial reasons. It is an undeniable fact that the high demand for children has created business networks that search and sell children without any regard to their well-being (Hubing, 2001). Other people eventually have children after they adopt, causing neglect of the adopted child once their biological child is born.

Secondly, there is no regulatory framework on rehoming. This can put an adopted person in a dangerous situation, as the rehoming process would be undertaken privately outside the state's or the courts' supervision to ensure that the process is indeed for the best interests of the child.

Rehoming is where a child is transferred from a home in which he or she was legally adopted to another home. The transaction can be easily effected by the granting of the power to transfer a custodial right and it is often without the involvement of a lawyer or authorized official to oversee the well-being of the child (Dahl, 2016).

Thirdly, there are some cases where the adoption process in court deviates from the established statutory procedures. This would certainly place the child in an adverse circumstance. The following are a number of inter country adoption cases that do not follow the applicable laws and regulations.

In the case of Sarah Mary Jane (Indonesian girl), the adoptive parents were granted to adopt Sarah Mary Jane who was under the care of her parents, by virtue of Court Decision No. 58/Pdt.P/2000/PN.Jakarta Selatan. Private adoption of a person who is still under the care of the parents is clearly against the applicable regulations set forth in the Regulation of the Minister of Social Affairs No. 13/HUK/1993. The strong social and economic status of the petitioners was the primary point of consideration for the judge in awarding adoption to the German couple.

The second case relates to the neglect of an adopted child by the parents who filed for adoption. The adoption was on Tristan Dowse, an Indonesian national who was adopted by a couple of Irish citizenship, Josep Nigel Dowse and Lala Dowse. Tristan Dowse, whose birth name was Erwin, was neglected by his adoptive parents when they had a child of their own. Josep and Lala filed for adoption with the South Jakarta District Court and the petition was granted by virtue of Court Decision No. 192/Pdt.P/PN Jakarta Selatan. Even though the adoption process was carried out through the court, many provisions of the law were violated in the process. These violations include: firstly, the length of marriage of the petitioners has not reached the mandatory minimum years of 5 years; secondly, the petitioners have not resided in Indonesia for the statutory minimum period of 2 consecutive years; thirdly, the petitioners have not cared for the child to be adopted for at least 6 months; fourthly, the petitioners filed for the adoption of a child who was still under the care of his biological parents; and fifthly, no approval was given by the Indonesian Ministry of Social Affairs.

The third case in 2010 indicated that the implementation of the regulations regarding inter country adoption is going better. In Milli Tesalonika case, the adoptive parents almost fulfilled all the requirements for the adoption. The case which was decided by South Jakarta District Court even though they did not meet the length of marriage of the adoptive parents. The judge on his Decision No. 103/Pdt.P/2010PN.Jkt.Sel did not consider the requirement that adoptive parents should have been married for at least 5 years. In fact, when the adoption process was conducted by adoptive parents, they had been married for 4.5 years.

4 CONCLUSION

Although the legislation regarding adoption, particularly inter country adoption, has undergone a number of improvements, the implementation of such legislation to ensure that the best interests of the child are served still encounters challenges. These challenges include the fact that the Hague Convention is still yet to be ratified by the Indonesian government, causing difficulties for the state to provide protection and supervision for the adopted children when they are taken by the adoptive parents to another country. Supervision and monitoring by the state is passive and greatly relies on the good faith of the adoptive parents to send regular reports on the development of the adopted child. This creates challenges in effectively providing protection for the child. The Indonesian legislation also does not provide for rehoming, which can place an adopted child in a dangerous situation. Furthermore, there are some cases which indicate unfamiliarity of the judges with the legislation governing adoption leading to an impediment in achieving the best interests of the child in inter country adoption process. The economic status of the adoptive parents often becomes the primary point of consideration for such judges in awarding adoption.

REFERENCES

Alam, A.S. & Fauzan, M. (2008) *Hukum Pengangkatan Anak Perspektif Islam*. Jakarta, Kencana.
Dahl, E.A. (2016) Re-homing: The underground market for adopted children and how current laws fail to protect the innocent. *Wake Forest Journal of Law & Policy*, 6 (2), 549–566.
Graff, N.B. (2000) Inter-country adoption and the convention on the rights of the child: Can the free market in children be controlled? *Syracuse Journal of International Law and Commerce*, 27 (2) 405–430.
Hague Conference on Private International Law (1993) *33 Convention on Protection of Children and Co-operation in Respect of Inter country Adoption*.
Hubing, B.M. (2001) International child adoptions: Who should decide what is in the best interests of the family. 15 *Notre Dame Journal of Law, Ethics & Public Policy*, 15 (2), 655–698.
Indonesia (1917) *Staatsblad No. 129 of 1917 regarding Civil and Commercial Laws for People of Chinese Descent in Indonesia*.
Indonesia (1979) *Surat Edaran Mahkamah Agung Republik Indonesia No. 2 Tahun 1979 tentang Pengangkatan Anak [Circular of the Supreme Court Republic of Indonesia Number 2 of 1979 regarding Child Adoption]*.
Indonesia (1979) *Undang-undang No. 4 Tahun 1979 tentang Kesejahteraan Anak [The Law Number 4 of 1979 regarding Child Walfare]*.
Indonesia (1983) *Surat Edaran Mahkamah Agung Republik Indonesia No. 6 Tahun 1983 tentang Penyempurnaan Surat Edaran No. 2 Tahun 1979 tentang Pengangkatan Anak [Circular of the Supreme Court Republic of Indonesia Number 6 of 1983 regarding Amendment of Circular Letter Number 2 of 1979 regarding Child Adoption]*.
Indonesia (1999) *Undang-undang No. 30 Tahun 1999 tentang Arbitrase dan Alternatif Penyelesaian Sengketa [The Law No. 30 of 1999 regarding Arbitration and Alternative Dispute Resolution]*.
Indonesia (2002) *Undang-undang No. 23 Tahun 2002 tentang Perlindungan Anak [The Law No. 23 of 2002 regarding Child Protection]*.
Indonesia (2005) *Edaran Mahkamah Agung Republik Indonesia No. 3 Tahun 2005 tentang Pengangkatan Anak [Circular of the Supreme Court Republic of Indonesia Number 3 of 2005 regarding Child Adoption]*.
Indonesia (2006) *Undang-undang No. 12 Tahun 2006 tentang Kewarganegaraan Indonesia [The Law No. 12 of 2006 regarding Indonesian Citizenship]*.
Indonesia (2007) *Peraturan Pemerintah No. 54 Tahun 2007 tentang Pelaksanaan Adopsi [Government Regulation No. 54 of 2007 regarding Adoption Implementation]*.
Kamil, A. & Fauzan, M. (2008). *Hukum Perlindungan dan Pengangkatan Anak di Indonesia*, Jakarta, Rajawali Press.
Kementrian Sosial [Social Affairs] Republik Indonesia (1984) Putusan [Decree] No. 41/HUK/KEP/VII/1984.
Kementrian Sosial [Social Affairs] Republik Indonesia (2010) Putusan [Decree] No. 37/HUK/2010.
Kementrian Sosial [Social Affairs] Republik Indonesia (1995) Putusan [Decree] No. 2/HUK/1995.
Kementrian Sosial [Social Affairs] Republik Indonesia (2009) Putusan [Decree] No. 13/HUK/1993.
Kementrian Sosial [Social Affairs] Republik Indonesia (2009) Putusan [Decree] No. 110/HUK/2009.
Misca, G. (2014) The "Quiet Migration": Is inter country adoption a successful intervention in the lives of vulnerable children? *Family Court Review* 52 (1), 60–68.
Office of The High Commissioner United Nation of Human Rights [OHCHR] (1989) *Convention on the Rights of the Child Adopted*. General Assembly Resolution Number 44/25 of 20 November 1989.
Olsen, L.J. (2004) Live or let die: Could inter country adoption make the difference? *Penn State International Law Review*, 22 (3), 483–525.
Pengadilan Negeri [District Court] Jakarta Selatan [South Jakarta] (2010) Putusan [Decision] No. 103/Pdt.P/2010/PN.Jkt.Sel.

Pengadilan Negeri [District Court] Jakarta Selatan [South Jakarta] (2001) Putusan [Decision] No. 192/Pdt.P/2001/PN.Jak.Sel.

Pengadilan Negeri [District Court] Jakarta Selatan [South Jakarta] (2000) Putusan [Decision] No. 58/Pdt.P/2000/ PN.Jak.Sel.

Presiden Republik Indonesia (1990) *Keputusan Presiden No. 36 Tahun 1990 tentang Pengesahan Konvensi tentang Hak-Hak Anak [Presidential Decree Number 36 of 1990 regarding Ratification Convention on the Rights of the Child]*.

Selman, P. (2009). The rise and fall of inter country adoption in the 21st century. *International Social Work,* 52 (5), 575–594.

Zaini, M. (1995) *Adopsi: Suatu Tinjauan dari Tiga Sistem Hukum*, Jakarta, Sinar Grafika.

Electronic identity management in ensuring national cyber security and resilience: Legal aspects of online identity and its secured transaction

E. Makarim
Faculty of Law, Universitas Indonesia, Depok, Indonesia

E.G. Pantouw
Center for Law and Technology Studies, Faculty of Law, Universitas Indonesia, Depok, Indonesia

ABSTRACT: Some prominent issues within the scope of today's Cyberlaw discussion are about Cybersecurity, Privacy, and Identity, particularly the National Electronic Identity Management in e-transactions. Identity has a great value as authentic information in representing personal data. In practice, there are many digital identities that may be used by people in the market. Even though there are many countries implementing a single credential system in their National e-ID Management, they actually cannot deny the existence of many online identities in the society that have already been used and accepted by the public in the market. Therefore, in order to support regional e-commerce, some countries inevitably should accommodate federated identity management in their policy. It may be classified as a critical public infrastructure in Cyber security. Through the normative research and methods of the qualitative data analysis by studying some lessons learned from the US-National Strategy for Trusted Identity in Cyberspace (NSTIC) and the European Regulation No. 910 of 2014 regarding e-Identification and Trust Services, this paper aims at recommending the National legal framework for trusted identity ecosystem and/or the e-Authentication System in Indonesia to support e-transaction in a regional context, ASEAN Economic Community.

1 INTRODUCTION

According to the latest developments in some discussions in various global forums on cyber law, two important issues have been raised, namely Privacy and Cyber Security. This is certainly understandable considering that the cyberspace has become a means for global e-commerce and access to e-public services/e-government. All legal relationships or e-transactions certainly require information security for the purpose of evidence, but there is clarity to the identity of the natural person responsible for the transactions.

On one hand, electronic systems require certainty of the identity of the parties intending to make transactions, but on the other hand, users also require certainty as to protection of the privacy of their personal data. Vulnerability and abuse of personal data may be the micro impacts on individuals and families, but they also pose a clear threat to the safety of the nation, so that such matters have been covered in discussions on cyber security. There are three keywords for the successful implementation of the Internet of things in the future: security, privacy, and trust consideration.

This paper aims at answering the following research questions; (i) what is the correlation between the National e-ID Management to Cyber security? and (ii) how can Indonesia and ASEAN, where the majority are using single credential system, consider a policy for accommodating many digital identities in order to support e-transactions in the region?

Looking at the global dynamics, there are some lessons learned from the results of research conducted by some international organizations (e.g. OECD and UNCITRAL), the

United States and also the European Union. It is a fact that many countries have a different policy regarding e-identity management in facilitating e-transactions, but they actually need an interoperability of those identities in the region. Indonesia needs to take some lessons learned in order to make a good policy on the management of digital identity and personal data, especially in order to face the MEA's challenges that require an interoperability of digital identities among ASEAN member countries to support the regional e-commerce transactions.

Technically, the authenticity of e-information cannot be separated from the secured electronic communication system principles, namely: confidentiality, integrity, authorization/ authenticity, and non-repudiation (Bajaj & Nag, 2005). The level of quality assurance of security will determine the level of reliability of the security system that will affect the probative value of evidence, from the lowest to the highest value that is not repudiated as evidence. Based on the functional equivalent approach, electronic information and/or electronic documents would be treated equally like written evidence if they meet at least three conditions, namely: pursuant to the UNCITRAL Model Law of e-commerce, the UNICTRAL Model Law of e-signatures, and the United Nations Convention for Electronic Communication in International Contract (UN-ECC), as follows:

1. the e-information is considered as "in writing" if it can be stored and retrieved; (availability)
2. the information is considered as 'original' if the information would not change when it is stored, retrieved, and displayed (integrity); and
3. the information is considered as having been "signed" if there is information that explains the existence of a legal subject responsible for it or there are reliable authentication methods and systems that determine, verify, and confirm the identity so that the authorization can be attributed to such party.

It should be noted that, in the distributed networks for communication, such as the Internet, any information exchanged for transactions should be secured in order to be used as evidence of the transactions. The method or technology for securing information and communication is commonly known as Electronic Signature that may have many types or differences in the market, including any identity provided by any government (e.g.: National ID cards) or provided by the private sector (private identity cards). It is undisputable that there is a need for a clear policy and a legal framework regarding the e-identification process (the process for identifying and acknowledging a set of attributes of data/identity) and the e-authentication process (the process for confirming the validity of digital identity).

2 CYBER SECURITY AND NATIONAL e-ID MANAGEMENT

There is a strong correlation between Cyber Security and National e-ID management because every e-transaction will use the identity and security in the exchange of information. In general, Cyber Security is often understood as the implementation of information security management in a corporation against the existence of the system and its assets by the application of ISO 27001 (International Organization for Standardization 2013). Meanwhile, ITU Global Cyber Security Agenda focuses more on non-war civilian interests by implementing 5 important things, Legal Measures, Technical and Procedural Measures, Organizational Structure, Capacity Building, and International Cooperation (2007). Meanwhile, NATO provides broader guidelines, namely the application of the protection of the national security in the use of cyberspace that covers the aspects of the national security and resilience and the possibility of retaliatory measures in order to maintain the interest of the nation (NATO, 2012). In line with NATO, the European Union also provides guidance and assessment to its member states regarding the importance of the National Cyber Security Strategy which includes the existence of the Critical Information Infrastructure in order to maintain regional security and stability (ENISA, 2012).

By combining various paradigms of Cyber Security that have been well developed by the ITU, NATO, and ENISA, it can be basically said that the coverage of Cyber Security actually does not only guarantee the aspects of Confidentiality, Integrity, and Availability of information but also provides (i) guarantee for every citizen to access the Internet as a representation of human rights to information and communication, (ii) secured, safe, and stable e-system while maintaining the interests of the National Security including the protection of public or personal data for the growth of the digital economy, and (iv) resilience or quick incident response and recovery from emergency situations (crisis management), which includes multi-stakeholders' participation and interests (coordination, cooperation, and collaboration) at the domestic, regional, and international levels as well as (v) cyber diplomacy and law enforcement.

With regard to National e-ID Management, based on the results of the UNCITRAL's research on Promoting Confidence in E-Commerce: Legal Issues on the International Use of Electronic Authentication and Signature Methods, UNCITRAL (2009) encourages interoperable application for the international harmonization of e-signature and/or e-authentication by benchmarking to the model of e-apostille, whereas at the national level, it includes the implementation of domestic models, such as Cyber-notary or Electronic-Notary. In the second part of the research's results, it prompts legal recognition by every country of e-authentication from other countries (legal recognition of foreign e-authentication and signature methods), the criteria of which include references to the country of origin, the principle of reciprocity and local validation, as well as the application of minimum requirements and/or the substantive functional equivalence approach.

In the Cyber Security Framework Guide, ITU highlights the Generic, and Universal Digital Identity System in an effort to secure digital identities across geographical boundaries as an important point contained in a Cyber Security Framework (2011). ITU recommends the application of the Secured Identity approach for the digital identity authentication process through an electronic system. The main focus is to ensure the application of schemes by building security protocols, standards, and accreditation software for authentication and authorization technology in the implementation identity schemes.

Meanwhile, OECD conducted research on (i) Digital Identity Management: enabling innovation and trust in the internet economy; and (ii) National Strategies and Policies for Digital Identity Management in OECD Countries (2011). Furthermore, in 2015, OECD made a report on the use of electronic identity in providing security and privacy in the implementation of the digital economy, including: (i) Rising Awareness of the Digital Identity, (ii) Furthering the Implementation of the Privacy Guidelines and Security Risk Recommendations, (iii) the Business Case for Digital Identity, (iv) Environmental Scan of Multi-Jurisdictional Approaches to Digital Identity, (v) Principles for Trust Frameworks, (vi) Exploring the Role of Public-Private Cooperation in Digital Identity Management (2015).

Such matter may be facilitated either in law or electronically, with clear policies and rules on accounting for the implementation of electronic signatures or digital signatures that are supported by electronic certification systems (digital certificates) which are orphaned (root) and based on Public Key Infrastructure (PKI). It should be noted that there are at least two models to be applied in its organization, i.e. orphaned or geared towards the government PKI (Government PKI applying the mechanism of National Root CA) and privately orphaned at each root (Private Root CA) that can be subsequently bridged by Government CA (Bridge CA). In the context of Indonesia, such policy may be accommodated pursuant to the Government Regulation Number 82 of 2012 (PP-PSTE), whereby CA may be provided by and rooted to the Government PKI (Government Root CA), while National CA or Bridging CA has not been determined yet.

With regard to e-ID Management, OECD indicates that in general there are four (4) models for determining policy model for e-ID Management, namely (i) Silo, (ii) Centralized, (iii) Federated, and (iv) User Centric (Table OECD (2011) Features of Technology Models for Id-Management Systems, pp.164–165 of the OECD's reference)

	SILO	**Centralized**	**Federated**	**User centric**
Method of Authentication	The user authenticates to each account when he wishes to use it.	The user authenticates to one main account.	The user authenticates to an identity provider, with this one authentication serving for the federation.	The user authenticates to identity providers, and service providers have to rely on that authentication.
Location of Identity Information	Identity information is stored in separate service provider accounts.	Identity information is stored in the one main account, a super account.	Service providers in the federation keep separate accounts in different locations. They may have agreements for sharing information.	Identity information is stored by identity providers chosen by the user. The user can help prevent the buildup of profiles that others hold about him.
Method of linking accounts/ learning if they belong to the same person	There is no linking between accounts and no information flow between them.	Linking between accounts is not applicable. (A user's full profile resides in that single place.)	The identity provider can indicate what identifiers for accounts with federation members correspond to the same person.	Uses of cryptography can prevent linkages between a user's different digital identities, leaving the user in control.
Trust Characteristics (who is dependent on whom, for what)	The user is reliant on the service provider to protect their information, even if limited. The absence of information sharing has privacy advantages.	The user is reliant on the service provider to maintain the privacy and security of all of his or her data.	Users have rights from contracts, but they may be unfamiliar with options. The federation has leverage as it is in possession of the user's information.	Users can keep accounts separate and still allow information to flow, but bear greater responsibility.
Convenience	Siloed accounts are inconvenient for users and service providers due to multiple authentications, redundant entry of information, and lack of data flow.	This arrangement is easy for the user since he or she only has to deal with one credential to call up the account and since he or she has to authenticate just once.	Other members of the federation avoid the burden of credential management. Organisations that provide services to a user can coordinate service delivery.	Users may be ill-equipped to manage their own data (also a vulnerability) and may need training and awareness raising.
Vulnerabilities	Siloed systems offer the advantage of having limited data on hand, thus creating less of an incentive for attack. They also have a better defined and stronger security boundary to keep attackers out and limit exposure from failures.	The central party controls the person's entire profile; other entities have little to check that profile against, and an insider could impersonate the person or alter data. Currently there is no way to safeguard data after it has been shared.	Users have little input into the business-partner agreements. Some service providers will set up federation systems to exploit users. Currently there is no way to safeguard data after it has been shared.	Concentration in the market for identity providers could leave them with much power. Currently there is no way to safeguard data after it has been shared.

OECD noted some important things worth learning from what is happening in the United States. On the other hand, the US delegation reminded all countries in the UNCITRAL forum in 2013 about the importance of clarity in e-ID Management for supporting global e-commerce. They themselves were developing and implementing National Strategies for Trusted Identity in Cyberspace (NSTIC) based on a basic policy on Federated Identity or Open Identity Management. The policy essentially indicated that the US Government prefers to create a Trusted Framework for the formation of an open yet reliable (trusted) Identity Ecosystem, so that they can integrate all existing Identities organized by the government or by the private sector. NSTIC provides four principles that must be applied in the Identity Ecosystem: Privacy-Enhancing, Voluntary, Secure, Interoperable, and easy to use (the United States of America—White House 2011). With these principles, NSTIC aims at providing guarantee for privacy to individuals and providing anonymity for entities requesting identity. To ensure the anonymity of a person's identity, NSTIC requires all state officials and businesses to utilize credential-based authentication, on the other hand, however, NSTIC also provides pathways for Policy-Based Access Authentication in the implementation of activities. Accordingly, there is less information to be released by individuals and the privacy of the individuals is guaranteed. Measures for preventing the use of Invisible Channels for attracting the disclosure of new information are reduced. Up to now, the US has not implemented any national identity card (single credential) system for its citizens. Although Social Security Number and Driving License systems have been implemented, they do not constitute a single card for all business transactions either to the government or to private parties. As the consequence of this policy, users can use their existing identities provided by social media network providers, for example, by login with Face book, Google, or any other soft identities issued by other providers as long as those identities meet the requirements of trusted framework.

In the latest development, European Union has revised its Directives on e-signature in the Regulation Number 910 of 2014 regarding e-Identification and Trust Services. The revised directives essentially set out an arrangement for a mutual recognition process of electronic identification among Member States (EU, 2014). The EU Regulation No. 910/2014 provides definitions of reliable identity identification and identification process in relation to parties making transactions. Such identification process uses unique data that distinguish every legal subject. Furthermore, such identification process is implemented by using the services of some reliable parties. Reliable Parties is defined as electronic service providers providing services for the creation, verification and confirmation of Electronic Signatures, Electronic Seals, Electronic Time Stampings, Electronic Registered Delivery Services, Electronic Documents and Website authentication.

The European Union has been harmonizing the needs of its Member States to have interoperability among identity systems in the future that will be inevitably applied in various sectors of life. EU has also been conducting STORK research to examine the level of trustworthiness (low level, substantial level, and high level) of digital certificates issued by each of its Member States, as well as their compliance with the agreed interoperability at the regional level. Basically, they are divided into four Quality Assurance Levels, ranging from minimum or lower level of assurance to high level of assurance.

Arrangement for the aforementioned four models worldwide has led to the application of federated identity, which essentially accommodates the presence of all existing digital identities for interoperability among the systems so that everyone can connect to the regional electronic system more easily. In fact, in addition to the identity of legal subjects provided by the government in the form of citizen identity cards, there are also several identities provided by companies, such as; membership numbers, credit card numbers, electronic payment transaction numbers, as well as other digital identities. All of those are categorized as soft-identities issued by private companies.

Indonesia can adopt some lessons learned from the above mentioned developments in the US and EU to improve its national legal system in pursuing regional e-commerce in the context of ASEAN Economic Community. ASEAN needs an interoperable system for electronic identification and authentication among ASEAN member states in order to access their public services and government administration.

3 INDONESIAN NATIONAL ELECTRONIC IDENTITY CARD (e-KTP)

In the perspective of cyber security in Indonesia, the Law Number 3 of Year 2002 regarding National Defence does not expressly define cyberspace as a defence territory (Indonesia, 2002). However, the Law Number 11 Year 2008 regarding Electronic Information and Transaction (ITE) sets out the foundation of security in cyberspace. The first basis is the provisions of Article 15 of the ITE Law regarding accountability for electronic system that must be organized in a reliable, safe, and responsible manner (Indonesia, 2008). The ITE Law requires e-system providers to bear responsibility or liability to the public (presumed liability), so that in a legal perspective, every operator must be held legally responsible/liable for the accountability for the electronic systems, except in the event of errors attributable to users (contributory negligence) or due to force majeure. The second basis is the provisions of Article 40 of the ITE Law setting out the possibility for the Government to actively protect the nation and to prevent anything that may be harmful to the national interest (Indonesia, 2008). Furthermore, the Indonesian Government has also issued the Presidential Regulation Number 82/2012 regarding PTSE for the application of IT Governance. The regulation also sets out that electronic public service systems, including electronic administrative systems, must be registered, and subject to feasibility certification based on assessment on the implementation of good IT governance (Indonesia, 2012). The PTSE Regulation also sets out mandatory use of Signature for Electronic transactions, which provides the strongest evidence of the use of TTE, with the support of an Electronic Certification rooted to the relevant Ministry of CA.

There were several issues related to the e-KTP development project in 2014, including the effective application of e-KTP, the procurement process, irregularities in the implementation of e-KTP project, as well as the use of e-KTP for obtaining public services and participating in general elections. There might also a problem if there was an issue of leakage of personal information data at the office of the General Election Commission.

In the context of the implementation of the National Single Identification Number (Nomor Induk Kependudukan/NIK) based on the aforementioned Law, electronic identity cards must be issued by referring to such National Single Identification Numbers (NIK), the specifications and formats of which are set out in Article 63 of the Law Number 23/2006. The special formats provide specific security system that provides official identity for the people (Indonesia 2006). Such security system is in the form of the use of security codes as a means for the identification of people's identity precisely and accurately as an authentication process, and for ensuring that the demographic documents belong to those people. In this case, e-KTP also includes the electronic recording of personal data that can be read electronically using a reader, as well as personal data security. By using e-KTP, Indonesia has adopted a single-user credentials, which requires all transactions to be made by using e-KTP. The e-KTP registration and issuance processes have reached more than 50%, but unfortunately the authentication system is still not running well. EDC machines have not been spread to all those requiring them and devices for authenticating identities.

The conventional paper-based ID card displayed the identity card number on its reverse side. The numbers were unique only to one document ID card, which corresponded to a set of data displayed on the document and a set of data stored in the administration system through the registration system by wards. Apart from the unique Single Identity Numbers, the ID card numbers guarantee the authenticity of the public documents using the set of data as displayed on the cards. There has been a question arising following the change to the system as ID card numbers are no longer displayed on the back side of e-KTP. On the electronic ID cards, the numbers are stored in a chip planted in the plastic cards, so that it is invisible to naked eyes. In this regard, there is an issue of security, not only related to possible budget leak, but also to the rights of the card holders to their personal data that are now vulnerable to possible abuse by unauthorized parties. There is also a possibility that some cards that have been issued do not reach the eligible persons and are used by other persons, or cards issued in duplicates but are not tracked properly. People are also not aware whether their ID cards have chips or not, while card reader systems for authenticating their cards are still unavailable. It is ironic that ID cards are the only proof of citizenship, but there is no guarantee of the security of data ensuring that the data displayed and stored on one card is different from another card. As the result, there is no assurance of the validity of identity and such issue may have a systemic impact on the life of the individual concerned.

In general, e-KTP does not have good reliability because it does not fully meet the provisions of Article 15 of the ITE Law and Presidential Regulation on PSTE, as e-KTP has not obtained any eligibility certification and it does not use any electronic signature pursuant to the Government PKI. The condition is getting even worse considering the less thorough dissemination of information, lack of goods news and non-satisfactory interaction, in addition to the fact that there people who have not received their e-KTP. Such condition leads to less trust on the implementation of the e-KTP project. In its development, in response to the expectations regarding interoperability with other agencies, a second generation of e-KTP cards will also be developed, which will include the inclusion of chip modules by the BPPT to allow cross-sectoral use of the ID Cards for accessing education and health services.

With regard to the people's needs for privacy, it is possible to have a domestic identity provider that may be treated as a provider of online identity or electronic signature.

4 ASEAN ECONOMIC COMMUNITY

Several ASEAN countries have implemented Identity Management and have applied CA for public services and government administration, but they are still not interoperable with each other. It seems that most of ASEAN countries select a single credential system. For Malaysia and Singapore, with a relatively smaller population, it is certainly easier to apply, so that they can run relatively well. In the regional context, a commitment will be required to mutually accept the existence of those identities, by at least applying a mutual recognition mechanism. However, it would be better if there is a mechanism that allows connectedness (Bridging CA) in more automated manner among the member states.

In view of the developments in the US and EU, ASEAN should consider the need for a regional approach by combining the paradigms set out in NSTIC and EU Regulation Number 910/2014 to provide space for private identity providers to interact with each other for the purpose of regional commerce. This is necessary for the distribution of legal digital products in ASEAN market. This approach shows cases of an identity ecosystem that consists of private parties and governments in a single environment for the provision of interoperable electronic identity mutual recognition (ASETF ASEAN 2015) namely Regional Authentication Framework.

Accordingly, a federated identity ecosystem can be created, where every ASEAN member state can provide facilities for the identities of their citizens. The assurance level of every identity made by the parties will determine whether or not the implementation of the ASEAN Economic Community would be effectively interoperable. The Intra factor will be a catalyst for e-commerce and earnings improvement in various sectors as well as improvement of convenience for parties to transactions in the ASEAN Single Market.

5 CONCLUSION

As the electronic identity system in Indonesia administered by the Government through the e-KTP Project does not have strong security, it is not easy to use or to access the system. The paradigm of e-KTP has been set, namely for using the identity system for the purpose of general elections. However, accordingly, it is difficult to apply the identity system as a means for ensuring every person's identity with a single credential system as intended in the Population Administration Law. Indonesia is, therefore, obligated to identify an Identity Framework to be developed for serving every citizen as mandated by the aforementioned law.

It is a challenge if Indonesia does not want to be in low level of trust, as e-KTP cannot be used for cross border e-commerce. However, it can be empowered by using the services of a private digital identity provider (such as privy.id) to fill the gap in the implementation of electronic identification. Cooperation between private parties and the government can increase the level of assurance of Indonesian reliability in the implementation of ASEAN Single Market. The Intra-ASEAN Secure Transaction Framework Report addressed three main components of e-authentication, namely: (i) Assurance Levels and Risk Assessments—The levels of assurance are defined so that different levels of importance of getting e-authentication right can be distinguished. For a transaction, risks are assessed and summarized in impact values of authentication failure associated with the types of potential impacts. Subsequently, an appropriate level of assurance is selected to reflect the potential impacts involved; (ii) Identity Proofing and Verification—For each level of assurance, an objective of authentication and a set of controls are defined. Details about identity proofing and verification methods are provided later for the registration process; (iii) Authentication Mechanism—Different token technologies are listed and mapped according to the levels of assurance. Moreover, recommendations are made as to how identity should be managed.

To ensure the implementation of Regional Authentication Framework, ASEAN should formulate a trusted digital identity framework as the basis for the ASEAN members in determining the implementation standard for various e-ID among ASEAN member states. The Framework used as the foundation for calculating the variables will be used by each member to examine the trusted identities of citizens of the member states. This framework will serve as a foundation for the implementation of electronic standards of authentication, identity proofing and privacy protection, as well as a foundation for level qualification or assurance against identity based on the capabilities of electronic system providers in each member state of ASEAN.

REFERENCES

ASEAN (2015) *Intra-ASEAN Secure Transaction Framework Report*.
ASEAN ICT *Master Plan 2020 (2015) Propelling ASEAN towards a digitally enabled economy that is secure, sustainable, and transformative; and enabling an innovative, inclusive and integrated ASEAN Community*.
Bajaj, K.K & Nag, D. (2005) *E-Commerce: The Cutting Edge of Business*. New Delhi, McGraw-Hill Education.
ENISA (2012) National Cyber Security Strategies. Heraklion, ENISA.
European Union (2014) EU Regulation No 910/2014 of the European Parliament and of the Council on Electronic Identification and Trust Services for Electronic Transactions in the Internal Market and Repealing Directive 1999/93/EC, OJ L 257.
Indonesia (2008) *Undang-Undang No. 11 Tahun 2008 tentang Informasi dan Transaksi Elektronik. Lembar Negara No. 58 Tahun 2008, Tambahan Lembar Negara No. 4843*.
Indonesia (2012) *Peraturan Pemerintah No. 82 Tahun 2012 tentang Penyelenggaraan Sistem dan Transaksi Elektronik. Lembar Negara No. 189 Tahun 2012, Tambahan Lembar Negara No. 5348*.
Indonesia (2013) *Undang-Undang No. 24 Tahun 2013 tentang Perubahan atas Undang-Undang No 23 Tahun 2006 tentang Administrasi Kependudukan. Lembar Negara No.232 Tahun 2013, Tambahan Lembar Negara No. 5475*.
Indonesia. 2006. *Undang-Undang No. 23 Tahun 2006 tentang Administrasi Kependudukan. Lembar Negara No.124 Tahun 2006, Tambahan Lembar Negara No. 4674*.
International Telecommunication Union *(2007) ITU Global Cyber Security Agenda (GCA), Framework for International Cooperation in Cyber Security*.
International Telecommunication Union (2011) ITU National Cyber Security Strategy Guide.
Makarim, E. (2010) *Tanggung Jawab Hukum Penyelenggara Sistem Elektronik*. Jakarta, Rajawali Press.
NATO Cooperative Cyber Defence Center of Excellence (2012) *National Cyber Security Framework Manual*. Tallinn, NATO CCD COE Publication.
OECD (2011) *Digital Identity Management, Enabling Innovation, and Trust in the Internet Economy*. OECD Publishing.
OECD (2015) *Working Party on Security and Privacy in the Digital Economy*. OECD Publishing.
United Nations (2009) *UNCITRAL Promoting Confidence in Electronic Commerce: Legal Issues on International use of Electronic Authentication and Signature Methods*. New York, United Nations Publication.
United States of America (2011) *National Strategy for Trusted Identities in Cyberspace, Enhancing Online Choice, Efficiency, Security, and Privacy*. Washington, White House.

The legal impact of the signing of the multilateral competent authority agreement on automatic exchange of financial account information: A banking law perspective

Rouli Anita Velentina Napitupulu
Faculty of Law, Universitas Indonesia, Depok, Indonesia

ABSTRACT: The Indonesian Minister of Finance signed the Multilateral Competent Authority Agreement (MCAA) on Automatic Exchange of Financial Account Information on 3 June 2015. The MCAA is a multilateral instrument that facilitates the implementation of the Automatic Exchange of Information (AEOI) of financial accounts globally using the Common Reporting Standard (CRS). The grounds for signing the MCAA were that thousands of trillions of rupiah belonging to Indonesian citizens had been indicated as being hidden in tax haven countries. Through the AEOI system, the treasures hidden in foreign countries can be tracked much more easily. However, Article 1 point 28 of Law Number 7 of 1992 concerning Banking as amended by Law Number 10 of 1998 (Republic of Indonesia, 1992, 1998) provides that details of a depositor's financial circumstances shall be included in the scope of bank secrecy. Disclosing any information within the scope of bank secrecy is a criminal offence. To protect banks from any liability, the Financial Services Authority (FSA) stipulates that customers should provide written letters of attorney that authorise banks to provide details of their financial circumstances in the context of AEOI using CRS. This causes a legal dilemma: What is the legal impact of the signing of the MCAA from the perspective of banking law? This legal research employs a qualitative doctrinal legal method. The provision that requires bank customers to give their prior consent to the banks in providing such financial information, in normal circumstances, is contrary to the code of the Banking Act. According to the code of the Banking Act, the consent should be given on the basis of the free will of the depositors. It would be unlawful to require the depositors to give their consent under any duress. Moreover, the enforceability of the signed MCAA instrument is lower than that of the Banking Act. As a result, this may cause the signing of the MCAA and the issuance of the FSA Regulation to be null and void.

1 INTRODUCTION

Indonesia, through its Competent Authority, the Minister of Finance of the Republic of Indonesia, signed the Multilateral Competent Authority Agreement (MCAA) on Automatic Exchange of Financial Account Information on 3 June 2015 (Ministry of Finance, 2015). The preamble of the MCAA declares that the jurisdictions of the signatories to the MCAA are the parties to, or those who have expressed their intention to be a signatory to, the Convention on Mutual Administrative Assistance in Tax Matters as amended by the Protocol amending this same Convention. Indonesia signed the Protocol amending the Convention on Mutual Administrative Assistance in Tax Matters ("the Convention") on 3 November 2011 (Ministry of Finance, 2015). The Convention was then to come into force on 1 May 2015 (OECD, 2016). The signing of the Convention has been ratified by the issuance of Presidential Regulation Number 159 of 2014 (Ministry of Finance, 2015).

The MCAA is one of the multilateral instruments that are created by the US as an effort to improve the compliance with its tax laws around the world (Snyder, 2015). International tax cooperation has been discussed by international organisations and individual countries in order to effectively resolve destructive tax practices. Under the international tax cooperation framework, both transparency and exchange of tax information are very important (Ferreira de Almeida, 2015). Therefore, on a global level, the MCAA would be a major initiative to tackle tax avoidance and evasion as well as to improve international tax compliance, for the preamble of the MCAA explicitly states that the MCAA intends to enable the participating countries to effectively activate the cross-border, automatic exchange of tax information. The preamble of the MCAA also specifies that the Automatic Exchange of financial account Information (AEOI) is carried out by using the Common Reporting Standard (CRS), which is based on the Convention. Article 6 of the Convention stipulates that two or more parties shall determine a certain procedure by mutual agreement concerning the automatic exchange of information. Therefore, the MCAA is determined by a mutual agreement. In relation to the information, Article 4 of the Convention provides that:

a. The information shall be foreseeably relevant for the administration or enforcement of domestic laws concerning the taxes.
b. The authorities may inform their residents or nationals before transmitting any information concerning them provided that the internal legislation regulates to do so.

Indonesia has also attempted to improve its tax compliance. Considering that thousands of trillions of rupiah belonging to Indonesian citizens, including government officials, are hidden in foreign countries, it is very important for Indonesia to get involved in the international tax cooperation. Through AEOI using CRS, it would be much easier to track the treasures hidden in foreign countries, especially in tax haven countries. Indonesia, therefore, decided to sign the MCAA. Nevertheless, any information concerning depositors and their deposits is included in the scope of bank secrecy under Indonesian banking law. This causes a legal dilemma: what is the legal impact of the signing of the Multilateral Competent Authority Agreement on Automatic Exchange of Financial Account Information from the perspective of banking law?

2 METHODOLOGY

Qualitative doctrinal legal method is the legal method used in this legal research. A doctrinal legal method, known as "black-letter law", aims to *"systematise, rectify, and clarify the law on any particular topic by a distinctive mode of analysis to authoritative texts that consist of primary and secondary source."* One of the assumptions is that *"the character of legal scholarship is derived from law itself"* (Rubin, 1997).

This research aimed to systematise, rectify, and clarify the signing of the MCAA in relation to the bank secrecy provisions. The primary legal materials that are used consist of: Law Number 7 of 1992 concerning Banking as amended by Law Number 10 of 1998 (Republic of Indonesia, 1992, 1998); Law Number 21 of 2008 concerning Sharia (Islamic) Banking (Republic of Indonesia, 2008); Law No.12 of 2011 concerning Formulation of Laws and Regulations (Republic of Indonesia, 2011a); Law No.21 of 2011 concerning the Financial Service Authority (Republic of Indonesia, 2011b); Bank Indonesia Regulation No.2/19/2000 concerning the Requirements and the Procedures for Granting Written Order or Permission to Waive Bank Secrecy (Bank Indonesia, 2009); and Financial Services Authority Regulation No.25/POJK.03/2015 concerning Submission of Information on Foreign Customers in relation to Taxation to a Partner Country or Partner Jurisdiction (Financial Services Authority, 2009). In addition, this research also uses legal material found using a review of the relevant literature, which consists of books, journals and research papers.

3 RESULTS AND ANALYSIS

The MCAA consists of eight sections and stipulates the details of the information (section 2); the period and procedure for information exchange (section 3); the compliance and enforcement collaboration (section 4); the personal data protection (section 5); and consultation between one or more other competent authorities (section 6).

Some participating members shall begin their information exchanges on September 2017; meanwhile, some others, including Indonesia, will begin in 2018 (OECD, 2106). However, it will not be a simple task to conduct this tax information exchange. The efficient flow of tax information could be threatened by several obstacles, including the bank secrecy laws, administrative procedure uncertainty, and the customer information recording and reporting systems of the concerned financial institutions (Sullivan & Cromwell, 2014).

From the above-mentioned barriers, tax authorities would find bank secrecy as the major barrier in conducting the exchange of tax information using CRS (Anamourlis & Nethercott, 2009). The scope of bank secrecy provision thus varies from one country to another. There is no bank secrecy provision which applies internationally (Soepraptomo, 2005).

3.1 *Bank secrecy in Indonesia and its exceptions*

In Indonesia, the tax authority is not permitted to automatically obtain information concerning depositors and their deposits. Under Law Number 7 of 1992 concerning Banking as amended by Law Number 10 of 1998 ("Banking Act") (Republic of Indonesia, 1992, 1998) and Law Number 21 of 2008 concerning Sharia (Islamic) Banking ("Sharia Banking Act") (Republic of Indonesia, 2008), anything related to information regarding any depositors and their deposits is included in the scope of bank secrecy. Pursuant to the Banking Act, the scope of bank secrecy is limited to information concerning "depositors" and "their deposits". Pursuant to Article 1 Number 28 of the Banking Act, bank secrecy is anything that is related to information concerning any depositor and his deposits. Meanwhile, under the Sharia Banking Act, the scope of bank secrecy includes information concerning "investors and their investments". Further, Article 1 Number 14 of the Sharia Banking Act stipulates that bank secrecy is anything that is related to information concerning any depositor and their deposits and any investor and their investments.

The obligation of banks to maintain bank secrecy is strictly regulated in both the Banking Act (Republic of Indonesia, 1992, 1998) and the Sharia Banking Act (Republic of Indonesia, 2008). Under the Banking Act, except those stipulated in Articles 41, 41A, 42, 43, 44 and 44A, the secrecy of information concerning any depositor and their deposits shall be maintained by banks and their affiliated parties (Article 40 (1) and (2) of the Banking Act). Thus, the exceptions to bank secrecy are limited in the Banking Act. Similarly, the Sharia Banking Act requires sharia banks and their affiliated parties to maintain secrecy of information concerning any depositor and their deposits and any investor and their investments (Article 41 of Sharia Banking Act).

From the above-mentioned articles, several things could be concluded. Firstly, the bank secrecy is related to information about depositors and their deposits. For a sharia bank, the bank secrecy includes not only deposits of its customers but also the investments of its customers. This means that the information concerning debtors and their debts is not included in the scope of bank secrecy. Secondly, all kinds of deposits are included in the scope of bank secrecy for both conventional banks and sharia banks. Thirdly, in both the Banking Act and the Sharia Banking Act, there are several exceptions to bank secrecy provisions. They are:

a. in the interest of taxation (Article 41 of the Banking Act; Article 42 of the Sharia Banking Act);
b. in the interest of the settlement of a bank's claims which have been handed over to the Agency for State Debt and the Auction Affair/State Debt Affair Committee (Article 41A of the Banking Act);
c. in the interest of criminal court proceedings in a criminal case (Article 42 of the Banking Act; Article 43 of the Sharia Banking Act);
d. in a civil suit between a bank and its customer (Article 43 of the Banking Act; Article 45 of the Sharia Banking Act);
e. for the purpose of information exchange between/among banks (Article 44 of the Banking Act; Article 46 of the Sharia Banking Act);

f. on the basis of a written request, consent, or letter of attorney from a depositor (Article 44A (1) of the Banking Act; Article 47 of the Sharia Banking Act); and
g. for the sake of the legal heirs of a depositor—in the case where the customer is deceased—to obtain information concerning deposits of the customer (Article 44A (2) of the Banking Act; Article 48 of the Sharia Banking Act).

They are two theories of bank secrecy, which are the theory of absolute bank secrecy and the theory of relative bank secrecy (Djumhana, 1993). According to the former, a bank has an absolute obligation to maintain bank secrecy under any circumstances. Meanwhile, based on the latter, bank secrets could be disclosed in the interest of the State (Djumhana, 1993). Based on the aforementioned provisions concerning bank secrecy exceptions, Indonesia effectively implements the theory of relative bank secrecy because in Indonesia bank secrecy can be waived for several limited reasons.

Based on the Torunier Case, bank secrecy could be waived on the basis of four reasons: the stipulation of certain regulations; the public interest; the bank's interest; consent from the customer(s) concerned (Husein, 2003). Torunier's case was between National Provincial and Union Bank of England. (Sjahdeini, 1999).

For conventional banks, the exception is stipulated in Article 41 of the Banking Act (Republic of Indonesia, 1992, 1998). Under this article, in the interest of taxation, the Chairman of Bank Indonesia, at the request of the Minister of Finance, has the authority to issue a written order to a bank to provide information and to disclose written evidence and document(s) as to a particular depositor's financial circumstances to a tax officer. Thus, a conventional bank is entitled to disclose its customer's financial circumstances to the tax authority on the basis of a written order from Bank Indonesia. The written order shall be in the interest of taxation, at the request of the Minister of Finance, and it should clearly mention the name of the tax officer and the name of the bank customer.

For sharia banks, not all kinds of "tax interests" could be classified as the exception to bank secrecy provisions. Article 42 (1) of the Sharia Banking Act clearly provides that the exception is merely in the interest of tax crime investigation. A written order could only be issued by the Chairman of Bank Indonesia to a bank to provide information and disclose written evidence and document(s) regarding the financial circumstances of a particular depositor or investor to a tax officer provided it is in the interest of tax crime investigation and at the request of the Minister of Finance. In the elucidation of this article, it is stated that "disclosing written evidence" includes providing an explanation and its copy. Consequently, a sharia bank shall be able to give information about the financial circumstances of a depositor when there is a written order from Bank Indonesia. The order shall be for the purpose of tax investigation and at the request of the Minister of Finance. Furthermore, Article 42 (2) of the Sharia Banking Act (Republic of Indonesia, 2008) stipulates that the written order shall mention the name of the tax officer, the name of the bank customer, and the tax case being investigated.

The provisions concerning bank secrecy are included in imperative provisions and contain sanctions in the form of punishment. In Article 47 of the Banking Act (Republic of Indonesia, 1992, 1998), whoever without a written order or permission from the Chairman of Bank Indonesia unlawfully compels a bank or its affiliated party to disclose information for the purpose of taxation (as stipulated in Article 40) shall be imprisoned for a period of two to four years and be sentenced to a penalty of between ten billion and two hundred billion rupiah. Furthermore, this article also provides that bank commissioners, bank directors, bank employees, or its other affiliated parties who unlawfully disclose confidential information—without a written order from the Chairman of Bank Indonesia—shall be imprisoned for a period of two to four years and be sentenced to a penalty of between four billion and eight billion rupiah. The term 'bank employee' includes all officials and employees of the bank (Article 47 (2) of the Banking Act). Moreover, the Banking Act clearly states that any offence to the disclosure of bank secrecy in the interest of taxation is a felony (Article 51).

Similarly to Article 47 of the Banking Act, Article 60 of the Sharia Banking Act (Republic of Indonesia, 2008) provides that whoever without a written order or permission from the

Chairman of Bank Indonesia unlawfully compels a sharia bank, syaria business unit *(unit usaha syariah)* UUS, or its affiliated parties to disclose confidential information shall be imprisoned for a period of two to four years and be sentenced to a penalty of between ten billion two hundred billion rupiah. Article 60 (2) of the Sharia Banking Act provides that bank commissioners, bank directors, bank employees, or its other affiliated parties who unlawfully disclose confidential information shall be imprisoned for a period of two to four years and be sentenced to a penalty of between four billion and eight billion rupiah.

Related to the issue of disclosing information under bank secrecy principles, Bank Indonesia issued *Peraturan Bank Indonesia* No.2/19/2000 *tentang Persyaratan dan Tata Cara Pemberian Perintah atau Izin Tertulis Membuka Rahasia Bank* – PBI No.2/19/2000 (Regulation of Bank Indonesia No.2/19/2000 concerning the Requirements and the Procedures for Granting Written Order or Permission to Waive Bank Secrecy) (Bank Indonesia, 2000). In relation to "bank secrecy" and "in the interest of taxation", PBI No.2/19/2000 provides that bank secrecy could only be waived on the basis of a written order from the Chairman of Bank Indonesia. The scope of the disclosure of bank secrecy is limited to what is mentioned in the written order (Article 8 of PBI No.2/19/2000 (Bank Indonesia, 2000)).

PBI No.2/19/2000 (Bank Indonesia, 2000) states that the written order shall be issued upon written request from the Minister of Finance, which shall be signed by the Minister of Finance (Article 4 (2)). The written request shall clearly point out five things, namely, the name of the tax officer; the name of the deposit customer (who is the taxpayer); the name of the bank office where the customer deposits their money; the information required; and the reason why the information is needed (Article 4 (3)).

3.2 Legal risks that may arise in relation to the signing of the MCAA

As explained above, the disclosure of bank secrets as stipulated in the Banking Act (Republic of Indonesia, 1992, 1998) and the Sharia Banking Act (Republic of Indonesia, 2008) of Indonesia shall be based on the written order from Bank Indonesia upon the written request from the Minister of Finance. Meanwhile, the AEOI of financial account globally using the CRS requires an automatic mechanism between contracting countries. Thus, AEOI using the CRS as stipulated in the MCAA is contrary to bank secrecy stipulated in the Banking Act and the Sharia Banking Act of Indonesia.

Under Article 7 (1) of the Law No. 12 of 2011 concerning Formulation of Rules and Regulations (*Undang-undang No. 12 Tahun 2011 tentang Pembentukan Peraturan Perundang-undangan*) (Republic of Indonesia, 2011a), the hierarchy of Indonesian rules consists of: the 1945 Constitution of the Republic of Indonesia; People's Consultative Council Decree; the Law/Government Regulation in Lieu of Law; the Government Regulation; the Presidential Regulation; the Province Regulation; and the Regency/Municipality Regulation. Then, it is stipulated in Article 7 (2) of this act (Republic of Indonesia, 2011a) that the power of rules is in accordance with the hierarchy. This provision is in line with the legal principle of *lex superior derogate legi inferior*, which means that a regulation of higher standing in the hierarchy overrides a regulation of lower standing.

The MCAA was signed by the Minister of Finance. The MCAA could thus be considered as a regulation issued by the Minister. It means that, in hierarchical terms, the MCAA is lower than the Banking Act and the Sharia Banking Act of Indonesia. Therefore, the MCAA would be overridden by the Banking and Sharia Banking Acts of Indonesia.

In December 2015, *Otoritas Jasa Keuangan* (OJK) (Financial Services Authority) issued *Peraturan OJK No.25/POJK.03/2015 tentang Penyampaian Informasi Nasabah Asing Terkait Perpajakan kepada Negara Mitra atau Yurisdiksi Mitra* (FSA Regulation No. 25/POJK.03/2015 concerning the Submission of Information on Foreign Customers in relation to Taxation to a Partner Country or Partner Jurisdiction) (Financial Services Authority, 2015). Pursuant to Article 3 of this regulation, financial service institutions could disclose information concerning foreign customers and/or prospective foreign customers to Indonesia's tax authority to be forwarded to the tax authority of a partner country or partner jurisdiction as a result of written consents, instructions, or letters of consent given by foreign customers and prospective foreign customers to financial services institutions. Furthermore, Article 7 (1) of FSA Regulation No. 25/POJK.03/2015 (Financial Services Authority, 2015) states that the submission of the report on the information of foreign customers shall be conducted within a maximum period of 60 days before the deadline to the tax authority of a partner country or partner jurisdiction.

Some people might argue that FSA Regulation No. 25/POJK.03/2015 (Financial Services Authority, 2015) is a way to force disclosure of bank secrets in the interest of taxation without a written order from the Chairman of Bank Indonesia. Article 3 of this regulation requires foreign customers or prospective customers to give written consent to banks to disclose bank secrets to the tax authority. However, it is important to review whether this provision is in line with Article 44A (1) of the Banking Act (Republic of Indonesia, 1992, 1998), which states that on the basis of a written request, consent, or letter of attorney from a depositor, information concerning deposits of a depositor in a bank shall be provided by the bank to any party specified by the depositor.

The provision that requires bank customers to authorise the bank to provide such financial information, in normal circumstances, is contrary to the code of the Banking Act. Under Article 44A (1) of the Banking Act, written consent shall be given on the basis of the free will of bank customers. However, under Article 3 of FSA Regulation No. 25/POJK.03/2015, the written consent is given under duress.

The authority of the FSA to issue its regulation is given by Article 31 of Law Number 21 of 2011 concerning the Financial Services Authority ("FSA Act") (Republic of Indonesia, 2011b). This article provides that further provisions regarding the protection of consumers and the community shall be regulated through the the FSA. In hierarchical terms, the FSA Act is equal to that of the Banking Act and the Sharia Banking Act of Indonesia. This means that the FSA Regulation's hierarchical standing is lower than that of the Banking and Sharia Banking Acts of Indonesia. Based on Article 7 (1) of Law No. 12 of 2011 concerning the establishment of laws and regulations (Republic of Indonesia, 2011a) and the legal principle of *lex superior derogate legi inferior* (a law that is higher in the hierarchy repeals one that is lower), the Banking Act and the Sharia Banking Act of Indonesia would override FSA Regulation No. 25/POJK.03/2015.

FSA Regulation No. 25/POJK.03/2015 would not be a valid legal basis to be an "implementing regulation" of the MCAA because the FSA Regulation would be overridden by the Banking Act and the Sharia Banking Act of Indonesia. Both the MCAA and FSA Regulation No. 25/POJK.03/2015 are contrary to bank secrecy provisions as stipulated in the Banking Act and the Sharia Banking Act of Indonesia. As a result, the signing of the MCAA and the issuance of FSA Regulation No. 25/POJK.03/2015 could be rendered null and void.

The General Elucidation of the FSA Act states that the protection of the financial services consumer, which was not as yet optimal, was one of the reasons for the establishment of FSA. Therefore, the obligation to make a contract on the basis of Article 3 of FSA Regulation No. 25/POJK.03/2015 would be contrary to the purpose of the establishment of the FSA. It would be unjust if depositors were required to give consent under duress because it is contrary to the code of the Banking and Sharia Banking Acts of Indonesia.

If a bank officer discloses any information concerning the financial circumstances of depositors, then the depositor could submit a civil lawsuit based on the perpetration of an unlawful act as stipulated in Article 1365 of the Indonesian Civil Code. Article 1365 of the Civil Code stipulates that every illegitimate act that causes any damage to third parties requires the party at fault to pay for the damage caused.

Moreover, pursuant to the Banking Act and the Sharia Banking Act of Indonesia, employees of a bank or its affiliated parties who unlawfully disclose confidential information shall be imprisoned for a period of two to four years and shall be sentenced to a penalty of between four billion and eight billion rupiah. As explained above, the provisions of the MCAA and FSA Regulation No. 25/POJK.03/2015 are contrary to bank secrecy provisions as stipulated in the Banking Act and the Sharia Banking Act of Indonesia. Thus could possibly cause the MCAA and FSA Regulation No. 25/POJK.03/2015 to be rendered null and void, or may, if a bank officer discloses any information concerning a deposit customer and his/her deposit, allow the customer to legitimately make a complaint to the police.

4 CONCLUSION

Based on the research findings and the data analysis in this article, it could be concluded that the provisions of the MCAA are contrary to the bank secrecy principles stipulated in the Banking Act and the Sharia Banking Act of Indonesia. As a result, the signing of the MCAA could possibly give rise to the following legal risks:

a. the MCAA could be rendered null and void;
b. a civil lawsuit could be submitted against the bank and/or employees of the bank or its other affiliated parties who disclose any information concerning the financial circumstances of any depositors if there is no written order from the Chairman of Bank Indonesia; and
c. a police report could be made against employees of the bank or its other affiliated parties who disclose any information concerning the financial circumstances of any depositors without a written order from the Chairman of Bank Indonesia.

REFERENCES

Anamourlis, T. & Nethercott, L. (2009). An overview of tax information exchange agreements and bank secrecy. *IBFD: Bulletin for International Taxation, 63*(12), 616–621.

Bank Indonesia. (2000). *Peraturan Bank Indonesia No. 2/19/PBI/2000 tentang Persyaratan dan Tata Cara Pemberian Perintah atau Izin Tertulis Membuka Rahasia Bank [Bank Indonesia Regulation No. 2/19/PBI/2000 regarding the requirements and the procedures for granting written order or permission to waive bank secrecy]*.

Djumhana, M. (1993). *Hukum Perbankan di Indonesia [Banking law in Indonesia]*. Bandung, Indonesia: Citra Adya Bakti.

Ferreira de Almeida, C.O. (2015). International tax cooperation, Taxpayers' rights and bank secrecy: Brazilian difficulties to fit global standards. *ExpressO*. Retrieved from http://works.bepress.com/carlos_almeida/1/.

Financial Services Authority. (2015). *Peraturan Otoritas Jasa Keuangan No. 25 Tahun 2015 tentang Penyampaian Informasi Nasabah Asing Terkait Perpajakan kepada Negara Mitra atau Yurisdiksi Mitra [Financial Services Authority Regulation No. 25 of 2015 regarding disclosure of foreign customers]*.

Husein, Y. (2003). *Rahasia Bank versus Kepentingan Umum [Bank secrecy versus public interest]*. Depok, Indonesia: Program Pascasarjana Fakultas Hukum Universitas Indonesia.

Ministry of Finance. (2016, June 8). The signing of multilateral competent authority agreement. Jakarta, Indonesia: Ministry of Finance, Republic of Indonesia. Available from http://www.kemenkeu.go.id/en/SP/signing-multilateral-competent-authority-agreement.

OECD. (2016). *Jurisdictions participating in the Convention on Mutual Administrative Assistance in Tax Matters Status, 26 September 2016*. Paris, France: Organization for Economic Co-operation and Development Retrieved from www.oecd.org/tax/exchange-of-tax-information/Status_of_convention.pdf.

Republic of Indonesia (2014). *Peraturan Presiden No. 159 Tahun 2014 tentang Pengesahan Konvensi tentang Bantuan Administratif Bersama di Bidang Perpajakan [Supreme Court Republic of Indonesia Regulation No. 1 of 2014 regarding Ratification Convention on Mutual Administrative Assistance in Tax Matters]*.

Republic of Indonesia. (1992). *Undang-undang No. 7 Tahun 1992 tentang Perbankan [Law No. 7 of 1992 concerning banking]*.

Republic of Indonesia. (1998). *Undang-undang No. 10 Tahun 1998 tentang Perubahan atas Undang-undangn No. 7 Tahun 1992 tentang Perbankan [Law No. 10 of 1998 concerning Amendment of Law No. 7 of 1992 concerning banking]*.

Republic of Indonesia. (2008). *Undang-undang No. 21 Tahun 2008 tentang Perbankan Syariah [Law No. 21 of 2008 concerning sharia banking]*.

Republic of Indonesia. (2011a). *Undang-undang No. 12 Tahun 2011 tentang Pembentukan Peraturan Perundang-undangan [Law No. 12 of 2011 concerning establishment of laws and regulations]*.

Republic of Indonesia. (2011b). *Undang-undang No. 21 Tahun 2011 tentang Otoritas Jasa Keuangan [Law No. 21 of 2011 concerning the Financial Services Authority]*.

Rubin, E.L. (1997). Law and the methodology of law. *Wisconsin Law Review*, 521–565.

Sjahdeini, S.R. (1999). Rahasia bank: Berbagai masalah di sekitarnya [Bank secrecy: Various surrounding problems]. *Jurnal Hukum Bisnis, 8*, 4–17.

Snyder, E.J. (2015). FATCA and the broader tax crackdown. *Trusts & Trustees, 21*(6), 596–604.

Soepraptomo, H. (2005). *Hukum Perbankan di Indonesia [Banking law in Indonesia]*. Jakarta, Indonesia: Raja Grafindo Persada.

Sullivan & Cromwell LLP. (2014). International tax cooperation: OECD standard for automatic exchange of financial account information. *S&C Memos*, 27 February 2014. Retrieved from https://www.sullcrom.com/international-tax-cooperation-oecd-standard-for-automatic-exchange-of-financial-account-information.

The unfair rules of intellectual property rights section of the trans-pacific partnership agreement

B.A. Prastyo & A. Sardjono
Faculty of Law, Universitas Indonesia, Depok, Indonesia

ABSTRACT: The fairness of Intellectual Property Rights' (IPR) provisions in the Trans Pacific Partnership Agreement (TPP) should be reconsidered. The primary objective of free trade is to foster competition in the supply of goods/services. That objective is difficult to be achieved completely, because natural monopoly practices can be triggered through the IPR's provisions. The objective of this paper is to provide an analysis to the IPR's section of the TPP to understand the future consequences of the rules for Indonesia. The methodology chosen is non-doctrinaire legal research methods. The result from this study shows that the TPP should be viewed as a multilateral free trade agreement that is fundamentally unfair for Indonesia. This paper's significance is an academic contribution that can be used by the government of Indonesia in the decision-making activities related to the TPP.

1 INTRODUCTION

TPP [Trans Pacific Partnership Agreement] and the entire free trade agreements aim to open up market access as much as possible, through a commitment to the elimination or reduction of tariffs and non-tariff trade barriers. Smith (2008), who was famous for suggesting free trade, argued that consumers will benefit if we "allow the most perfect freedom to the trade of all such mercantile nations", because free trade will create a competition that is forcing manufacturers to act more efficiently and ultimately makes the price of goods to be cheap. However, Smith apparently did not consider IPR [intellectual property rights] aspects when stating the argument. Although tariffs do not exist, a restriction on imports does not exist, and there are other manufacturers capable of producing such products, the price of some IPR's protected products is still expensive because the protection of IPR has created natural monopoly practices for some products.

Needless to say, it should be understood that the IPR ownership does not necessarily give birth to monopolistic practices. For example, a copyrighted song does not have specific consumers. Therefore, no song's author can be a monopolist in the music market. However, there are several types of products that have a specific market in which the protection of IPR may create natural monopoly practices. For example, pharmaceuticals and pesticides. Each pharmaceutical has specific uses as medication to treat certain diseases, so the market for each pharmaceutical is specific. Pesticides also have the same characteristics as a pharmaceutical because of its uses in dealing with specific pests of plants. Granting patents on pharmaceuticals or certain pesticides may make the owner of the patent becomes a natural monopolist.

Based on those thoughts, the benefit of the IPR regime is basically very clear for the producers, or in this context, IPR owners, but it is not clear for the consumers. IPR regime has mostly functioned as means for producers who have excelled in engineering, technology, and science to prevent strong competitors in the same market and to sell their products at high prices to consumers.

Considering the relative huge gap in technological capability and money among developed and developing countries, the insertion of IPR's provisions in any free trade agreement should be assessed to determine whether those provisions have been formulated under the circumstances, as Rawls (1999) states, "behind the veil of ignorance." The transparency and equality of the negotiation should also be taken into account because those will determine the fairness of the free trade agreement. Fairness is an essential issue in the field of law because fairness reflects justice (Rawls, 1999).

> "Justice is the first virtue of social institutions, as truth is of systems of thought... Being first virtues of human activities, truth and justice are uncompromising...The natural distribution is neither just nor unjust; nor is it unjust that persons are born into society at some particular position. These are simply natural facts. What is just and unjust is the way that institutions deal with these facts...The social system is not an unchangeable order beyond human control but a pattern of human action. In justice as fairness men agree to avail themselves of the accidents of nature and social circumstance only when doing so is for the common benefit. The two principles are a fair way of meeting the arbitrariness of fortune; and while no doubt imperfect in other ways, the institutions which satisfy these principles are just." (where is the citation?)

Based on the notion of justice as fairness that is stated by Rawls, this article aims to assess whether the rules in the IPR section of the TPP can be considered just or unjust. Furthermore, this article will answer three questions. First, were the IPR's rules of the TPP Agreement drafted in accordance to "behind the veil of ignorance" situation? Second, do the procedures of accession of the TPP Agreement accept a new proposal to amend the rules in the IPR section? Third, should Indonesia join the TPP Agreement?

In order to answer those questions, the non-doctrinaire type of legal research methodology was applied. The laws assessed include the IPR rules of the TPP Agreement. In conducting the assessment, a conceptual approach was applied, in which John Rawls' theory of justice and Adam Smith's concept of free trade were used as the analytical knife. Conclusions were drawn by applying a deductive logical scheme, whereas the major premise used is that all laws regulating international trade are just when equality and inequality of rights and duties of the IPR owners, and the consumers are for the advantage of all.

2 PREVIOUS CRITICISM RELATED TO THE IPR SECTION OF THE TPP AGREEMENT

Several authors have commented on the IPR Section of the TPP Agreement. Correa (2015), Baker (2016), Bitton (2014), and Silverman (2014) concluded that the IPR Section of the TPP Agreement will more likely bring harm to the public in the form of the increased prices of pharmaceutical products. Cox (2014) also shares similar negative view on that agreement, but she focuses more on difficulties that will be faced by industries in developing nations. Weatherall (2016) states a strong criticism to the TPP Agreement, as he explains that "far from being a 21st Century trade agreement, it seems that the TPP looks backwards. The Asia-Pacific deserved better."

This article will join those previous articles that criticize the IPR Section of the TPP Agreement. The difference that this article makes is on two aspects. First, this article tries to elaborate the TPP rules in a legal theoretical context, especially the theory of justice. Second, this article tries to predict the impact of the rules in the IPR Section of the TPP Agreement to the context of Indonesia's consumers, small industries, tax payers, and the government.

3 THEORY OF JUSTICE AT A GLANCE

Referring to the theory of justice put forward by John Rawls (1999), justice consists of two principles. The first one is equality principle: "each person is to have an equal right to the most extensive scheme of equal basic liberties compatible with a similar scheme of liberties for others." For the second principle, Rawls (1999) indicates that law makers may make a rule that regulates inequality, but that rule must be arranged so that it will "reasonably expected to be to everyone's advantage". Furthermore, Rawls also stated that a rule or agreement is not fair if the procedures in the rule-making or the agreement-making does not reflect the equality of the parties involved. Equality of the parties is indicated when equal opportunity to submit proposals regarding mutual rights and obligations to be regulated is given to all parties.

Among the various obligations that the IPR section set forth in the TPP Agreement, 4 (four) aspects will be observed in this article, namely: a) the confidentiality of the data related to the marketing authorization chemical products agriculture, b) the extension of patents for pharmaceutical products, c) the authority to detain or release of goods in the customs territory suspected to contain elements of IPR infringement, and d) an obligation for internet service providers to reveal alleged IPR's infringers. Those aspects were chosen because the future consequences of the rules in those aspects have indicated to be not to everyone's advantage.

4 DATA EXCLUSIVITY IN AGRO-CHEMICAL PRODUCTS

Along with the increase in world population and the reduced availability of agricultural land, the food resources become very important for any country in the world. Therefore, the agricultural sector is one of the most highly commercial sectors. Agricultural chemical products, such as fertilizers and pesticides, are vital for farmers. The bigger the scale of the production, the more agro-chemical products will be needed.

As an illustration, the value of imports of fertilizers to Indonesia in 2014 (BPS, 2015) reached US$ 1.8 Billion. The Government of Indonesia in the fiscal year of 2016 (Ditjen Anggaran Kementerian Keuangan Republik Indonesia, 2016) allocated the subsidy for the fertilizers at Rp 30.1 trillion (more than US$ 30 billion). The huge amount of that subsidy shows a strong commitment from the President Jokowi's government to enlarge the scale of Indonesia's agricultural production. However, in addition to that, it also shows that Indonesia is a big market for agro-chemical products.

If Indonesia ratifies the TPP, based on Article 18.47, the government will be restrained to issue market authorization to any other parties or allow any other parties to market the same or similar product when a party has obtained market authorization for a new product, unless the owner of that new product gives permission to the government to do so. The time of that restraining will last for ten years starting from the date of the given market authorization. Fertilizers and pesticides are products in which are included in that provision.

If that provision is implemented, the owner of a new agro-chemical product will obtain the most benefits since she will have the opportunity to be the sole supplier of the product to the market for ten years. Noticing the fact that the meaning of "new product" in that provision does not exclusively mean "product patented in Indonesia", that provision has substantially reintroduced the old-time European patent law which existed in the era of mercantilism and colonialism in the 18th and 19th centuries. Comparing to WTO's TRIPS Agreement, that provision will jeopardize the opportunity of generic agro-chemical producers to supply the similar products to the market with a cheaper price. Based on the circumstances that most fertilizers and pesticides used in Indonesia and any other developing countries are imported from developed countries, it is reasonable to think that this provision is promulgated to serve the interests of the exporting producers without considering the burden of the consumers in the importing countries.

5 EXTENSION OF THE PATENT TERM FOR PHARMACEUTICAL PRODUCTS

Pharmaceutical products may include pharmaceutics and certain substances for the manufacture of pharmaceutics. As we all know, the effect of consuming pharmaceutics on the quality of health of the individual is very large. Therefore, it is important that pharmaceutics can be consumed by any individual who needs it regardless of the level of income. However, in the reality, many people cannot obtain the necessary pharmaceutics for their medication because the price is too expensive for them. Because of that, many people lose their wealth in large quantities to pay for the medication. This circumstances should be the primary concern of the government, because the government is responsible for its people's welfare.

As is the case in the context of the marketing of agro-chemical products, the major competitors of the patented-pharmaceutical industry are the generic pharmaceutical industry. Based on one of the disputes in the WTO, currently the implementation of the experiments and the use of the invention of a patented pharmaceutics without the permission of the patent holder for the purpose of obtaining a marketing authorization are not in conflict with the WTO Agreement.

However, if Indonesia ratified the TPP, the Article 18.48.2 would require the government to provide the extension of the term of patent protection to the owner of patented pharmaceutics. Furthermore, Article 18.50.1 would oblige the government to ban the circulation or marketing of pharmaceutics that is identical or similar to the new pharmaceutics for 5 years starting from the new pharmaceutics obtained marketing authorization. In Article 18:52, new pharmaceutical products (including pharmaceutics) are defined as pharmaceutical products that do not contain chemicals that have previously been approved by the government for distribution. That means, "a new pharmaceutical product" does not exclusively mean "pharmaceutical patented in Indonesia".

Those provisions about "new pharmaceutical products" in the TPP greatly advance the interests of non-generic pharmaceutical industry. The patent owner will get the extension of the term of patent protection. It means that the owner of the new patented-pharmaceutics will enjoy a longer time to act in natural monopoly. The owner of an "off-patent but new pharmaceutical product" will also obtain the most benefit of those provisions because he/she can get "exclusive distribution license" for 5 years. Therefore, it is reasonable to think that the interest of the consumers or the patients, especially those classified as poor, is not addressed at all in those provisions.

6 BORDER PROTECTION RELATED TO THE ALLEGATION OF IPR'S INFRINGEMENT

The role of customs officials in IPR enforcement is quite significant. The owner of IPR may ask the customs officials to prevent goods that allegedly infringe its IPR from entering the domestic territory. Although this rule may affect the flow of the goods in international trade, it is not considered as a barrier to free trade. However, learning from the cases in the Netherlands and Germany, the exporters from developing countries should be aware to the possibility of misapplication of this rule.

In 2008 and 2009, at least 20 cargoes containing generic pharmaceutics from India to be sent to Brazil or Nigeria were detained by customs officials in the Netherlands and Germany. The cargoes were in the status of transit and were in the process to be transferred to other modes of transport to their final destination. Based on that status, it was supposed to be clear that the goods within the cargoes were not meant to be distributed in the Netherlands and Germany's domestic territory. However, based on the request of the owner of registered trademark, GlaxoSmithKline, Germany's customs officials detained some cargoes. The company alleged that the cargoes contained goods that infringed its registered trademark, namely the "Amoxil". That allegation was ridiculous because they knew that the goods within the cargoes were generic pharmaceutics with a generic name called "Amoxicillin".

The other cargoes were detained by the Netherlands' customs officials based on the allegation that goods within the cargoes were counterfeit pharmaceutics. Later, it was known that the cargoes were "Losartan Potassium", a kind of medicine that is not patented either in the exporting country (India) or in the importing country (Brazil), and were not meant to be distributed within the Netherlands or any other area of European Union. Other cargoes that were detained were containing "Abacavir Sulfate," a generic medicine used for HIV/AIDS medication that is not counterfeit pharmaceutics nor violating any intellectual property rights in the exporting country (India) or in the importing country (Nigeria), and was not meant to be distributed within the Netherlands or any other area of European Union. The exporters and importers have suffered huge losses due to the actions of those customs officials because some of the goods detained were destroyed, sent back to India, or sent to Brazil after a long time on hold.

India filed a request for consultation for those cases to the WTO. Later on, India and European Union reached a mutual understanding that the European Union's countries will not intercept in-transit generic medicines anymore, unless there is adequate evidence that the goods are likely to be diverted to enter and be distributed in a domestic area of European Union. In addition, the European Union has promised to issue a guideline to its members to clarify how an IPR's enforcement should be conducted in the border area. Although that understanding has resolved the issue of in-transit generic medicine, there is no guarantee that the treatment will be the same to other kinds of manufactured products exported from developing countries.

If Indonesia ratifies the TPP, Indonesia must perform the duty set out in the Article 18.76.4. (B) that requires customs officials of Indonesia to inform the owner of the registered trademark and/or copyright-related several kinds of information. The information includes the identity of the sender/exporter/receiver/importers, description of goods, and, if known, the country of origin of the product manufacturer that has been detained by the customs authorities.

That kind of provision will make Indonesian customs officials serve the interest of IPR's owners. The custom officials will be acting as if they are the employees of the owners of registered trademarks and/or copyrights in collecting information that could be used for the preparation of their IPR's enforcement. Seeing from the perspective of distributive justice, it should be clear that that provision does not uphold the principle of the balance of interest. The owner of IPR receives so many legal rights that may hugely benefit them, but almost no IPR's duty nor obligation is assigned to them to make contribution to the consumers. Furthermore, we can say that that provision was designed to help the owner of IPR enjoy the full benefit of their exclusive rights using the public's money to pay for the border protection's expenses and coercive authority.

7 DUTY OF INTERNET SERVICE PROVIDER

If Indonesia ratifies the TPP, based on Article 18.82.7, Internet service providers will have the duty or obligation to help the owner of IPR to control IPR infringement occurring on their service and become an informant for the owner of IPR. The costs or expenses to perform such obligations are paid by the Internet service provider itself. After receiving information from the Internet service providers, the IPR owner may send a letter of summons to the individuals who are on the list, asking them to halt the infringement and pay a sum of money as compensation for the losses suffered by the owners of IPR. If the letter of summons is not complied with, the owner of IPR can report them to the police or sue them in the court.

If the Internet service provider does the duty, then as the "legal incentives" they do not need to be responsible for in compensating financially from IPR infringement which they do not control. However, if the Internet service provider does not carry out this obligation, they should not be considered guilty of an offense, and nothing can prevent the owner of IPR to report the Internet service provider to the police or to sue them based on the allegation that they are participating in IPR infringement. TPP's provision is much more burdensome for Internet service providers, even when compared with the rules in force in the United States.

In the United States, the lawsuit filed by the owners of IPRs to thousands of Internet users have occurred several times. In 2012 Cintel Films sued 1,052 BitTorrent users and Nu Image sued 4.165 Internet users of various Internet service providers. Under the United States' law, Cintel and Nu Image were forced to file a lawsuit against the thousands of people with the name of fiction, i.e. Doe, because they cannot obtain information from the Internet service providers regarding the identity of the individual. Through a court order, they could ask the relevant Internet service providers to provide information about users of its services. However, Internet service providers may only provide such information if their users agree to provide personal information. Comparing that United States' law and the TPP's provisions, it is reasonable to think that the TPP provisions were designed to prevent the global spreading of the privacy protection model of the United States in the Internet service provider business.

However, although the TPP drafters do not like the United States' privacy protection model, the Copyright Alert System of the United States is likely to be duplicated in TPP member countries. The system is built based on the agreement between the owner of IPR and Internet service providers. The Internet service provider is obliged to monitor the use of its services for controlling IPR infringement. If the Internet service providers find users who are suspected of conducting IPR infringement, then the user will receive a notice or warning for several times. If the user in question does not need the letter, the Internet service provider is expected to slow down the speed of the user's Internet access or apply any other "mitigation measures".

8 CONCLUSION

Rules of IPR in the TPP Agreement observed in this article were not chosen as if they had been behind a veil of ignorance, but rather designed by those who wish to provide maximum protection for IPR-owned producers, regardless of the burden faced by the small industries and consumers. With that conclusion, the authors share a similar view to Correa (2015) who states that "the TPP provides a paradigmatic example of international law making led by the interests of a business group, adopted by a government as an essential component of its own agenda." Such provisions, some of which are in conflict with the principles of the law applied in Indonesia, increase the systemic risk of import-dependency, may create too much burden to domestic Internet service providers, and may force the consumers to pay higher prices for the products. All of those prospects indicate that the rights of the consumers, the domestic infant industries, and the Indonesian tax payers have not been adequately taken into account in the TPP provisions.

Noticing that the TPP negotiations have been completed, it will be difficult for Indonesia to submit its own proposal of provisions. It means that Indonesia will not get an equal chance to propose for the rights and obligations which can be advantageous for its own. Therefore, it is reasonable to conclude that the TPP Agreement is fundamentally unfair for Indonesia. The Indonesian government is expected not to join the membership of TPP.

REFERENCES

Badan Pusat Statistik (2015) *BPS: Impor Pupuk Menurut Negara Asal Utama*, 2000–2014. [Online] Retrieved from https://www.bps.go.id/linkTabelStatis/view/id/1044.

Baker, B.K. (2016) Trans-Pacific Partnership provisions in intellectual property, transparency, and investment chapters threaten access to medicines in the US and elsewhere. *PLoS Medicine* 13 (3), e1001970. [Online] Available from: doi:10.1371/journal.pmed.1001970.

Bitton, M. (2014) Examining the Trans-Pacific Partnership agreement. *Journal of Internet Law*, 17 (9), 25–38.

Correa, C.M. (2015) *Intellectual Property in the Trans-Pacific Partnership: Increasing the Barriers for the Access to Affordable Medicines*. Geneva, South Centre. Research Paper 62, September 2015. [Online] Available from: https://www.researchgate.net/publication/304576192.

Cox, K.L. (2014) The intellectual property chapter of the Trans Pacific Partnership agreement and investment in developing nations. *University of Pennsylvania Journal of International Law*, 35 (4), 1045–1059.

Rawls, J. (1999) *A Theory of Justice.* Revised edition. *Cambridge.* Cambridge, The Belknap Press of Harvard University Press.

Silverman, M.E. (2014) The case for flexible intellectual property protections in the Trans-Pacific Partnership. *Journal of Law and Health*, 27 (2), 215–231.

Smith, A. (2008) *An Inquiry into the Nature and Causes of the Wealth of Nations*. Oxford, Oxford University Press.

Trans-Pacific Partnership Full Text. (n.d). Retrieved from https://ustr.gov/sites/default/files/TPP-Final-Text-Intellectual-Property.pdf.

Weatherall, K.G. (2016) Intellectual property in the TPP: Not 'the New TRIPS'. *Melbourne Journal of International Law*, 17 (2), 1–29.

Author index

Abdullah, Z. 187
Afriana, A. 167
Agustina, R. 55, 71
Anam, S. 181
Anindita, S.L. 71
Arinanto, S. 85
Asa, S. 101
Asshiddiqie, J. 63, 181

Bakry, M.R. 45
Barlinti, Y.S. 109

Cahyono, A.B. 241

Darmawan, I. 229
Dewaranu, T.A. 221
Dewi, G. 187
Dewi, R.I. 203
Dewi, Y.K. 235

Elda, E. 79
Erliyana, A. 45, 93

Fakhriah, E.L. 9, 167
Farida, I. 85
Fitriasih, S. 101, 173

Gabor, M.M. 23

Harahap, M.Y. 127
Harkrisnowo, H. 1, 229
Hasanah, U. 109, 127, 187
Hayati, T. 119

Indrawati, Y. 93

Lita, H.N. 109

Makarim, E. 249
Mannas, Y.A. 9
Marlyna, H. 15
Munthe, A.K. 141

Napitupulu, R.A.V. 257
Nelson, F.M. 39

Nuning, W.P. 195
Nurdin, A.R. 211

Pangaribuan, A. 149
Pantouw, E.G. 249
Prastyo, B.A. 265
Prihatini, F. 141

Rianarizkiwati, N. 63
Ridwan, F.H. 55
Rizal, J. 55, 85

Saibih, J. 31
Santoso, T. 79
Sardjono, A. 15, 23, 265
Setiawati, W. 157
Sofian, A. 173
Sunarti, E.S. 119

Wibisana, A.G. 195, 221
Wirdyaningsih 133